Hilbert Space

Edited by Paul F. Kisak

Contents

Chapter 1

Hilbert space

For the Hilbert space-filling curve, see Hilbert curve.

The mathematical concept of a **Hilbert space**, named after

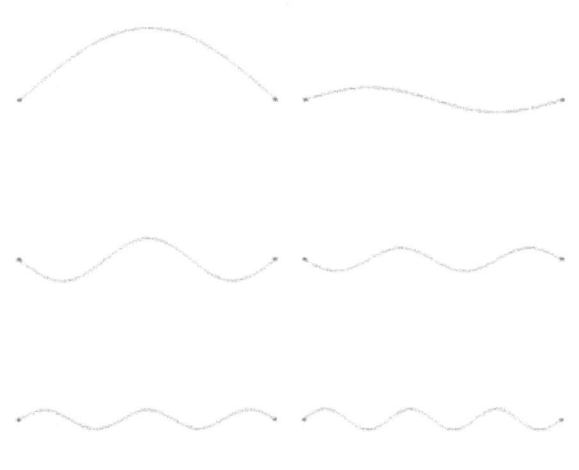

The state of a vibrating string can be modeled as a point in a Hilbert space. The decomposition of a vibrating string into its vibrations in distinct overtones is given by the projection of the point onto the coordinate axes in the space.

David Hilbert, generalizes the notion of Euclidean space. It extends the methods of vector algebra and calculus from the two-dimensional Euclidean plane and three-dimensional space to spaces with any finite or infinite number of dimensions. A Hilbert space is an abstract vector space possessing the structure of an inner product that allows length and angle to be measured. Furthermore, Hilbert spaces are complete: there are enough limits in the space to allow the techniques of calculus to be used.

Hilbert spaces arise naturally and frequently in mathematics and physics, typically as infinite-dimensional function spaces. The earliest Hilbert spaces were studied from this point of view in the first decade of the 20th century by David Hilbert, Erhard Schmidt, and Frigyes Riesz. They are indispensable tools in the theories of partial differential equations, quantum mechanics, Fourier analysis (which includes applications to signal processing and heat transfer)—and ergodic theory, which forms the mathematical underpinning of thermodynamics. John von Neumann coined the term *Hilbert space* for the abstract concept that underlies many of these diverse applications. The success of Hilbert space methods ushered in a very fruitful era for functional analysis. Apart from the classical Euclidean spaces, examples of Hilbert spaces include spaces of square-integrable functions, spaces of sequences, Sobolev spaces consisting of generalized functions, and Hardy spaces of holomorphic functions.

Geometric intuition plays an important role in many aspects of Hilbert space theory. Exact analogs of the Pythagorean theorem and parallelogram law hold in a Hilbert space. At a deeper level, perpendicular projection onto a subspace (the analog of "dropping the altitude" of a triangle) plays a significant role in optimization problems and other aspects of the theory. An element of a Hilbert space can be uniquely specified by its coordinates with respect to a set of coordinate axes (an orthonormal basis), in analogy with Cartesian coordinates in the plane. When that set of axes is countably infinite, this means that the Hilbert space can also usefully be thought of in terms of infinite sequences that are square-summable. Linear operators on a Hilbert space are likewise fairly concrete objects: in good cases, they are simply transformations that stretch the space by different factors in mutually perpendicular directions in a sense that is made precise by the study of their spectrum.

1.1 Definition and illustration

1.1.1 Motivating example: Euclidean space

One of the most familiar examples of a Hilbert space is the Euclidean space consisting of three-dimensional vectors, denoted by $\mathbb{R}^3$, and equipped with the dot product. The dot product takes two vectors $\mathbf{x}$ and $\mathbf{y}$, and produces a real number $\mathbf{x} \cdot \mathbf{y}$. If $\mathbf{x}$ and $\mathbf{y}$ are represented in Cartesian coordinates, then the dot product is defined by

$$(x_1, x_2, x_3) \cdot (y_1, y_2, y_3) = x_1 y_1 + x_2 y_2 + x_3 y_3.$$

The dot product satisfies the properties:

1. It is symmetric in $\mathbf{x}$ and $\mathbf{y}$: $\mathbf{x} \cdot \mathbf{y} = \mathbf{y} \cdot \mathbf{x}$.

2. It is linear in its first argument: $(a\mathbf{x}_1 + b\mathbf{x}_2) \cdot \mathbf{y} = a\mathbf{x}_1 \cdot \mathbf{y} + b\mathbf{x}_2 \cdot \mathbf{y}$ for any scalars a, b, and vectors $\mathbf{x}_1$, $\mathbf{x}_2$, and $\mathbf{y}$.

3. It is positive definite: for all vectors $\mathbf{x}$, $\mathbf{x} \cdot \mathbf{x} \geq 0$, with equality if and only if $\mathbf{x} = 0$.

An operation on pairs of vectors that, like the dot product, satisfies these three properties is known as a (real) inner product. A vector space equipped with such an inner product is known as a (real) inner product space. Every finite-dimensional inner product space is also a Hilbert space. The basic feature of the dot product that connects it with Euclidean geometry is that it is related to both the length (or norm) of a vector, denoted ‖x‖, and to the angle θ between two vectors $\mathbf{x}$ and $\mathbf{y}$ by means of the formula

$$\mathbf{x} \cdot \mathbf{y} = \|\mathbf{x}\| \, \|\mathbf{y}\| \, \cos \theta.$$

Multivariable calculus in Euclidean space relies on the abil-

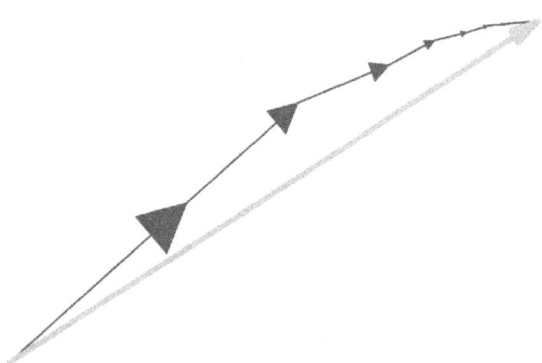

Completeness means that if a particle moves along the broken path (in blue) travelling a finite total distance, then the particle has a well-defined net displacement (in orange).

ity to compute limits, and to have useful criteria for concluding that limits exist. A mathematical series

$$\sum_{n=0}^{\infty} \mathbf{x}_n$$

consisting of vectors in $\mathbb{R}^3$ is absolutely convergent provided that the sum of the lengths converges as an ordinary series of real numbers:[1]

$$\sum_{k=0}^{\infty} \|\mathbf{x}_k\| < \infty.$$

Just as with a series of scalars, a series of vectors that converges absolutely also converges to some limit vector $\mathbf{L}$ in the Euclidean space, in the sense that

$$\left\| \mathbf{L} - \sum_{k=0}^{N} \mathbf{x}_k \right\| \to 0 \quad \text{as} N \to \infty.$$

This property expresses the *completeness* of Euclidean space: that a series that converges absolutely also converges in the ordinary sense.

Hilbert spaces are often taken over the complex numbers. The complex plane denoted by $\mathbb{C}$ is equipped with a notion of magnitude, the complex modulus |z| which is defined as the square root of the product of z with its complex conjugate:

$$|z|^2 = z\overline{z}.$$

If $z = x + iy$ is a decomposition of z into its real and imaginary parts, then the modulus is the usual Euclidean two-dimensional length:

$$|z| = \sqrt{x^2 + y^2}.$$

The inner product of a pair of complex numbers z and w is the product of z with the complex conjugate of w:

$$\langle z, w \rangle = z\overline{w}.$$

This is complex-valued. The real part of $\langle z, w \rangle$ gives the usual two-dimensional Euclidean dot product.

A second example is the space $\mathbb{C}^2$ whose elements are pairs of complex numbers $z = (z_1, z_2)$. Then the inner product of z with another such vector $w = (w_1, w_2)$ is given by

$$\langle z, w \rangle = z_1 \overline{w}_1 + z_2 \overline{w}_2.$$

The real part of $\langle z, w \rangle$ is then the four-dimensional Euclidean dot product. This inner product is *Hermitian* symmetric, which means that the result of interchanging z and w is the complex conjugate:

$$\langle w, z \rangle = \overline{\langle z, w \rangle}.$$

1.1.2 Definition

A **Hilbert space** H is a real or complex inner product space that is also a complete metric space with respect to the distance function induced by the inner product.[2] To say that H is a complex inner product space means that H is a complex vector space on which there is an inner product $\langle x, y \rangle$ associating a complex number to each pair of elements x, y of H that satisfies the following properties:

- The inner product of a pair of elements is equal to the complex conjugate of the inner product of the swapped elements:

$$\langle y, x \rangle = \overline{\langle x, y \rangle}.$$

- The inner product is linear in its first argument.[3] For all complex numbers a and b,

$$\langle ax_1 + bx_2, y \rangle = a\langle x_1, y \rangle + b\langle x_2, y \rangle.$$

- The inner product of an element with itself is positive definite:

$$\langle x, x \rangle \geq 0$$

where the case of equality holds precisely when $x = 0$.

It follows from properties 1 and 2 that a complex inner product is antilinear in its second argument, meaning that

$$\langle x, ay_1 + by_2 \rangle = \bar{a}\langle x, y_1 \rangle + \bar{b}\langle x, y_2 \rangle.$$

A real inner product space is defined in the same way, except that H is a real vector space and the inner product takes real values. Such an inner product will be bilinear: that is, linear in each argument.

The norm is the real-valued function

$$\|x\| = \sqrt{\langle x, x \rangle},$$

and the distance d between two points x, y in H is defined in terms of the norm by

$$d(x, y) = \|x - y\| = \sqrt{\langle x - y, x - y \rangle}.$$

That this function is a distance function means (1) that it is symmetric in x and y, (2) that the distance between x and itself is zero, and otherwise the distance between x and y must be positive, and (3) that the triangle inequality holds, meaning that the length of one leg of a triangle xyz cannot exceed the sum of the lengths of the other two legs:

$$d(x, z) \leq d(x, y) + d(y, z).$$

This last property is ultimately a consequence of the more

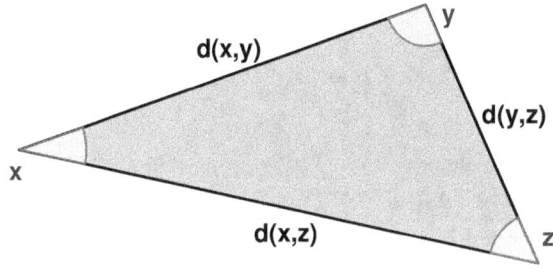

fundamental Cauchy–Schwarz inequality, which asserts

$$|\langle x, y \rangle| \leq \|x\| \, \|y\|$$

with equality if and only if x and y are linearly dependent.

Relative to a distance function defined in this way, any inner product space is a metric space, and sometimes is known as a **pre-Hilbert space**.[4] Any pre-Hilbert space that is additionally also a complete space is a Hilbert space. Completeness is expressed using a form of the Cauchy criterion for sequences in H: a pre-Hilbert space H is complete if every Cauchy sequence converges with respect to this norm to an element in the space. Completeness can be characterized by the following equivalent condition: if a series of vectors $\sum_{k=0}^{\infty} u_k$ converges absolutely in the sense that

$$\sum_{k=0}^{\infty} \|u_k\| < \infty,$$

then the series converges in H, in the sense that the partial sums converge to an element of H.

As a complete normed space, Hilbert spaces are by definition also Banach spaces. As such they are topological vector spaces, in which topological notions like the openness and closedness of subsets are well-defined. Of special importance is the notion of a closed linear subspace of a Hilbert space that, with the inner product induced by restriction, is also complete (being a closed set in a complete metric space) and therefore a Hilbert space in its own right.

1.1.3   Second example: sequence spaces

The sequence space ℓ^2 consists of all infinite sequences $\mathbf{z} = (z1, z_2, ...)$ of complex numbers such that the series

$$\sum_{n=1}^{\infty} |z_n|^2$$

converges. The inner product on ℓ^2 is defined by

$$\langle \mathbf{z}, \mathbf{w} \rangle = \sum_{n=1}^{\infty} z_n \overline{w_n},$$

with the latter series converging as a consequence of the Cauchy–Schwarz inequality.

Completeness of the space holds provided that whenever a series of elements from ℓ^2 converges absolutely (in norm), then it converges to an element of ℓ^2. The proof is basic in mathematical analysis, and permits mathematical series of elements of the space to be manipulated with the same ease as series of complex numbers (or vectors in a finite-dimensional Euclidean space).[5]

1.2   History

Prior to the development of Hilbert spaces, other generalizations of Euclidean spaces were known to mathematicians and physicists. In particular, the idea of an abstract linear space had gained some traction towards the end of the 19th century:[6] this is a space whose elements can be added together and multiplied by scalars (such as real or complex numbers) without necessarily identifying these elements with "geometric" vectors, such as position and momentum vectors in physical systems. Other objects studied by mathematicians at the turn of the 20th century, in particular spaces of sequences (including series) and spaces of functions,[7] can naturally be thought of as linear spaces. Functions, for instance, can be added together or multiplied by constant scalars, and these operations obey the algebraic laws satisfied by addition and scalar multiplication of spatial vectors.

In the first decade of the 20th century, parallel developments led to the introduction of Hilbert spaces. The first of these was the observation, which arose during David Hilbert and Erhard Schmidt's study of integral equations,[8] that two

David Hilbert

square-integrable real-valued functions f and g on an interval [a,b] have an *inner product*

$$\langle f, g \rangle = \int_a^b f(x) g(x) \, dx$$

which has many of the familiar properties of the Euclidean dot product. In particular, the idea of an orthogonal family of functions has meaning. Schmidt exploited the similarity of this inner product with the usual dot product to prove an analog of the spectral decomposition for an operator of the form

$$f(x) \mapsto \int_a^b K(x, y) f(y) \, dy$$

where K is a continuous function symmetric in x and y. The resulting eigenfunction expansion expresses the function K as a series of the form

$$K(x, y) = \sum_n \lambda_n \varphi_n(x) \varphi_n(y)$$

where the functions φ_n are orthogonal in the sense that $\langle \varphi n, \varphi m \rangle = 0$ for all $n \neq m$. The individual terms in this series are sometimes referred to as elementary product solutions. However, there are eigenfunction expansions that fail to converge in a suitable sense to a square-integrable function: the missing ingredient, which ensures convergence, is completeness.[9]

The second development was the Lebesgue integral, an alternative to the Riemann integral introduced by Henri Lebesgue in 1904.[10] The Lebesgue integral made it possible to integrate a much broader class of functions. In 1907, Frigyes Riesz and Ernst Sigismund Fischer independently proved that the space L^2 of square Lebesgue-integrable functions is a complete metric space.[11] As a consequence of the interplay between geometry and completeness, the 19th century results of Joseph Fourier, Friedrich Bessel and Marc-Antoine Parseval on trigonometric series easily carried over to these more general spaces, resulting in a geometrical and analytical apparatus now usually known as the Riesz–Fischer theorem.[12]

Further basic results were proved in the early 20th century. For example, the Riesz representation theorem was independently established by Maurice Fréchet and Frigyes Riesz in 1907.[13] John von Neumann coined the term *abstract Hilbert space* in his work on unbounded Hermitian operators.[14] Although other mathematicians such as Hermann Weyl and Norbert Wiener had already studied particular Hilbert spaces in great detail, often from a physically motivated point of view, von Neumann gave the first complete and axiomatic treatment of them.[15] Von Neumann later used them in his seminal work on the foundations of quantum mechanics,[16] and in his continued work with Eugene Wigner. The name "Hilbert space" was soon adopted by others, for example by Hermann Weyl in his book on quantum mechanics and the theory of groups.[17]

The significance of the concept of a Hilbert space was underlined with the realization that it offers one of the best mathematical formulations of quantum mechanics.[18] In short, the states of a quantum mechanical system are vectors in a certain Hilbert space, the observables are hermitian operators on that space, the symmetries of the system are unitary operators, and measurements are orthogonal projections. The relation between quantum mechanical symmetries and unitary operators provided an impetus for the development of the unitary representation theory of groups, initiated in the 1928 work of Hermann Weyl.[17] On the other hand, in the early 1930s it became clear that classical mechanics can be described in terms of Hilbert space (Koopman–von Neumann classical mechanics) and that certain properties of classical dynamical systems can be analyzed using Hilbert space techniques in the framework of ergodic theory.[19]

The algebra of observables in quantum mechanics is naturally an algebra of operators defined on a Hilbert space, according to Werner Heisenberg's matrix mechanics formulation of quantum theory. Von Neumann began investigating operator algebras in the 1930s, as rings of operators on a Hilbert space. The kind of algebras studied by von Neumann and his contemporaries are now known as von Neumann algebras. In the 1940s, Israel Gelfand, Mark Naimark and Irving Segal gave a definition of a kind of operator algebras called C*-algebras that on the one hand made no reference to an underlying Hilbert space, and on the other extrapolated many of the useful features of the operator algebras that had previously been studied. The spectral theorem for self-adjoint operators in particular that underlies much of the existing Hilbert space theory was generalized to C*-algebras. These techniques are now basic in abstract harmonic analysis and representation theory.

1.3 Examples

1.3.1 Lebesgue spaces

Main article: L^p space

Lebesgue spaces are function spaces associated to measure spaces (X, M, μ), where X is a set, M is a σ-algebra of subsets of X, and μ is a countably additive measure on M. Let $L^2(X, \mu)$ be the space of those complex-valued measurable functions on X for which the Lebesgue integral of the square of the absolute value of the function is finite, i.e., for a function f in $L^2(X, \mu)$,

$$\int_X |f|^2 d\mu < \infty,$$

and where functions are identified if and only if they differ only on a set of measure zero.

The inner product of functions f and g in $L^2(X, \mu)$ is then defined as

$$\langle f, g \rangle = \int_X f(t)\overline{g(t)}\, d\mu(t).$$

For f and g in L^2, this integral exists because of the Cauchy–Schwarz inequality, and defines an inner product on the space. Equipped with this inner product, L^2 is in fact complete.[20] The Lebesgue integral is essential to ensure completeness: on domains of real numbers, for instance, not enough functions are Riemann integrable.[21]

The Lebesgue spaces appear in many natural settings. The spaces $L^2(\mathbf{R})$ and $L^2([0,1])$ of square-integrable functions with respect to the Lebesgue measure on the real line and unit interval, respectively, are natural domains on which to define the Fourier transform and Fourier series. In other situations, the measure may be something other than the ordinary Lebesgue measure on the real line. For instance, if w is any positive measurable function, the space of all measurable functions f on the interval [0, 1] satisfying

$$\int_0^1 |f(t)|^2 w(t)\, dt < \infty$$

is called the weighted L^2 space $L2$ $w([0,1])$, and w is called the weight function. The inner product is defined by

$$\langle f, g \rangle = \int_0^1 f(t)\overline{g(t)}w(t)\, dt.$$

The weighted space $L2$ $w([0,1])$ is identical with the Hilbert space $L^2([0,1],\mu)$ where the measure μ of a Lebesgue-measurable set A is defined by

$$\mu(A) = \int_A w(t)\, dt.$$

Weighted L^2 spaces like this are frequently used to study orthogonal polynomials, because different families of orthogonal polynomials are orthogonal with respect to different weighting functions.

1.3.2   Sobolev spaces

Sobolev spaces, denoted by H^s or $W^{s,2}$, are Hilbert spaces. These are a special kind of function space in which differentiation may be performed, but that (unlike other Banach spaces such as the Hölder spaces) support the structure of an inner product. Because differentiation is permitted, Sobolev spaces are a convenient setting for the theory of partial differential equations.[22] They also form the basis of the theory of direct methods in the calculus of variations.[23]

For s a non-negative integer and $\Omega \subset \mathbf{R}^n$, the Sobolev space $H^s(\Omega)$ contains L^2 functions whose weak derivatives of order up to s are also L^2. The inner product in $H^s(\Omega)$ is

$$\langle f, g \rangle = \int_\Omega f(x)\bar{g}(x)\, dx + \int_\Omega Df(x)\cdot D\bar{g}(x)\, dx +$$

$$\cdots + \int_\Omega D^s f(x)\cdot D^s \bar{g}(x)\, dx$$

where the dot indicates the dot product in the Euclidean space of partial derivatives of each order. Sobolev spaces can also be defined when s is not an integer.

Sobolev spaces are also studied from the point of view of spectral theory, relying more specifically on the Hilbert space structure. If Ω is a suitable domain, then one can define the Sobolev space $H^s(\Omega)$ as the space of Bessel potentials;[24] roughly,

$$H^s(\Omega) = \{(1 - \Delta)^{-s/2}f | f \in L^2(\Omega)\}.$$

Here Δ is the Laplacian and $(1 - \Delta)^{-s/2}$ is understood in terms of the spectral mapping theorem. Apart from providing a workable definition of Sobolev spaces for non-integer s, this definition also has particularly desirable properties under the Fourier transform that make it ideal for the study of pseudodifferential operators. Using these methods on a compact Riemannian manifold, one can obtain for instance the Hodge decomposition, which is the basis of Hodge theory.[25]

1.3.3   Spaces of holomorphic functions

Hardy spaces

The Hardy spaces are function spaces, arising in complex analysis and harmonic analysis, whose elements are certain holomorphic functions in a complex domain.[26] Let U denote the unit disc in the complex plane. Then the Hardy space $H^2(U)$ is defined as the space of holomorphic functions f on U such that the means

$$M_r(f) = \frac{1}{2\pi}\int_0^{2\pi} |f(re^{i\theta})|^2\, d\theta$$

remain bounded for $r < 1$. The norm on this Hardy space is defined by

$$\|f\|_2 = \lim_{r \to 1} \sqrt{M_r(f)}.$$

Hardy spaces in the disc are related to Fourier series. A function f is in $H^2(U)$ if and only if

$$f(z) = \sum_{n=0}^{\infty} a_n z^n$$

where

$$\sum_{n=0}^{\infty} |a_n|^2 < \infty.$$

Thus $H^2(U)$ consists of those functions that are L^2 on the circle, and whose negative frequency Fourier coefficients vanish.

Bergman spaces

The Bergman spaces are another family of Hilbert spaces of holomorphic functions.[27] Let D be a bounded open set in the complex plane (or a higher-dimensional complex space) and let $L^{2,h}(D)$ be the space of holomorphic functions f in D that are also in $L^2(D)$ in the sense that

$$\|f\|^2 = \int_D |f(z)|^2 \, d\mu(z) < \infty,$$

where the integral is taken with respect to the Lebesgue measure in D. Clearly $L^{2,h}(D)$ is a subspace of $L^2(D)$; in fact, it is a closed subspace, and so a Hilbert space in its own right. This is a consequence of the estimate, valid on compact subsets K of D, that

$$\sup_{z \in K} |f(z)| \leq C_K \|f\|_2,$$

which in turn follows from Cauchy's integral formula. Thus convergence of a sequence of holomorphic functions in $L^2(D)$ implies also compact convergence, and so the limit function is also holomorphic. Another consequence of this inequality is that the linear functional that evaluates a function f at a point of D is actually continuous on $L^{2,h}(D)$. The Riesz representation theorem implies that the evaluation functional can be represented as an element of $L^{2,h}(D)$. Thus, for every $z \in D$, there is a function $\eta_z \in L^{2,h}(D)$ such that

$$f(z) = \int_D f(\zeta)\overline{\eta_z(\zeta)} \, d\mu(\zeta)$$

for all $f \in L^{2,h}(D)$. The integrand

$$K(\zeta, z) = \overline{\eta_z(\zeta)}$$

is known as the Bergman kernel of D. This integral kernel satisfies a reproducing property

$$f(z) = \int_D f(\zeta)K(\zeta, z) \, d\mu(\zeta).$$

A Bergman space is an example of a reproducing kernel Hilbert space, which is a Hilbert space of functions along with a kernel $K(\zeta,z)$ that verifies a reproducing property analogous to this one. The Hardy space $H^2(D)$ also admits a reproducing kernel, known as the Szegő kernel.[28] Reproducing kernels are common in other areas of mathematics

as well. For instance, in harmonic analysis the Poisson kernel is a reproducing kernel for the Hilbert space of square-integrable harmonic functions in the unit ball. That the latter is a Hilbert space at all is a consequence of the mean value theorem for harmonic functions.

1.4 Applications

Many of the applications of Hilbert spaces exploit the fact that Hilbert spaces support generalizations of simple geometric concepts like projection and change of basis from their usual finite dimensional setting. In particular, the spectral theory of continuous self-adjoint linear operators on a Hilbert space generalizes the usual spectral decomposition of a matrix, and this often plays a major role in applications of the theory to other areas of mathematics and physics.

1.4.1 Sturm–Liouville theory

Main articles: Sturm–Liouville theory and Spectral theory of ordinary differential equations

In the theory of ordinary differential equations, spectral

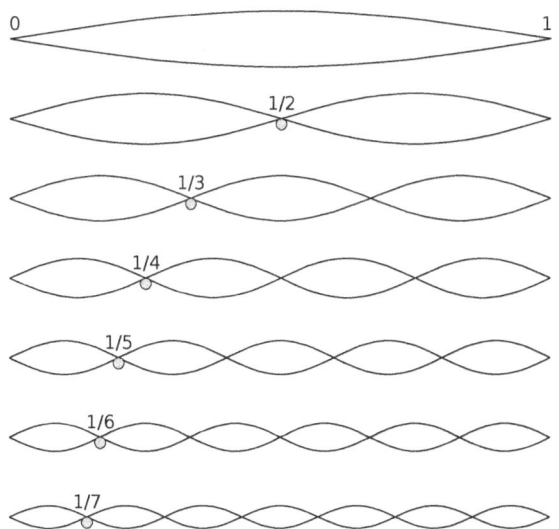

The overtones of a vibrating string. These are eigenfunctions of an associated Sturm–Liouville problem. The eigenvalues 1,1/2,1/3,... form the (musical) harmonic series.

methods on a suitable Hilbert space are used to study the behavior of eigenvalues and eigenfunctions of differential equations. For example, the Sturm–Liouville problem arises in the study of the harmonics of waves in a violin string or a drum, and is a central problem in ordinary differential equations.[29] The problem is a differential equation of the form

$$-\frac{d}{dx}\left[p(x)\frac{dy}{dx}\right] + q(x)y = \lambda w(x)y$$

for an unknown function y on an interval $[a,b]$, satisfying general homogeneous Robin boundary conditions

$$\begin{cases} \alpha y(a) + \alpha' y'(a) = 0 \\ \beta y(b) + \beta' y'(b) = 0. \end{cases}$$

The functions p, q, and w are given in advance, and the problem is to find the function y and constants λ for which the equation has a solution. The problem only has solutions for certain values of λ, called eigenvalues of the system, and this is a consequence of the spectral theorem for compact operators applied to the integral operator defined by the Green's function for the system. Furthermore, another consequence of this general result is that the eigenvalues λ of the system can be arranged in an increasing sequence tending to infinity.[30]

1.4.2 Partial differential equations

Hilbert spaces form a basic tool in the study of partial differential equations.[22] For many classes of partial differential equations, such as linear elliptic equations, it is possible to consider a generalized solution (known as a weak solution) by enlarging the class of functions. Many weak formulations involve the class of Sobolev functions, which is a Hilbert space. A suitable weak formulation reduces to a geometrical problem the analytic problem of finding a solution or, often what is more important, showing that a solution exists and is unique for given boundary data. For linear elliptic equations, one geometrical result that ensures unique solvability for a large class of problems is the Lax–Milgram theorem. This strategy forms the rudiment of the Galerkin method (a finite element method) for numerical solution of partial differential equations.[31]

A typical example is the Poisson equation $-\Delta u = g$ with Dirichlet boundary conditions in a bounded domain Ω in $\mathbf{R}^2$. The weak formulation consists of finding a function u such that, for all continuously differentiable functions v in Ω vanishing on the boundary:

$$\int_\Omega \nabla u \cdot \nabla v = \int_\Omega gv.$$

This can be recast in terms of the Hilbert space $H1$ $0(\Omega)$ consisting of functions u such that u, along with its weak partial derivatives, are square integrable on Ω, and vanish on the boundary. The question then reduces to finding u in this space such that for all v in this space

$$a(u, v) = b(v)$$

where a is a continuous bilinear form, and b is a continuous linear functional, given respectively by

$$a(u, v) = \int_\Omega \nabla u \cdot \nabla v, \quad b(v) = \int_\Omega gv.$$

Since the Poisson equation is elliptic, it follows from Poincaré's inequality that the bilinear form a is coercive. The Lax–Milgram theorem then ensures the existence and uniqueness of solutions of this equation.

Hilbert spaces allow for many elliptic partial differential equations to be formulated in a similar way, and the Lax–Milgram theorem is then a basic tool in their analysis. With suitable modifications, similar techniques can be applied to parabolic partial differential equations and certain hyperbolic partial differential equations.

1.4.3 Ergodic theory

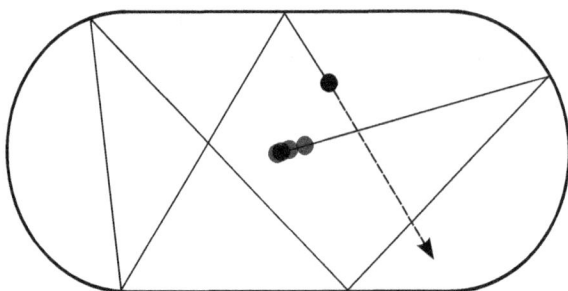

The path of a billiard ball in the Bunimovich stadium is described by an ergodic dynamical system.

The field of ergodic theory is the study of the long-term behavior of chaotic dynamical systems. The protypical case of a field that ergodic theory applies to is thermodynamics, in which—though the microscopic state of a system is extremely complicated (it is impossible to understand the ensemble of individual collisions between particles of matter)—the average behavior over sufficiently long time intervals is tractable. The laws of thermodynamics are assertions about such average behavior. In particular, one formulation of the zeroth law of thermodynamics asserts that over sufficiently long timescales, the only functionally independent measurement that one can make of a thermodynamic system in equilibrium is its total energy, in the form of temperature.

An ergodic dynamical system is one for which, apart from the energy—measured by the Hamiltonian—there are no other functionally independent conserved quantities on the phase space. More explicitly, suppose that the energy E is fixed, and let ΩE be the subset of the phase space consisting of all states of energy E (an energy surface), and let Tt denote the evolution operator on the phase space. The dynamical system is ergodic if there are no continuous non-constant functions on ΩE such that

$$f(T_t w) = f(w)$$

for all w on ΩE and all time t. Liouville's theorem implies that there exists a measure μ on the energy surface that is invariant under the time translation. As a result, time translation is a unitary transformation of the Hilbert space $L^2(\Omega E, \mu)$ consisting of square-integrable functions on the energy surface ΩE with respect to the inner product

$$\langle f, g \rangle_{L^2(\Omega_E, \mu)} = \int_E f \bar{g} \, d\mu.$$

The von Neumann mean ergodic theorem[19] states the following:

- If Ut is a (strongly continuous) one-parameter semi-group of unitary operators on a Hilbert space H, and P is the orthogonal projection onto the space of common fixed points of Ut, $\{x \in H \mid Utx = x \text{ for all } t > 0\}$, then

$$Px = \lim_{T \to \infty} \frac{1}{T} \int_0^T U_t x \, dt.$$

For an ergodic system, the fixed set of the time evolution consists only of the constant functions, so the ergodic theorem implies the following:[32] for any function $f \in L^2(\Omega E, \mu)$,

$$L^2 - \lim_{T \to \infty} \frac{1}{T} \int_0^T f(T_t w) \, dt = \int_{\Omega_E} f(y) \, d\mu(y).$$

That is, the long time average of an observable f is equal to its expectation value over an energy surface.

1.4.4 Fourier analysis

One of the basic goals of Fourier analysis is to decompose a function into a (possibly infinite) linear combination of

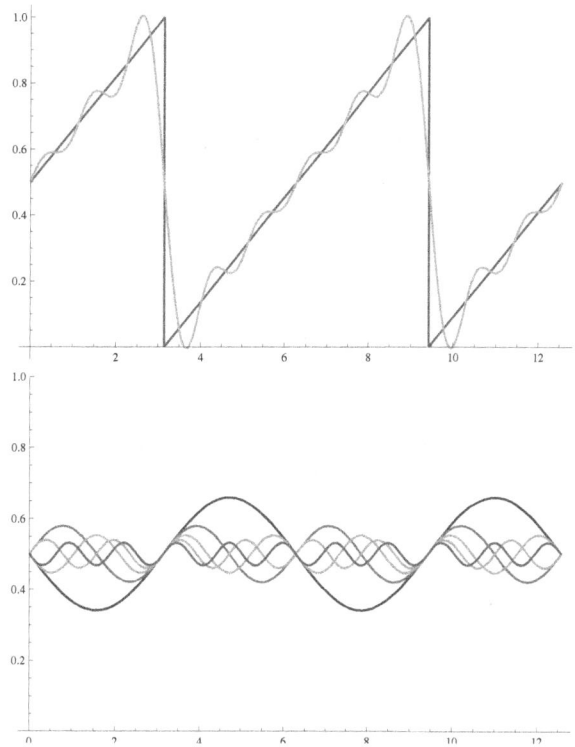

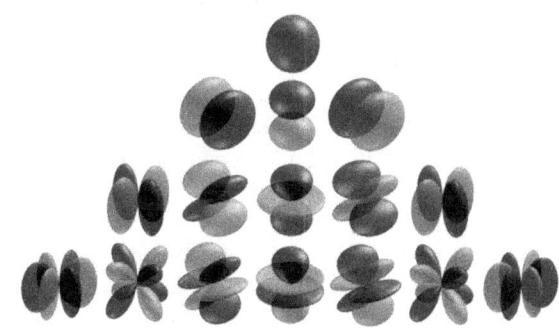

Superposition of sinusoidal wave basis functions (bottom) to form a sawtooth wave (top)

Spherical harmonics, an orthonormal basis for the Hilbert space of square-integrable functions on the sphere, shown graphed along the radial direction

given basis functions: the associated Fourier series. The classical Fourier series associated to a function f defined on the interval $[0, 1]$ is a series of the form

$$\sum_{n=-\infty}^{\infty} a_n e^{2\pi i n \theta}$$

where

$$a_n = \int_0^1 f(\theta)e^{-2\pi in\theta}\,d\theta.$$

The example of adding up the first few terms in a Fourier series for a sawtooth function is shown in the figure. The basis functions are sine waves with wavelengths λ/n (n=integer) shorter than the wavelength λ of the sawtooth itself (except for n=1, the *fundamental* wave). All basis functions have nodes at the nodes of the sawtooth, but all but the fundamental have additional nodes. The oscillation of the summed terms about the sawtooth is called the Gibbs phenomenon.

A significant problem in classical Fourier series asks in what sense the Fourier series converges, if at all, to the function f. Hilbert space methods provide one possible answer to this question.[33] The functions $en(\theta) = e^{2\pi in\theta}$ form an orthogonal basis of the Hilbert space $L^2([0,1])$. Consequently, any square-integrable function can be expressed as a series

$$f(\theta) = \sum_n a_n e_n(\theta), \quad a_n = \langle f, e_n \rangle$$

and, moreover, this series converges in the Hilbert space sense (that is, in the L^2 mean).

The problem can also be studied from the abstract point of view: every Hilbert space has an orthonormal basis, and every element of the Hilbert space can be written in a unique way as a sum of multiples of these basis elements. The coefficients appearing on these basis elements are sometimes known abstractly as the Fourier coefficients of the element of the space.[34] The abstraction is especially useful when it is more natural to use different basis functions for a space such as $L^2([0,1])$. In many circumstances, it is desirable not to decompose a function into trigonometric functions, but rather into orthogonal polynomials or wavelets for instance,[35] and in higher dimensions into spherical harmonics.[36]

For instance, if en are any orthonormal basis functions of $L^2[0,1]$, then a given function in $L^2[0,1]$ can be approximated as a finite linear combination[37]

$$f(x) \approx f_n(x) = a_1 e_1(x) + a_2 e_2(x) + \cdots + a_n e_n(x).$$

The coefficients $\{aj\}$ are selected to make the magnitude of the difference $\|f - fn\|^2$ as small as possible. Geometrically, the best approximation is the orthogonal projection of f onto the subspace consisting of all linear combinations of the $\{ej\}$, and can be calculated by[38]

$$a_j = \int_0^1 \overline{e_j(x)} f(x)\,dx.$$

That this formula minimizes the difference $\|f - fn\|^2$ is a consequence of Bessel's inequality and Parseval's formula.

In various applications to physical problems, a function can be decomposed into physically meaningful eigenfunctions of a differential operator (typically the Laplace operator): this forms the foundation for the spectral study of functions, in reference to the spectrum of the differential operator.[39] A concrete physical application involves the problem of hearing the shape of a drum: given the fundamental modes of vibration that a drumhead is capable of producing, can one infer the shape of the drum itself?[40] The mathematical formulation of this question involves the Dirichlet eigenvalues of the Laplace equation in the plane, that represent the fundamental modes of vibration in direct analogy with the integers that represent the fundamental modes of vibration of the violin string.

Spectral theory also underlies certain aspects of the Fourier transform of a function. Whereas Fourier analysis decomposes a function defined on a compact set into the discrete spectrum of the Laplacian (which corresponds to the vibrations of a violin string or drum), the Fourier transform of a function is the decomposition of a function defined on all of Euclidean space into its components in the continuous spectrum of the Laplacian. The Fourier transformation is also geometrical, in a sense made precise by the Plancherel theorem, that asserts that it is an isometry of one Hilbert space (the "time domain") with another (the "frequency domain"). This isometry property of the Fourier transformation is a recurring theme in abstract harmonic analysis, as evidenced for instance by the Plancherel theorem for spherical functions occurring in noncommutative harmonic analysis.

1.4.5   Quantum mechanics

In the mathematically rigorous formulation of quantum mechanics, developed by John von Neumann,[41] the possible states (more precisely, the pure states) of a quantum mechanical system are represented by unit vectors (called *state vectors*) residing in a complex separable Hilbert space, known as the state space, well defined up to a complex number of norm 1 (the phase factor). In other words, the possible states are points in the projectivization of a Hilbert space, usually called the complex projective space. The exact nature of this Hilbert space is dependent on the system; for example, the position and momentum states for a single non-relativistic spin zero particle is the space of all square-integrable functions, while the states for the spin of a single proton are unit elements of the two-dimensional complex Hilbert space of spinors. Each observable is represented by a self-adjoint linear operator acting on the state space. Each eigenstate of an observable corresponds to an eigenvector of the operator, and the associated eigenvalue corresponds to

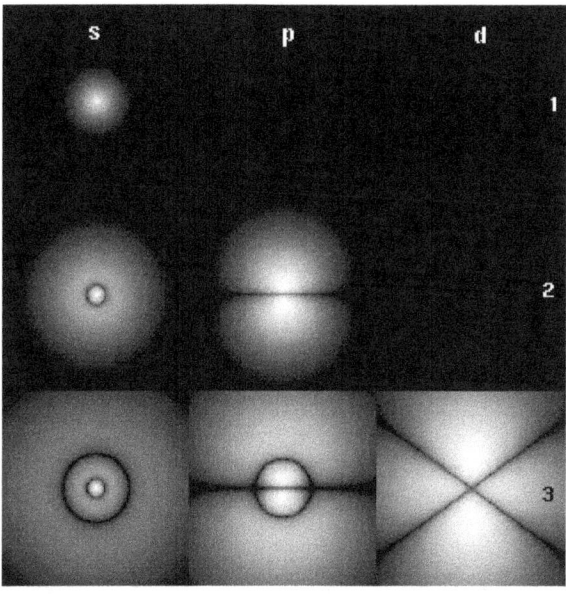

The orbitals of an electron in a hydrogen atom are eigenfunctions of the energy.

the value of the observable in that eigenstate.

The inner product between two state vectors is a complex number known as a probability amplitude. During an ideal measurement of a quantum mechanical system, the probability that a system collapses from a given initial state to a particular eigenstate is given by the square of the absolute value of the probability amplitudes between the initial and final states. The possible results of a measurement are the eigenvalues of the operator—which explains the choice of self-adjoint operators, for all the eigenvalues must be real. The probability distribution of an observable in a given state can be found by computing the spectral decomposition of the corresponding operator.

For a general system, states are typically not pure, but instead are represented as statistical mixtures of pure states, or mixed states, given by density matrices: self-adjoint operators of trace one on a Hilbert space. Moreover, for general quantum mechanical systems, the effects of a single measurement can influence other parts of a system in a manner that is described instead by a positive operator valued measure. Thus the structure both of the states and observables in the general theory is considerably more complicated than the idealization for pure states.

1.5 Properties

1.5.1 Pythagorean identity

Two vectors u and v in a Hilbert space H are orthogonal when $\langle u, v \rangle = 0$. The notation for this is $u \perp v$. More generally, when S is a subset in H, the notation $u \perp S$ means that u is orthogonal to every element from S.

When u and v are orthogonal, one has

$$\|u+v\|^2 = \langle u+v, u+v \rangle = \langle u,u \rangle + 2\,\mathrm{Re}\langle u,v \rangle + \langle v,v \rangle$$

$$= \|u\|^2 + \|v\|^2.$$

By induction on n, this is extended to any family $u_1,...,u_n$ of n orthogonal vectors,

$$\|u_1 + \cdots + u_n\|^2 = \|u_1\|^2 + \cdots + \|u_n\|^2.$$

Whereas the Pythagorean identity as stated is valid in any inner product space, completeness is required for the extension of the Pythagorean identity to series. A series $\Sigma\, u_k$ of *orthogonal* vectors converges in H if and only if the series of squares of norms converges, and

$$\left\| \sum_{k=0}^{\infty} u_k \right\|^2 = \sum_{k=0}^{\infty} \|u_k\|^2.$$

Furthermore, the sum of a series of orthogonal vectors is independent of the order in which it is taken.

1.5.2 Parallelogram identity and polarization

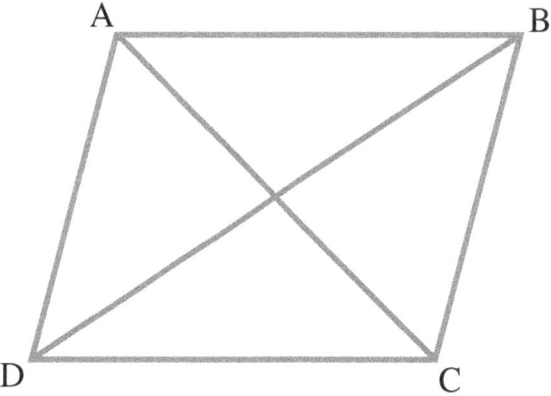

Geometrically, the parallelogram identity asserts that $AC^2 + BD^2 = 2(AB^2 + AD^2)$. In words, the sum of the squares of the diagonals is twice the sum of the squares of any two adjacent sides.

By definition, every Hilbert space is also a Banach space. Furthermore, in every Hilbert space the following parallelogram identity holds:

$$\|u+v\|^2 + \|u-v\|^2 = 2(\|u\|^2 + \|v\|^2).$$

Conversely, every Banach space in which the parallelogram identity holds is a Hilbert space, and the inner product is uniquely determined by the norm by the polarization identity.[42] For real Hilbert spaces, the polarization identity is

$$\langle u, v \rangle = \frac{1}{4} \left(\|u+v\|^2 - \|u-v\|^2 \right).$$

For complex Hilbert spaces, it is

$$\langle u, v \rangle$$
$$= \frac{1}{4} \left(\|u+v\|^2 - \|u-v\|^2 + i\|u+iv\|^2 - i\|u-iv\|^2 \right).$$

The parallelogram law implies that any Hilbert space is a uniformly convex Banach space.[43]

1.5.3 Best approximation

This subsection employs the Hilbert projection theorem. If C is a non-empty closed convex subset of a Hilbert space H and x a point in H, there exists a unique point $y \in C$ that minimizes the distance between x and points in C,[44]

$$y \in C, \quad \|x-y\| = \text{dist}(x, C) = \min\{\|x-z\| : z \in C\}.$$

This is equivalent to saying that there is a point with minimal norm in the translated convex set $D = C - x$. The proof consists in showing that every minimizing sequence $(dn) \subset D$ is Cauchy (using the parallelogram identity) hence converges (using completeness) to a point in D that has minimal norm. More generally, this holds in any uniformly convex Banach space.[45]

When this result is applied to a closed subspace F of H, it can be shown that the point $y \in F$ closest to x is characterized by[46]

$$y \in F, \quad x - y \perp F.$$

This point y is the *orthogonal projection* of x onto F, and the mapping $PF : x \to y$ is linear (see Orthogonal complements and projections). This result is especially significant in applied mathematics, especially numerical analysis, where it forms the basis of least squares methods.[47]

In particular, when F is not equal to H, one can find a non-zero vector v orthogonal to F (select x not in F and $v = x - y$). A very useful criterion is obtained by applying this observation to the closed subspace F generated by a subset S of H.

A subset S of H spans a dense vector subspace if (and only if) the vector 0 is the sole vector $v \in H$ orthogonal to S.

1.5.4 Duality

The dual space H^* is the space of all continuous linear functions from the space H into the base field. It carries a natural norm, defined by

$$\|\varphi\| = \sup_{\|x\|=1, x \in H} |\varphi(x)|.$$

This norm satisfies the parallelogram law, and so the dual space is also an inner product space. The dual space is also complete, and so it is a Hilbert space in its own right.

The Riesz representation theorem affords a convenient description of the dual. To every element u of H, there is a unique element φu of H^*, defined by

$$\varphi_u(x) = \langle x, u \rangle.$$

The mapping $u \mapsto \varphi_u$ is an antilinear mapping from H to H^*. The Riesz representation theorem states that this mapping is an antilinear isomorphism.[48] Thus to every element φ of the dual H^* there exists one and only one u_φ in H such that

$$\langle x, u_\varphi \rangle = \varphi(x)$$

for all $x \in H$. The inner product on the dual space H^* satisfies

$$\langle \varphi, \psi \rangle = \langle u_\psi, u_\varphi \rangle.$$

The reversal of order on the right-hand side restores linearity in φ from the antilinearity of u_φ. In the real case, the antilinear isomorphism from H to its dual is actually an isomorphism, and so real Hilbert spaces are naturally isomorphic to their own duals.

The representing vector u_φ is obtained in the following way. When $\varphi \neq 0$, the kernel $F = \text{Ker}(\varphi)$ is a closed vector subspace of H, not equal to H, hence there exists a non-zero vector v orthogonal to F. The vector u is a suitable scalar multiple λv of v. The requirement that $\varphi(v) = \langle v, u \rangle$ yields

$$u = \langle v, v \rangle^{-1} \overline{\varphi(v)} \, v.$$

This correspondence $\varphi \leftrightarrow u$ is exploited by the bra-ket notation popular in physics. It is common in physics to assume that the inner product, denoted by $\langle x|y \rangle$, is linear on the right,

$\langle x|y\rangle = \langle y, x\rangle.$

The result $\langle x|y\rangle$ can be seen as the action of the linear functional $\langle x|$ (the *bra*) on the vector $|y\rangle$ (the *ket*).

The Riesz representation theorem relies fundamentally not just on the presence of an inner product, but also on the completeness of the space. In fact, the theorem implies that the topological dual of any inner product space can be identified with its completion. An immediate consequence of the Riesz representation theorem is also that a Hilbert space H is reflexive, meaning that the natural map from H into its double dual space is an isomorphism.

1.5.5 Weakly convergent sequences

Main article: Weak convergence (Hilbert space)

In a Hilbert space H, a sequence $\{xn\}$ is weakly convergent to a vector $x \in H$ when

$$\lim_n \langle x_n, v\rangle = \langle x, v\rangle$$

for every $v \in H$.

For example, any orthonormal sequence $\{fn\}$ converges weakly to 0, as a consequence of Bessel's inequality. Every weakly convergent sequence $\{xn\}$ is bounded, by the uniform boundedness principle.

Conversely, every bounded sequence in a Hilbert space admits weakly convergent subsequences (Alaoglu's theorem).[49] This fact may be used to prove minimization results for continuous convex functionals, in the same way that the Bolzano–Weierstrass theorem is used for continuous functions on $\mathbf{R}^d$. Among several variants, one simple statement is as follows:[50]

If $f \colon H \to \mathbf{R}$ is a convex continuous function such that $f(x)$ tends to $+\infty$ when $\|x\|$ tends to ∞, then f admits a minimum at some point $x_0 \in H$.

This fact (and its various generalizations) are fundamental for direct methods in the calculus of variations. Minimization results for convex functionals are also a direct consequence of the slightly more abstract fact that closed bounded convex subsets in a Hilbert space H are weakly compact, since H is reflexive. The existence of weakly convergent subsequences is a special case of the Eberlein–Šmulian theorem.

1.5.6 Banach space properties

Any general property of Banach spaces continues to hold for Hilbert spaces. The open mapping theorem states that a continuous surjective linear transformation from one Banach space to another is an open mapping meaning that it sends open sets to open sets. A corollary is the bounded inverse theorem, that a continuous and bijective linear function from one Banach space to another is an isomorphism (that is, a continuous linear map whose inverse is also continuous). This theorem is considerably simpler to prove in the case of Hilbert spaces than in general Banach spaces.[51] The open mapping theorem is equivalent to the closed graph theorem, which asserts that a function from one Banach space to another is continuous if and only if its graph is a closed set.[52] In the case of Hilbert spaces, this is basic in the study of unbounded operators (see closed operator).

The (geometrical) Hahn–Banach theorem asserts that a closed convex set can be separated from any point outside it by means of a hyperplane of the Hilbert space. This is an immediate consequence of the best approximation property: if y is the element of a closed convex set F closest to x, then the separating hyperplane is the plane perpendicular to the segment xy passing through its midpoint.[53]

1.6 Operators on Hilbert spaces

1.6.1 Bounded operators

The continuous linear operators $A \colon H_1 \to H_2$ from a Hilbert space H_1 to a second Hilbert space H_2 are *bounded* in the sense that they map bounded sets to bounded sets. Conversely, if an operator is bounded, then it is continuous. The space of such bounded linear operators has a norm, the operator norm given by

$$\|A\| = \sup \{ \|Ax\| : \|x\| \le 1 \}.$$

The sum and the composite of two bounded linear operators is again bounded and linear. For y in H_2, the map that sends $x \in H_1$ to $\langle Ax, y\rangle$ is linear and continuous, and according to the Riesz representation theorem can therefore be represented in the form

$$\langle x, A^*y\rangle = \langle Ax, y\rangle$$

for some vector $A^* y$ in H_1. This defines another bounded linear operator $A^* \colon H_2 \to H_1$, the adjoint of A. One can see that $A^{**} = A$.

The set B(H) of all bounded linear operators on H, together with the addition and composition operations, the norm and the adjoint operation, is a C*-algebra, which is a type of operator algebra.

An element A of B(H) is called *self-adjoint* or *Hermitian* if $A* = A$. If A is Hermitian and $\langle Ax, x \rangle \geq 0$ for every x, then A is called *non-negative*, written $A \geq 0$; if equality holds only when $x = 0$, then A is called *positive*. The set of self adjoint operators admits a partial order, in which $A \geq B$ if $A - B \geq 0$. If A has the form $B^* B$ for some B, then A is non-negative; if B is invertible, then A is positive. A converse is also true in the sense that, for a non-negative operator A, there exists a unique non-negative square root B such that

$A = B^2 = B^*B.$

In a sense made precise by the spectral theorem, self-adjoint operators can usefully be thought of as operators that are "real". An element A of B(H) is called *normal* if $A* A = A A*$. Normal operators decompose into the sum of a self-adjoint operators and an imaginary multiple of a self adjoint operator

$A = \dfrac{A + A^*}{2} + i\dfrac{A - A^*}{2i}$

that commute with each other. Normal operators can also usefully be thought of in terms of their real and imaginary parts.

An element U of B(H) is called unitary if U is invertible and its inverse is given by $U*$. This can also be expressed by requiring that U be onto and $\langle Ux, Uy \rangle = \langle x, y \rangle$ for all x and y in H. The unitary operators form a group under composition, which is the isometry group of H.

An element of B(H) is compact if it sends bounded sets to relatively compact sets. Equivalently, a bounded operator T is compact if, for any bounded sequence $\{xk\}$, the sequence $\{Txk\}$ has a convergent subsequence. Many integral operators are compact, and in fact define a special class of operators known as Hilbert–Schmidt operators that are especially important in the study of integral equations. Fredholm operators differ from a compact operator by a multiple of the identity, and are equivalently characterized as operators with a finite dimensional kernel and cokernel. The index of a Fredholm operator T is defined by

index $T = \dim \ker T - \dim \operatorname{coker} T.$

The index is homotopy invariant, and plays a deep role in differential geometry via the Atiyah–Singer index theorem.

1.6.2 Unbounded operators

Unbounded operators are also tractable in Hilbert spaces,

and have important applications to quantum mechanics.[54] An unbounded operator T on a Hilbert space H is defined as a linear operator whose domain $D(T)$ is a linear subspace of H. Often the domain $D(T)$ is a dense subspace of H, in which case T is known as a densely defined operator.

The adjoint of a densely defined unbounded operator is defined in essentially the same manner as for bounded operators. Self-adjoint unbounded operators play the role of the *observables* in the mathematical formulation of quantum mechanics. Examples of self-adjoint unbounded operators on the Hilbert space $L^2(\mathbf{R})$ are:[55]

- A suitable extension of the differential operator

$(Af)(x) = -i\dfrac{d}{dx}f(x),$

where i is the imaginary unit and f is a differentiable function of compact support.

- The multiplication-by-x operator:

$(Bf)(x) = xf(x).$

These correspond to the momentum and position observables, respectively. Note that neither A nor B is defined on all of H, since in the case of A the derivative need not exist, and in the case of B the product function need not be square integrable. In both cases, the set of possible arguments form dense subspaces of $L^2(\mathbf{R})$.

1.7 Constructions

1.7.1 Direct sums

Two Hilbert spaces H_1 and H_2 can be combined into another Hilbert space, called the (orthogonal) direct sum,[56] and denoted

$H_1 \oplus H_2,$

consisting of the set of all ordered pairs (x_1, x_2) where $xi \in Hi$, $i = 1,2$, and inner product defined by

$\langle (x_1, x_2), (y_1, y_2) \rangle_{H_1 \oplus H_2} = \langle x_1, y_1 \rangle_{H_1} + \langle x_2, y_2 \rangle_{H_2}.$

More generally, if H_i is a family of Hilbert spaces indexed by $i \in I$, then the direct sum of the H_i, denoted

$$\bigoplus_{i \in I} H_i$$

consists of the set of all indexed families

$$x = (x_i \in H_i | i \in I) \in \prod_{i \in I} H_i$$

in the Cartesian product of the H_i such that

$$\sum_{i \in I} \|x_i\|^2 < \infty.$$

The inner product is defined by

$$\langle x, y \rangle = \sum_{i \in I} \langle x_i, y_i \rangle_{H_i}.$$

Each of the H_i is included as a closed subspace in the direct sum of all of the H_i. Moreover, the H_i are pairwise orthogonal. Conversely, if there is a system of closed subspaces, $V_i, i \in I$, in a Hilbert space H, that are pairwise orthogonal and whose union is dense in H, then H is canonically isomorphic to the direct sum of V_i. In this case, H is called the internal direct sum of the V_i. A direct sum (internal or external) is also equipped with a family of orthogonal projections E_i onto the ith direct summand H_i. These projections are bounded, self-adjoint, idempotent operators that satisfy the orthogonality condition

$$E_i E_j = 0, \quad i \neq j.$$

The spectral theorem for compact self-adjoint operators on a Hilbert space H states that H splits into an orthogonal direct sum of the eigenspaces of an operator, and also gives an explicit decomposition of the operator as a sum of projections onto the eigenspaces. The direct sum of Hilbert spaces also appears in quantum mechanics as the Fock space of a system containing a variable number of particles, where each Hilbert space in the direct sum corresponds to an additional degree of freedom for the quantum mechanical system. In representation theory, the Peter–Weyl theorem guarantees that any unitary representation of a compact group on a Hilbert space splits as the direct sum of finite-dimensional representations.

1.7.2 Tensor products

Main article: Tensor product of Hilbert spaces

If H_1 and H_2, then one defines an inner product on the (ordinary) tensor product as follows. On simple tensors, let

$$\langle x_1 \otimes x_2, y_1 \otimes y_2 \rangle = \langle x_1, y_1 \rangle \langle x_2, y_2 \rangle.$$

This formula then extends by sesquilinearity to an inner product on $H_1 \otimes H_2$. The Hilbertian tensor product of H_1 and H_2, sometimes denoted by $H_1 \widehat{\otimes} H_2$, is the Hilbert space obtained by completing $H_1 \otimes H_2$ for the metric associated to this inner product.[57]

An example is provided by the Hilbert space $L^2([0, 1])$. The Hilbertian tensor product of two copies of $L^2([0, 1])$ is isometrically and linearly isomorphic to the space $L^2([0, 1]^2)$ of square-integrable functions on the square $[0, 1]^2$. This isomorphism sends a simple tensor $f_1 \otimes f_2$ to the function

$$(s, t) \mapsto f_1(s) f_2(t)$$

on the square.

This example is typical in the following sense.[58] Associated to every simple tensor product $x_1 \otimes x_2$ is the rank one operator from H_*
1 to H_2 that maps a given $x^* \in H_1^*$ as

$$x^* \mapsto x^*(x_1) x_2.$$

This mapping defined on simple tensors extends to a linear identification between $H_1 \otimes H_2$ and the space of finite rank operators from H_1^* to H_2. This extends to a linear isometry of the Hilbertian tensor product $H_1 \widehat{\otimes} H_2$ with the Hilbert space $HS(H_1^*, H_2)$ of Hilbert–Schmidt operators from H_1^* to H_2.

1.8 Orthonormal bases

The notion of an orthonormal basis from linear algebra generalizes over to the case of Hilbert spaces.[59] In a Hilbert space H, an orthonormal basis is a family $\{e_k\}k \in B$ of elements of H satisfying the conditions:

1. *Orthogonality*: Every two different elements of B are orthogonal: $\langle e_k, e_j \rangle = 0$ for all k, j in B with $k \neq j$.

2. *Normalization*: Every element of the family has norm 1: $\|e_k\| = 1$ for all k in B.

3. *Completeness*: The linear span of the family e_k, $k \in B$, is dense in H.

A system of vectors satisfying the first two conditions basis is called an orthonormal system or an orthonormal set (or an orthonormal sequence if B is countable). Such a system is always linearly independent. Completeness of an orthonormal system of vectors of a Hilbert space can be equivalently restated as:

if $\langle v, e_k \rangle = 0$ for all $k \in B$ and some $v \in H$ then $v = \mathbf{0}$.

This is related to the fact that the only vector orthogonal to a dense linear subspace is the zero vector, for if S is any orthonormal set and v is orthogonal to S, then v is orthogonal to the closure of the linear span of S, which is the whole space.

Examples of orthonormal bases include:

- the set $\{(1,0,0), (0,1,0), (0,0,1)\}$ forms an orthonormal basis of $\mathbf{R}^3$ with the dot product;

- the sequence $\{f_n : n \in \mathbf{Z}\}$ with $f_n(x) = \exp(2\pi inx)$ forms an orthonormal basis of the complex space $L^2([0,1])$;

In the infinite-dimensional case, an orthonormal basis will not be a basis in the sense of linear algebra; to distinguish the two, the latter basis is also called a Hamel basis. That the span of the basis vectors is dense implies that every vector in the space can be written as the sum of an infinite series, and the orthogonality implies that this decomposition is unique.

1.8.1   Sequence spaces

The space ℓ^2 of square-summable sequences of complex numbers is the set of infinite sequences

$$(c_1, c_2, c_3, \dots)$$

of complex numbers such that

$$|c_1|^2 + |c_2|^2 + |c_3|^2 + \cdots < \infty.$$

This space has an orthonormal basis:

$$e_1 = (1, 0, 0, \dots)$$
$$e_2 = (0, 1, 0, \dots)$$
$$\vdots$$

More generally, if B is any set, then one can form a Hilbert space of sequences with index set B, defined by

$$\ell^2(B) = \left\{ x : B \xrightarrow{x} \mathbb{C} \mid \sum_{b \in B} |x(b)|^2 < \infty \right\}.$$

The summation over B is here defined by

$$\sum_{b \in B} |x(b)|^2 = \sup \sum_{n=1}^{N} |x(b_n)|^2$$

the supremum being taken over all finite subsets of B. It follows that, for this sum to be finite, every element of $\ell^2(B)$ has only countably many nonzero terms. This space becomes a Hilbert space with the inner product

$$\langle x, y \rangle = \sum_{b \in B} x(b)\overline{y(b)}$$

for all x and y in $\ell^2(B)$. Here the sum also has only countably many nonzero terms, and is unconditionally convergent by the Cauchy–Schwarz inequality.

An orthonormal basis of $\ell^2(B)$ is indexed by the set B, given by

$$e_b(b') = \begin{cases} 1 & \text{if } b = b' \\ 0 & \text{otherwise.} \end{cases}$$

1.8.2   Bessel's inequality and Parseval's formula

Let $f_1, \dots, f_n$ be a finite orthonormal system in H. For an arbitrary vector x in H, let

$$y = \sum_{j=1}^{n} \langle x, f_j \rangle f_j.$$

Then $\langle x, f_k \rangle = \langle y, f_k \rangle$ for every $k = 1, \dots, n$. It follows that $x - y$ is orthogonal to each f_k, hence $x - y$ is orthogonal to y. Using the Pythagorean identity twice, it follows that

$$\|x\|^2 = \|x - y\|^2 + \|y\|^2 \geq \|y\|^2 = \sum_{j=1}^{n} |\langle x, f_j \rangle|^2.$$

Let $\{f_i\}$, $i \in I$, be an arbitrary orthonormal system in H. Applying the preceding inequality to every finite subset J of I gives the *Bessel inequality*[60]

$$\sum_{i \in I} |\langle x, f_i \rangle|^2 \le \|x\|^2, \quad x \in H$$

(according to the definition of the sum of an arbitrary family of non-negative real numbers).

Geometrically, Bessel's inequality implies that the orthogonal projection of x onto the linear subspace spanned by the fi has norm that does not exceed that of x. In two dimensions, this is the assertion that the length of the leg of a right triangle may not exceed the length of the hypotenuse.

Bessel's inequality is a stepping stone to the more powerful Parseval identity, which governs the case when Bessel's inequality is actually an equality. If $\{ek\}k \in B$ is an orthonormal basis of H, then every element x of H may be written as

$$x = \sum_{k \in B} \langle x, e_k \rangle e_k.$$

Even if B is uncountable, Bessel's inequality guarantees that the expression is well-defined and consists only of countably many nonzero terms. This sum is called the *Fourier expansion* of x, and the individual coefficients $\langle x, ek \rangle$ are the *Fourier coefficients* of x. Parseval's formula is then

$$\|x\|^2 = \sum_{k \in B} |\langle x, e_k \rangle|^2.$$

Conversely, if $\{ek\}$ is an orthonormal set such that Parseval's identity holds for every x, then $\{ek\}$ is an orthonormal basis.

1.8.3 Hilbert dimension

As a consequence of Zorn's lemma, *every* Hilbert space admits an orthonormal basis; furthermore, any two orthonormal bases of the same space have the same cardinality, called the Hilbert dimension of the space.[61] For instance, since $\ell^2(B)$ has an orthonormal basis indexed by B, its Hilbert dimension is the cardinality of B (which may be a finite integer, or a countable or uncountable cardinal number).

As a consequence of Parseval's identity, if $\{ek\}k \in B$ is an orthonormal basis of H, then the map $\Phi : H \to \ell^2(B)$ defined by $\Phi(x) = (\langle x, ek \rangle)k \in B$ is an isometric isomorphism of Hilbert spaces: it is a bijective linear mapping such that

$$\langle \Phi(x), \Phi(y) \rangle_{\ell^2(B)} = \langle x, y \rangle_H$$

for all x and y in H. The cardinal number of B is the Hilbert dimension of H. Thus every Hilbert space is isometrically isomorphic to a sequence space $\ell^2(B)$ for some set B.

1.8.4 Separable spaces

A Hilbert space is separable if and only if it admits a countable orthonormal basis. All infinite-dimensional separable Hilbert spaces are therefore isometrically isomorphic to ℓ^2.

In the past, Hilbert spaces were often required to be separable as part of the definition.[62] Most spaces used in physics are separable, and since these are all isomorphic to each other, one often refers to any infinite-dimensional separable Hilbert space as "*the* Hilbert space" or just "Hilbert space".[63] Even in quantum field theory, most of the Hilbert spaces are in fact separable, as stipulated by the Wightman axioms. However, it is sometimes argued that non-separable Hilbert spaces are also important in quantum field theory, roughly because the systems in the theory possess an infinite number of degrees of freedom and any infinite Hilbert tensor product (of spaces of dimension greater than one) is non-separable.[64] For instance, a bosonic field can be naturally thought of as an element of a tensor product whose factors represent harmonic oscillators at each point of space. From this perspective, the natural state space of a boson might seem to be a non-separable space.[64] However, it is only a small separable subspace of the full tensor product that can contain physically meaningful fields (on which the observables can be defined). Another non-separable Hilbert space models the state of an infinite collection of particles in an unbounded region of space. An orthonormal basis of the space is indexed by the density of the particles, a continuous parameter, and since the set of possible densities is uncountable, the basis is not countable.[64]

1.9 Orthogonal complements and projections

If S is a subset of a Hilbert space H, the set of vectors orthogonal to S is defined by

$$S^\perp = \{x \in H : \langle x, s \rangle = 0 \ \forall s \in S\}.$$

$S^\perp$ is a closed subspace of H (can be proved easily using the linearity and continuity of the inner product) and so forms itself a Hilbert space. If V is a closed subspace of H, then $V^\perp$ is called the *orthogonal complement* of V. In fact, every x in H can then be written uniquely as $x = v + w$, with v in

V and w in $V^\perp$. Therefore, H is the internal Hilbert direct sum of V and $V^\perp$.

The linear operator $PV : H \to H$ that maps x to v is called the *orthogonal projection* onto V. There is a natural one-to-one correspondence between the set of all closed subspaces of H and the set of all bounded self-adjoint operators P such that $P^2 = P$. Specifically,

> **Theorem**. The orthogonal projection PV is a self-adjoint linear operator on H of norm ≤ 1 with the property $P^2 V = PV$. Moreover, any self-adjoint linear operator E such that $E^2 = E$ is of the form PV, where V is the range of E. For every x in H, $PV(x)$ is the unique element v of V, which minimizes the distance $\|x - v\|$.

This provides the geometrical interpretation of $PV(x)$: it is the best approximation to x by elements of V.[65]

Projections PU and PV are called mutually orthogonal if $PUPV = 0$. This is equivalent to U and V being orthogonal as subspaces of H. The sum of the two projections PU and PV is a projection only if U and V are orthogonal to each other, and in that case $PU + PV = PU_+V$. The composite $PUPV$ is generally not a projection; in fact, the composite is a projection if and only if the two projections commute, and in that case $PUPV = PU_{\cap}V$.

By restricting the codomain to the Hilbert space V, the orthogonal projection PV gives rise to a projection mapping $\pi: H \to V$; it is the adjoint of the inclusion mapping

$$i : V \to H,$$

meaning that

$$\langle ix, y \rangle_H = \langle x, \pi y \rangle_V$$

for all $x \in V$ and $y \in H$.

The operator norm of the orthogonal projection PV onto a non-zero closed subspace V is equal to one:

$$\|P_V\| = \sup_{x \in H, x \neq 0} \frac{\|P_V x\|}{\|x\|} = 1.$$

Every closed subspace V of a Hilbert space is therefore the image of an operator P of norm one such that $P^2 = P$. The property of possessing appropriate projection operators characterizes Hilbert spaces:[66]

- A Banach space of dimension higher than 2 is (isometrically) a Hilbert space if and only if, for every

closed subspace V, there is an operator PV of norm one whose image is V such that $P_V^2 = P_V$.

While this result characterizes the metric structure of a Hilbert space, the structure of a Hilbert space as a topological vector space can itself be characterized in terms of the presence of complementary subspaces:[67]

- A Banach space X is topologically and linearly isomorphic to a Hilbert space if and only if, to every closed subspace V, there is a closed subspace W such that X is equal to the internal direct sum $V \oplus W$.

The orthogonal complement satisfies some more elementary results. It is a monotone function in the sense that if $U \subset V$, then $V^\perp \subseteq U^\perp$ with equality holding if and only if V is contained in the closure of U. This result is a special case of the Hahn–Banach theorem. The closure of a subspace can be completely characterized in terms of the orthogonal complement: If V is a subspace of H, then the closure of V is equal to $V^{\perp\perp}$. The orthogonal complement is thus a Galois connection on the partial order of subspaces of a Hilbert space. In general, the orthogonal complement of a sum of subspaces is the intersection of the orthogonal complements:[68] $\left(\sum_i V_i \right)^\perp = \bigcap_i V_i^\perp$. If the V_i are in addition closed, then $\overline{\sum_i V_i^\perp} = \left(\bigcap_i V_i \right)^\perp$.

1.10   Spectral theory

There is a well-developed spectral theory for self-adjoint operators in a Hilbert space, that is roughly analogous to the study of symmetric matrices over the reals or self-adjoint matrices over the complex numbers.[69] In the same sense, one can obtain a "diagonalization" of a self-adjoint operator as a suitable sum (actually an integral) of orthogonal projection operators.

The spectrum of an operator T, denoted $\sigma(T)$ is the set of complex numbers λ such that $T - \lambda$ lacks a continuous inverse. If T is bounded, then the spectrum is always a compact set in the complex plane, and lies inside the disc $|z| \leq \|T\|$. If T is self-adjoint, then the spectrum is real. In fact, it is contained in the interval $[m, M]$ where

$$m = \inf_{\|x\|=1} \langle Tx, x \rangle, \quad M = \sup_{\|x\|=1} \langle Tx, x \rangle.$$

Moreover, m and M are both actually contained within the spectrum.

The eigenspaces of an operator T are given by

$$H_\lambda = \ker(T - \lambda).$$

Unlike with finite matrices, not every element of the spectrum of T must be an eigenvalue: the linear operator $T - \lambda$ may only lack an inverse because it is not surjective. Elements of the spectrum of an operator in the general sense are known as *spectral values*. Since spectral values need not be eigenvalues, the spectral decomposition is often more subtle than in finite dimensions.

However, the spectral theorem of a self-adjoint operator T takes a particularly simple form if, in addition, T is assumed to be a compact operator. The spectral theorem for compact self-adjoint operators states:[70]

- A compact self-adjoint operator T has only countably (or finitely) many spectral values. The spectrum of T has no limit point in the complex plane except possibly zero. The eigenspaces of T decompose H into an orthogonal direct sum:

$$H = \bigoplus_{\lambda \in \sigma(T)} H_\lambda.$$

  Moreover, if $E\lambda$ denotes the orthogonal projection onto the eigenspace $H\lambda$, then

$$T = \sum_{\lambda \in \sigma(T)} \lambda E_\lambda,$$

  where the sum converges with respect to the norm on B(H).

This theorem plays a fundamental role in the theory of integral equations, as many integral operators are compact, in particular those that arise from Hilbert–Schmidt operators.

The general spectral theorem for self-adjoint operators involves a kind of operator-valued Riemann–Stieltjes integral, rather than an infinite summation.[71] The *spectral family* associated to T associates to each real number λ an operator $E\lambda$, which is the projection onto the nullspace of the operator $(T - \lambda)^+$, where the positive part of a self-adjoint operator is defined by

$$A^+ = \frac{1}{2} \left(\sqrt{A^2} + A \right).$$

The operators $E\lambda$ are monotone increasing relative to the partial order defined on self-adjoint operators; the eigenvalues correspond precisely to the jump discontinuities. One has the spectral theorem, which asserts

$$T = \int_\mathbb{R} \lambda \, dE_\lambda.$$

The integral is understood as a Riemann–Stieltjes integral, convergent with respect to the norm on B(H). In particular, one has the ordinary scalar-valued integral representation

$$\langle Tx, y \rangle = \int_\mathbb{R} \lambda \, d\langle E_\lambda x, y \rangle.$$

A somewhat similar spectral decomposition holds for normal operators, although because the spectrum may now contain non-real complex numbers, the operator-valued Stieltjes measure $dE\lambda$ must instead be replaced by a resolution of the identity.

A major application of spectral methods is the spectral mapping theorem, which allows one to apply to a self-adjoint operator T any continuous complex function f defined on the spectrum of T by forming the integral

$$f(T) = \int_{\sigma(T)} f(\lambda) \, dE_\lambda.$$

The resulting continuous functional calculus has applications in particular to pseudodifferential operators.[72]

The spectral theory of *unbounded* self-adjoint operators is only marginally more difficult than for bounded operators. The spectrum of an unbounded operator is defined in precisely the same way as for bounded operators: λ is a spectral value if the resolvent operator

$$R_\lambda = (T - \lambda)^{-1}$$

fails to be a well-defined continuous operator. The self-adjointness of T still guarantees that the spectrum is real. Thus the essential idea of working with unbounded operators is to look instead at the resolvent $R\lambda$ where λ is non-real. This is a *bounded* normal operator, which admits a spectral representation that can then be transferred to a spectral representation of T itself. A similar strategy is used, for instance, to study the spectrum of the Laplace operator: rather than address the operator directly, one instead looks as an associated resolvent such as a Riesz potential or Bessel potential.

A precise version of the spectral theorem in this case is:[73]

Given a densely defined self-adjoint operator T on a Hilbert space H, there corresponds a unique resolution of the identity E on the Borel sets of **R**, such that

$$\langle Tx, y \rangle = \int_\mathbb{R} \lambda \, dE_{x,y}(\lambda)$$

for all $x \in D(T)$ and $y \in H$. The spectral measure E is concentrated on the spectrum of T.

There is also a version of the spectral theorem that applies to unbounded normal operators.

1.11   See also

- Hadamard space

- Hilbert algebra

- Hilbert C*-module

- Hilbert manifold

- Operator theory

- Operator topologies

- Rigged Hilbert space

1.12   Notes

[1] Marsden 1974, §2.8

[2] The mathematical material in this section can be found in any good textbook on functional analysis, such as Dieudonné (1960), Hewitt & Stromberg (1965), Reed & Simon (1980) or Rudin (1980).

[3] In some conventions, inner products are linear in their second arguments instead.

[4] Dieudonné 1960, §6.2

[5] Dieudonné 1960

[6] Largely from the work of Hermann Grassmann, at the urging of August Ferdinand Möbius (Boyer & Merzbach 1991, pp. 584–586). The first modern axiomatic account of abstract vector spaces ultimately appeared in Giuseppe Peano's 1888 account (Grattan-Guinness 2000, §5.2.2; O'Connor & Robertson 1996).

[7] A detailed account of the history of Hilbert spaces can be found in Bourbaki 1987.

[8] Schmidt 1908

[9] Titchmarsh 1946, §IX.1

[10] Lebesgue 1904. Further details on the history of integration theory can be found in Bourbaki (1987) and Saks (2005).

[11] Bourbaki 1987.

[12] Dunford & Schwartz 1958, §IV.16

[13] In Dunford & Schwartz (1958, §IV.16), the result that every linear functional on $L^2[0,1]$ is represented by integration is jointly attributed to Fréchet (1907) and Riesz (1907). The general result, that the dual of a Hilbert space is identified with the Hilbert space itself, can be found in Riesz (1934).

[14] von Neumann 1929.

[15] Kline 1972, p. 1092

[16] Hilbert, Nordheim & von Neumann 1927.

[17] Weyl 1931.

[18] Prugovečki 1981, pp. 1–10.

[19] von Neumann 1932

[20] Halmos 1957, Section 42.

[21] Hewitt & Stromberg 1965.

[22] Bers, John & Schechter 1981.

[23] Giusti 2003.

[24] Stein 1970

[25] Details can be found in Warner (1983).

[26] A general reference on Hardy spaces is the book Duren (1970).

[27] Krantz 2002, §1.4

[28] Krantz 2002, §1.5

[29] Young 1988, Chapter 9.

[30] The eigenvalues of the Fredholm kernel are $1/\lambda$, which tend to zero.

[31] More detail on finite element methods from this point of view can be found in Brenner & Scott (2005).

[32] Reed & Simon 1980

[33] A treatment of Fourier series from this point of view is available, for instance, in Rudin (1987) or Folland (2009).

[34] Halmos 1957, §5

[35] Bachman, Narici & Beckenstein 2000

[36] Stein & Weiss 1971, §IV.2.

[37] Lancos 1988, pp. 212–213

[38] Lanczos 1988, Equation 4-3.10

[39] The classic reference for spectral methods is Courant & Hilbert 1953. A more up-to-date account is Reed & Simon 1975.

[40] Kac 1966

[41] von Neumann 1955

[42] Young 1988, p. 23.

[43] Clarkson 1936.

[44] Rudin 1987, Theorem 4.10

[45] Dunford & Schwartz 1958, II.4.29

[46] Rudin 1987, Theorem 4.11

[47] Blanchet, Gérard; Charbit, Maurice (2014). *Digital Signal and Image Processing Using MATLAB*. Digital Signal and Image Processing **1** (Second ed.). New Jersey: Wiley. pp. 349–360. ISBN 978-1848216402.

[48] Weidmann 1980, Theorem 4.8

[49] Weidmann 1980, §4.5

[50] Buttazzo, Giaquinta & Hildebrandt 1998, Theorem 5.17

[51] Halmos 1982, Problem 52, 58

[52] Rudin 1973

[53] Trèves 1967, Chapter 18

[54] See Prugovečki (1981), Reed & Simon (1980, Chapter VIII) and Folland (1989).

[55] Prugovečki 1981, III, §1.4

[56] Dunford & Schwartz 1958, IV.4.17-18

[57] Weidmann 1980, §3.4

[58] Kadison & Ringrose 1983, Theorem 2.6.4

[59] Dunford & Schwartz 1958, §IV.4.

[60] For the case of finite index sets, see, for instance, Halmos 1957, §5. For infinite index sets, see Weidmann 1980, Theorem 3.6.

[61] Levitan 2001. Many authors, such as Dunford & Schwartz (1958, §IV.4), refer to this just as the dimension. Unless the Hilbert space is finite dimensional, this is not the same thing as its dimension as a linear space (the cardinality of a Hamel basis).

[62] Prugovečki 1981, I, §4.2

[63] von Neumann (1955) defines a Hilbert space via a countable Hilbert basis, which amounts to an isometric isomorphism with ℓ^2. The convention still persists in most rigorous treatments of quantum mechanics; see for instance Sobrino 1996, Appendix B.

[64] Streater & Wightman 1964, pp. 86–87

[65] Young 1988, Theorem 15.3

[66] Kakutani 1939

[67] Lindenstrauss & Tzafriri 1971

[68] Halmos 1957, §12

[69] A general account of spectral theory in Hilbert spaces can be found in Riesz & Sz Nagy (1990). A more sophisticated account in the language of C*-algebras is in Rudin (1973) or Kadison & Ringrose (1997)

[70] See, for instance, Riesz & Sz Nagy (1990, Chapter VI) or Weidmann 1980, Chapter 7. This result was already known to Schmidt (1907) in the case of operators arising from integral kernels.

[71] Riesz & Sz Nagy 1990, §§107–108

[72] Shubin 1987

[73] Rudin 1973, Theorem 13.30.

1.13 References

- Bachman, George; Narici, Lawrence; Beckenstein, Edward (2000), *Fourier and wavelet analysis*, Universitext, Berlin, New York: Springer-Verlag, ISBN 978-0-387-98899-3, MR 1729490.

- Bers, Lipman; John, Fritz; Schechter, Martin (1981), *Partial differential equations*, American Mathematical Society, ISBN 0-8218-0049-3.

- Bourbaki, Nicolas (1986), *Spectral theories*, Elements of mathematics, Berlin: Springer-Verlag, ISBN 0-201-00767-3.

- Bourbaki, Nicolas (1987), *Topological vector spaces*, Elements of mathematics, Berlin: Springer-Verlag, ISBN 978-3-540-13627-9.

- Boyer, Carl Benjamin; Merzbach, Uta C (1991), *A History of Mathematics* (2nd ed.), John Wiley & Sons, Inc., ISBN 0-471-54397-7.

- Brenner, S.; Scott, R. L. (2005), *The Mathematical Theory of Finite Element Methods* (2nd ed.), Springer, ISBN 0-387-95451-1.

- Buttazzo, Giuseppe; Giaquinta, Mariano; Hildebrandt, Stefan (1998), *One-dimensional variational problems*, Oxford Lecture Series in Mathematics and its Applications **15**, The Clarendon Press Oxford University Press, ISBN 978-0-19-850465-8, MR 1694383.

- Clarkson, J. A. (1936), "Uniformly convex spaces", *Trans. Amer. Math. Soc.* **40** (3): 396–414, doi:10.2307/1989630, JSTOR 1989630.

- Courant, Richard; Hilbert, David (1953), *Methods of Mathematical Physics, Vol. I*, Interscience.

- Dieudonné, Jean (1960), *Foundations of Modern Analysis*, Academic Press.

- Dirac, P.A.M. (1930), *Principles of Quantum Mechanics*, Oxford: Clarendon Press.

- Dunford, N.; Schwartz, J.T. (1958), *Linear operators, Parts I and II*, Wiley-Interscience.

- Duren, P. (1970), *Theory of H^p-Spaces*, New York: Academic Press.

- Folland, Gerald B. (2009), *Fourier analysis and its application* (Reprint of Wadsworth and Brooks/Cole 1992 ed.), American Mathematical Society Bookstore, ISBN 0-8218-4790-2.

- Folland, Gerald B. (1989), *Harmonic analysis in phase space*, Annals of Mathematics Studies **122**, Princeton University Press, ISBN 0-691-08527-7.

- Fréchet, Maurice (1907), "Sur les ensembles de fonctions et les opérations linéaires", *C. R. Acad. Sci. Paris* **144**: 1414–1416.

- Fréchet, Maurice (1904–1907), *Sur les opérations linéaires*.

- Giusti, Enrico (2003), *Direct Methods in the Calculus of Variations*, World Scientific, ISBN 981-238-043-4.

- Grattan-Guinness, Ivor (2000), *The search for mathematical roots, 1870–1940*, Princeton Paperbacks, Princeton University Press, ISBN 978-0-691-05858-0, MR 1807717.

- Halmos, Paul (1957), *Introduction to Hilbert Space and the Theory of Spectral Multiplicity*, Chelsea Pub. Co

- Halmos, Paul (1982), *A Hilbert Space Problem Book*, Springer-Verlag, ISBN 0-387-90685-1.

- Hewitt, Edwin; Stromberg, Karl (1965), *Real and Abstract Analysis*, New York: Springer-Verlag.

- Hilbert, David; Nordheim, Lothar (Wolfgang); von Neumann, John (1927), "Über die Grundlagen der Quantenmechanik", *Mathematische Annalen* **98**: 1–30, doi:10.1007/BF01451579.

- Kac, Mark (1966), "Can one hear the shape of a drum?", *American Mathematical Monthly* **73** (4, part 2): 1–23, doi:10.2307/2313748, JSTOR 2313748.

- Kadison, Richard V.; Ringrose, John R. (1997), *Fundamentals of the theory of operator algebras. Vol. I*, Graduate Studies in Mathematics **15**, Providence, R.I.: American Mathematical Society, ISBN 978-0-8218-0819-1, MR 1468229.

- Kakutani, Shizuo (1939), "Some characterizations of Euclidean space", *Japanese Journal of Mathematics* **16**: 93–97, MR 0000895.

- Kline, Morris (1972), *Mathematical thought from ancient to modern times, Volume 3* (3rd ed.), Oxford University Press (published 1990), ISBN 978-0-19-506137-6.

- Kolmogorov, Andrey; Fomin, Sergei V. (1970), *Introductory Real Analysis* (Revised English edition, trans. by Richard A. Silverman (1975) ed.), Dover Press, ISBN 0-486-61226-0.

- Krantz, Steven G. (2002), *Function Theory of Several Complex Variables*, Providence, R.I.: American Mathematical Society, ISBN 978-0-8218-2724-6.

- Lanczos, Cornelius (1988), *Applied analysis* (Reprint of 1956 Prentice-Hall ed.), Dover Publications, ISBN 0-486-65656-X.

- Lindenstrauss, J.; Tzafriri, L. (1971), "On the complemented subspaces problem", *Israel Journal of Mathematics* **9** (2): 263–269, doi:10.1007/BF02771592, ISSN 0021-2172, MR 0276734.

- O'Connor, John J.; Robertson, Edmund F. (1996), "Abstract linear spaces", *MacTutor History of Mathematics archive*, University of St Andrews..

- Lebesgue, Henri (1904), *Leçons sur l'intégration et la recherche des fonctions primitives*, Gauthier-Villars.

- B.M. Levitan (2001), "Hilbert space", in Hazewinkel, Michiel, *Encyclopedia of Mathematics*, Springer, ISBN 978-1-55608-010-4.

- Marsden, Jerrold E. (1974), *Elementary classical analysis*, W. H. Freeman and Co., MR 0357693.

- von Neumann, John (1929), "Allgemeine Eigenwerttheorie Hermitescher Funktionaloperatoren", *Mathematische Annalen* **102**: 49–131, doi:10.1007/BF01782338.

- von Neumann, John (1932), "Physical Applications of the Ergodic Hypothesis", *Proc Natl Acad Sci USA* **18** (3): 263–266, Bibcode:1932PNAS...18..263N, doi:10.1073/pnas.18.3.263, JSTOR 86260, PMC 1076204, PMID 16587674.

- von Neumann, John (1932), *Mathematical Foundations of Quantum Mechanics*, Princeton Landmarks in Mathematics, Princeton University Press (published 1996), ISBN 978-0-691-02893-4, MR 1435976.

- Prugovečki, Eduard (1981), *Quantum mechanics in Hilbert space* (2nd ed.), Dover (published 2006), ISBN 978-0-486-45327-9.

- Reed, Michael; Simon, Barry (1980), *Functional Analysis*, Methods of Modern Mathematical Physics, Academic Press, ISBN 0-12-585050-6.

- Reed, Michael; Simon, Barry (1975), *Fourier Analysis, Self-Adjointness*, Methods of Modern Mathematical Physics, Academic Press, ISBN 9780125850025.

- Riesz, Frigyes (1907), "Sur une espèce de Géométrie analytique des systèmes de fonctions sommables", *C. R. Acad. Sci. Paris* **144**: 1409–1411.

- Riesz, Frigyes (1934), "Zur Theorie des Hilbertschen Raumes", *Acta Sci. Math. Szeged* **7**: 34–38.

- Riesz, Frigyes; Sz.-Nagy, Béla (1990), *Functional analysis*, Dover, ISBN 0-486-66289-6.

- Rudin, Walter (1973), *Functional analysis*, Tata MacGraw-Hill.

- Rudin, Walter (1987), *Real and Complex Analysis*, McGraw-Hill, ISBN 0-07-100276-6.

- Saks, Stanisław (2005), *Theory of the integral* (2nd Dover ed.), Dover, ISBN 978-0-486-44648-6; originally published *Monografje Matematyczne*, vol. 7, Warszawa, 1937.

- Schmidt, Erhard (1908), "Über die Auflösung linearer Gleichungen mit unendlich vielen Unbekannten", *Rend. Circ. Mat. Palermo* **25**: 63–77, doi:10.1007/BF03029116.

- Shubin, M. A. (1987), *Pseudodifferential operators and spectral theory*, Springer Series in Soviet Mathematics, Berlin, New York: Springer-Verlag, ISBN 978-3-540-13621-7, MR 883081.

- Sobrino, Luis (1996), *Elements of non-relativistic quantum mechanics*, River Edge, New Jersey: World Scientific Publishing Co. Inc., ISBN 978-981-02-2386-1, MR 1626401.

- Stewart, James (2006), *Calculus: Concepts and Contexts* (3rd ed.), Thomson/Brooks/Cole.

- Stein, E (1970), *Singular Integrals and Differentiability Properties of Functions*, Princeton Univ. Press, ISBN 0-691-08079-8.

- Stein, Elias; Weiss, Guido (1971), *Introduction to Fourier Analysis on Euclidean Spaces*, Princeton, N.J.: Princeton University Press, ISBN 978-0-691-08078-9.

- Streater, Ray; Wightman, Arthur (1964), *PCT, Spin and Statistics and All That*, W. A. Benjamin, Inc.

- Teschl, Gerald (2009). *Mathematical Methods in Quantum Mechanics; With Applications to Schrödinger Operators*. Providence: American Mathematical Society. ISBN 978-0-8218-4660-5..

- Titchmarsh, Edward Charles (1946), *Eigenfunction expansions, part 1*, Oxford University: Clarendon Press.

- Trèves, François (1967), *Topological Vector Spaces, Distributions and Kernels*, Academic Press.

- Warner, Frank (1983), *Foundations of Differentiable Manifolds and Lie Groups*, Berlin, New York: Springer-Verlag, ISBN 978-0-387-90894-6.

- Weidmann, Joachim (1980), *Linear operators in Hilbert spaces*, Graduate Texts in Mathematics **68**, Berlin, New York: Springer-Verlag, ISBN 978-0-387-90427-6, MR 566954.

- Weyl, Hermann (1931), *The Theory of Groups and Quantum Mechanics* (English 1950 ed.), Dover Press, ISBN 0-486-60269-9.

- Young, Nicholas (1988), *An introduction to Hilbert space*, Cambridge University Press, ISBN 0-521-33071-8, Zbl 0645.46024.

1.14 External links

- Hazewinkel, Michiel, ed. (2001), "Hilbert space", *Encyclopedia of Mathematics*, Springer, ISBN 978-1-55608-010-4

- Hilbert space at Mathworld

- 245B, notes 5: Hilbert spaces by Terence Tao

Chapter 2

Hilbert curve

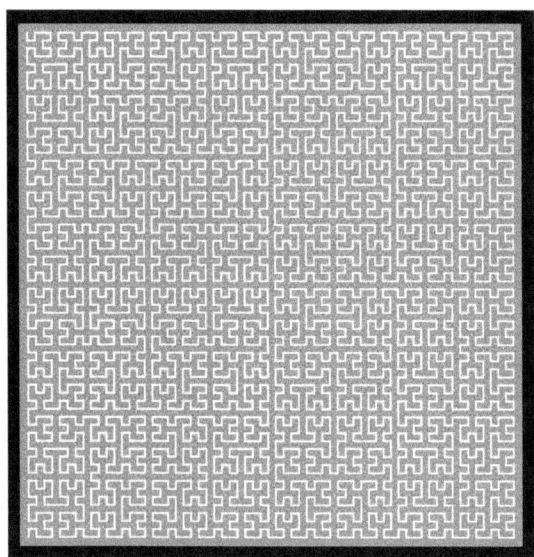

First steps toward building the Hilbert curve

A **Hilbert curve** (also known as a **Hilbert space-filling curve**) is a continuous fractal space-filling curve first described by the German mathematician David Hilbert in 1891,[1] as a variant of the space-filling Peano curves discovered by Giuseppe Peano in 1890.[2]

Because it is space-filling, its Hausdorff dimension is 2 (precisely, its image is the unit square, whose dimension is 2 in any definition of dimension; its graph is a compact set homeomorphic to the closed unit interval, with Hausdorff dimension 2).

H_n is the n th approximation to the limiting curve. The Euclidean length of H_n is $2^n - \frac{1}{2^n}$, i.e., it grows exponentially with n , while at the same time always being bounded by a square with a finite area.

2.1 Images

- Hilbert curve, first order

- Hilbert curves, first and second orders

- Hilbert curves, first to third orders

- String art

- Hilbert curve, construction color-coded

- A Hilbert curve in three dimensions

- A 3-D Hilbert curve with color showing progression

- This GIF file displays an animation of circles traveling along the path of a Hilbert Space filling Curve.

2.2 Applications and mapping algorithms

Both the true Hilbert curve and its discrete approximations are useful because they give a mapping between 1D and 2D space that fairly well preserves locality.[3] If (x, y) are the coordinates of a point within the unit square, and d is the distance along the curve when it reaches that point, then points that have nearby d values will also have nearby (x, y) values. The converse can't always be true. There will sometimes be points where the (x, y) coordinates are close but their d values are far apart.

Because of this locality property, the Hilbert curve is widely used in computer science. For example, the range of IP addresses used by computers can be mapped into a picture using the Hilbert curve. Code to generate the image would map from 2D to 1D to find the color of each pixel, and the Hilbert curve is sometimes used because it keeps nearby IP addresses close to each other in the picture. A grayscale photograph can be converted to a dithered black-and-white image using thresholding, with the leftover amount from each pixel added to the next pixel along the Hilbert curve. Code to do this would map from 1D to 2D, and the Hilbert curve is sometimes used because it does not create the distracting patterns that would be visible to the eye if the order were simply left to right across each row of pixels. Hilbert curves in higher dimensions are an instance of a generalization of Gray codes, and are sometimes used for similar purposes, for similar reasons. For multidimensional databases, Hilbert order has been proposed to be used instead of Z order because it has better locality-preserving behavior. For example, Hilbert curves have been used to compress and accelerate R-tree indexes[4] (see Hilbert R-tree). They have also been used to help compress data warehouses.[5][6]

Given the variety of applications, it is useful to have algorithms to map in both directions. In many languages, these are better if implemented with iteration rather than recursion. The following C code performs the mappings in both directions, using iteration and bit operations rather than recursion. It assumes a square divided into n by n cells, for n a power of 2, with integer coordinates, with (0,0) in the lower left corner, $(n-1,n-1)$ in the upper right corner, and a distance d that starts at 0 in the lower left corner and goes to $n^2 - 1$ in the lower-right corner.

```
//convert (x,y) to d int xy2d (int n, int x, int y) { int rx, ry,
s, d=0; for (s=n/2; s>0; s/=2) { rx = (x & s) > 0; ry = (y
& s) > 0; d += s * s * ((3 * rx) ^ ry); rot(s, &x, &y, rx,
ry); } return d; } //convert d to (x,y) void d2xy(int n, int
d, int *x, int *y) { int rx, ry, s, t=d; *x = *y = 0; for (s=1;
s<n; s*=2) { rx = 1 & (t/2); ry = 1 & (t ^ rx); rot(s, x, y,
rx, ry); *x += s * rx; *y += s * ry; t /= 4; } } //rotate/flip
a quadrant appropriately void rot(int n, int *x, int *y, int
rx, int ry) { if (ry == 0) { if (rx == 1) { *x = n-1 - *x;
*y = n-1 - *y; } //Swap x and y int t = *x; *x = *y; *y = t; } }
```

These use the C conventions: the & symbol is a bitwise AND, the ^ symbol is a bitwise XOR, the += operator adds onto a variable, and the /= operator divides a variable. The handling of booleans in C means that in xy2d, the variable rx is set to 0 or 1 to match bit s of x, and similarly for ry.

The xy2d function works top down, starting with the most significant bits of x and y, and building up the most significant bits of d first. The d2xy function works in the opposite order, starting with the least significant bits of d, and building up x and y starting with the least significant bits. Both functions use the rotation function to rotate and flip the (x,y) coordinate system appropriately.

The two mapping algorithms work in similar ways. The entire square is viewed as composed of 4 regions, arranged 2 by 2. Each region is composed of 4 smaller regions, and so on, for a number of levels. At level s, each region is s by s cells. There is a single FOR loop that iterates through levels. On each iteration, an amount is added to d or to x and y, determined by which of the 4 regions it is in at the current level. The current region out of the 4 is (rx,ry), where rx and ry are each 0 or 1. So it consumes 2 input bits, (either 2 from d or 1 each from x and y), and generates two

output bits. It also calls the rotation function so that (x,y) will be appropriate for the next level, on the next iteration. For xy2d, it starts at the top level of the entire square, and works its way down to the lowest level of individual cells. For d2xy, it starts at the bottom with cells, and works up to include the entire square.

It is possible to implement Hilbert curves efficiently even when the data space does not form a square.[7] Moreover, there are several possible generalizations of Hilbert curves to higher dimensions.[8][9]

2.3 Representation as Lindenmayer system

The Hilbert Curve can be expressed by a rewrite system (L-system).

Hilbert Curve at its sixth iteration

Alphabet : A, B

Constants : F + −

Axiom : A

Production rules:

$$A \rightarrow - B F + A F A + F B -$$
$$B \rightarrow + A F - B F B - F A +$$

Here, "F" means "draw forward", "−" means "turn left 90°", "+" means "turn right 90°" (see turtle graphics), and "A" and "B" are ignored during drawing.

2.4 Other implementations

Arthur Butz[10] provided an algorithm for calculating the Hilbert curve in multidimensions.

Graphics Gems II[11] discusses Hilbert Curve coherency, and provides implementation.

2.5 See also

- Hilbert curve scheduling

- Hilbert R-tree

- Sierpiński curve

- Moore curve

- Space-filling curves

- List of fractals by Hausdorff dimension

2.6 Notes

[1] D. Hilbert: Über die stetige Abbildung einer Linie auf ein Flächenstück. Mathematische Annalen 38 (1891), 459–460.

[2] G.Peano: Sur une courbe, qui remplit toute une aire plane. Mathematische Annalen 36 (1890), 157–160.

[3] Moon, B.; Jagadish, H.V.; Faloutsos, C.; Saltz, J.H. (2001), "Analysis of the clustering properties of the Hilbert space-filling curve", IEEE Transactions on Knowledge and Data Engineering 13 (1): 124–141, doi:10.1109/69.908985.

[4] I. Kamel, C. Faloutsos, Hilbert R-tree: An improved R-tree using fractals, in: Proceedings of the 20th International Conference on Very Large Data Bases, Morgan Kaufmann Publishers Inc., San Francisco, CA, USA, 1994, pp. 500–509.

[5] T. Eavis, D. Cueva, A Hilbert space compression architecture for data warehouse environments, Lecture Notes in Computer Science 4654 (2007) 1–12.

[6] Daniel Lemire and Owen Kaser, Reordering Columns for Smaller Indexes, Information Sciences 181 (12), 2011.

[7] C. H. Hamilton, A. Rau-Chaplin, Compact Hilbert indices: Space-filling curves for domains with unequal side lengths, Information Processing Letters 105 (5) (2007) 155–163.

[8] J. Alber, R. Niedermeier, On multidimensional curves with Hilbert property, Theory of Computing Systems 33 (4) (2000) 295–312.

[9] H. J. Haverkort, F. van Walderveen, Four-dimensional Hilbert curves for R-trees, in: Proceedings of the Eleventh Workshop on Algorithm Engineering and Experiments, 2009, pp. 63–73.

[10] A.R. Butz (April 1971). "Alternative algorithm for Hilbert's space filling curve.". IEEE Trans. On Computers, 20: 424–42. doi:10.1109/T-C.1971.223258.

[11] Voorhies, Douglas: Space-Filling Curves and a Measure of Coherence, p. 26-30, Graphics Gems II.

2.7 External links

- Dynamic Hilbert curve with JSXGraph

- Three.js WebGL 3D Hilbert curve demo

- XKCD cartoon using the locality properties of the Hilbert curve to create a "map of the internet"

- Gcode generator for Hilbert curve

Chapter 3

Lp space

In mathematics, the **L^p spaces** are function spaces defined using a natural generalization of the p-norm for finite-dimensional vector spaces. They are sometimes called **Lebesgue spaces**, named after Henri Lebesgue (Dunford & Schwartz 1958, III.3), although according to the Bourbaki group (Bourbaki 1987) they were first introduced by Frigyes Riesz (Riesz 1910). **L^p spaces** form an important class of Banach spaces in functional analysis, and of topological vector spaces. Lebesgue spaces have applications in physics, statistics, finance, engineering, and other disciplines.

3.1 The p-norm in finite dimensions

The length of a vector $x = (x_1, x_2, ..., x_n)$ in the n-dimensional real vector space $\mathbf{R}^n$ is usually given by the Euclidean norm:

$$\|x\|_2 = \left(x_1^2 + x_2^2 + \cdots + x_n^2\right)^{\frac{1}{2}}$$

The Euclidean distance between two points x and y is the length $\|x - y\|_2$ of the straight line between the two points. In many situations, the Euclidean distance is insufficient for capturing the actual distances in a given space. An analogy to this can be found in Manhattan taxi drivers who should measure distance not in terms of the length of the straight line to their destination, but in terms of the Manhattan distance, which takes into account that streets are either orthogonal or parallel to each other. The class of p-norms generalizes these two examples and has an abundance of applications in many parts of mathematics, physics, and computer science.

3.1.1 Definition

For a real number $p \geq 1$, the **p-norm** or **L^p-norm** of x is defined by

$$\|x\|_p = \left(|x_1|^p + |x_2|^p + \cdots + |x_n|^p\right)^{\frac{1}{p}}$$

(The absolute value bars are unnecessary if p is a rational number with even numerator and odd denominator.)

The Euclidean norm from above falls into this class and is the 2-norm, and the 1-norm is the norm that corresponds to the Manhattan distance.

The **L^∞-norm** or maximum norm (or uniform norm) is the limit of the L^p-norms for $p \to \infty$. It turns out that this limit is equivalent to the following definition:

$$\|x\|_\infty = \max\left\{|x_1|, |x_2|, \ldots, |x_n|\right\}$$

See L-infinity.

For all $p \geq 1$, the p-norms and maximum norm as defined above indeed satisfy the properties of a "length function" (or norm), which are that:

- only the zero vector has zero length,

- the length of the vector is positive homogeneous with respect to multiplication by a scalar (positive homogeneity), and

- the length of the sum of two vectors is no larger than the sum of lengths of the vectors (triangle inequality).

Abstractly speaking, this means that $\mathbf{R}^n$ together with the p-norm is a Banach space. This Banach space is the **L^p-space** over $\mathbf{R}^n$.

Relations between p-norms

The grid distance ("Manhattan distance") between two points is never shorter than the length of the line segment between them (the Euclidean or "as the crow flies" distance). Formally, this means that the Euclidean norm of any vector is bounded by its 1-norm:

$$\|x\|_2 \leq \|x\|_1$$

This fact generalizes to p-norms in that the p-norm ‖x‖p of any given vector x does not grow with p:

> ‖x‖p+a ≤ ‖x‖p for any vector x and real numbers $p \geq 1$ and $a \geq 0$. (In fact this remains true for $0 < p < 1$ and $a \geq 0$.)

For the opposite direction, the following relation between the 1-norm and the 2-norm is known:

$$\|x\|_1 \leq \sqrt{n}\|x\|_2$$

This inequality depends on the dimension n of the underlying vector space and follows directly from the Cauchy–Schwarz inequality.

In general, for vectors in $\mathbf{C}^n$ where $0 < r < p$:

$$\|x\|_p \leq \|x\|_r \leq n^{\left(\frac{1}{r} - \frac{1}{p}\right)}\|x\|_p$$

3.1.2   When $0 < p < 1$

In $\mathbf{R}^n$ for $n > 1$, the formula

$$\|x\|_p = \left(|x_1|^p + |x_2|^p + \cdots + |x_n|^p\right)^{\frac{1}{p}}$$

defines an absolutely homogeneous function of degree 1 for $0 < p < 1$; however, the resulting function does not define an F-norm, because it is not subadditive. In $\mathbf{R}^n$ for $n > 1$, the formula for $0 < p < 1$

$$|x_1|^p + |x_2|^p + \cdots + |x_n|^p$$

defines a subadditive function, which does define an F-norm. This F-norm is homogeneous of degree p.

However, the function

$$d_p(x, y) = \sum_{i=1}^{n} |x_i - y_i|^p$$

defines a metric. The metric space $(\mathbf{R}^n, dp)$ is denoted by ℓn^p.

Although the p-unit ball Bn^p around the origin in this metric is "concave", the topology defined on $\mathbf{R}^n$ by the metric

dp is the usual vector space topology of $\mathbf{R}^n$, hence ℓn^p is a locally convex topological vector space. Beyond this qualitative statement, a quantitative way to measure the lack of convexity of ℓn^p is to denote by $Cp(n)$ the smallest constant C such that the multiple $C \, Bn^p$ of the p-unit ball contains the convex hull of Bn^p, equal to Bn^1. The fact that for fixed $p < 1$ we have

$$C_p(n) = n^{\frac{1}{p} - 1} \to \infty, \qquad \text{as } n \to \infty,$$

shows that the infinite-dimensional sequence space ℓ^p defined below, is no longer locally convex.

3.1.3   When $p = 0$

There is one ℓ_0 norm and another function called the ℓ_0 "norm" (with quotation marks).

The mathematical definition of the ℓ_0 norm was established by Banach's *Theory of Linear Operations*. The space of sequences has a complete metric topology provided by the F-norm

$$(x_n) \mapsto \sum_n 2^{-n} \frac{|x_n|}{1 + |x_n|},$$

which is discussed by Stefan Rolewicz in *Metric Linear Spaces*.[1] The ℓ_0-normed space is studied in functional analysis, probability theory, and harmonic analysis.

Another function was called the ℓ_0 "norm" by David Donoho — whose quotation marks warn that this function is not a proper norm — is the number of non-zero entries of the vector x. Many authors abuse terminology by omitting the quotation marks. Defining $0^0 = 0$, the zero "norm" of x is equal to

$$|x_1|^0 + |x_2|^0 + \cdots + |x_n|^0.$$

This is not a norm (B-norm, with "B" for Banach) because it is not homogeneous. Despite these defects as a mathematical norm, the non-zero counting "norm" has uses in scientific computing, information theory, and statistics – notably in compressed sensing in signal processing and computational harmonic analysis.

3.2   The p-norm in countably infinite dimensions and ℓ^p spaces

For more details on this topic, see Sequence space.

The p-norm can be extended to vectors that have an infinite number of components, which yields the space ℓ^p. This contains as special cases:

- ℓ^1, the space of sequences whose series is absolutely convergent,

- ℓ^2, the space of **square-summable** sequences, which is a Hilbert space, and

- ℓ^∞, the space of bounded sequences.

The space of sequences has a natural vector space structure by applying addition and scalar multiplication coordinate by coordinate. Explicitly, the vector sum and the scalar action for infinite sequences of real (or complex) numbers are given by:

$$(x_1, x_2, \ldots, x_n, x_{n+1}, \ldots) + (y_1, y_2, \ldots, y_n, y_{n+1}, \ldots) =$$
$$(x1+y1, x2+y2, \ldots, xn+y$$
$$\lambda \cdot (x_1, x_2, \ldots, x_n, x_{n+1}, \ldots) =$$
$$(\lambda x1, \lambda x2, \ldots, \lambda xn, \lambda xn+1, \ldots).$$

Define the p-norm:

$$\|x\|_p = (|x_1|^p + |x_2|^p + \cdots + |x_n|^p + |x_{n+1}|^p + \cdots)^{\frac{1}{p}}$$

Here, a complication arises, namely that the series on the right is not always convergent, so for example, the sequence made up of only ones, (1, 1, 1, ...), will have an infinite p-norm for $1 \le p < \infty$. The space ℓ^p is then defined as the set of all infinite sequences of real (or complex) numbers such that the p-norm is finite.

One can check that as p increases, the set ℓ^p grows larger. For example, the sequence

$$\left(1, \frac{1}{2}, \ldots, \frac{1}{n}, \frac{1}{n+1}, \ldots\right)$$

is not in ℓ^1, but it is in ℓ^p for $p > 1$, as the series

$$1^p + \frac{1}{2^p} + \cdots + \frac{1}{n^p} + \frac{1}{(n+1)^p} + \cdots,$$

diverges for $p = 1$ (the harmonic series), but is convergent for $p > 1$.

One also defines the ∞-norm using the supremum:

$$\|x\|_\infty = \sup(|x_1|, |x_2|, \ldots, |x_n|, |x_{n+1}|, \ldots)$$

and the corresponding space ℓ^∞ of all bounded sequences. It turns out that[2]

$$\|x\|_\infty = \lim_{p \to \infty} \|x\|_p$$

if the right-hand side is finite, or the left-hand side is infinite. Thus, we will consider ℓ^p spaces for $1 \le p \le \infty$.

The p-norm thus defined on ℓ^p is indeed a norm, and ℓ^p together with this norm is a Banach space. The fully general L^p space is obtained — as seen below — by considering vectors, not only with finitely or countably-infinitely many components, but with "*arbitrarily many components*"; in other words, functions. An integral instead of a sum is used to define the p-norm.

3.3 L^p spaces

An L^p space may be defined as a space of functions for which the *p*-th power of the absolute value is Lebesgue integrable.[3] More generally, let $1 \le p < \infty$ and (S, Σ, μ) be a measure space. Consider the set of all measurable functions from S to **C** or **R** whose absolute value raised to the p-th power has finite integral, or equivalently, that

$$\|f\|_p \equiv \left(\int_S |f|^p \, \mathrm{d}\mu\right)^{\frac{1}{p}} < \infty$$

The set of such functions forms a vector space, with the following natural operations:

$$(f + g)(x) = f(x) + g(x),$$
$$(\lambda f)(x) = \lambda f(x)$$

for every scalar λ.

That the sum of two p-th power integrable functions is again p-th power integrable follows from the inequality

$$\|f + g\|_p^p \le 2^{p-1} \left(\|f\|_p^p + \|g\|_p^p\right).$$

(This comes from the convexity of $t \mapsto t^p$ for $p \ge 1$.)

In fact, more is true. *Minkowski's inequality* says the triangle inequality holds for $\| \cdot \|p$. Thus the set of p-th power integrable functions, together with the function $\| \cdot \|p$, is a seminormed vector space, which is denoted by $\mathcal{L}^p(S, \mu)$.

This can be made into a normed vector space in a standard way; one simply takes the quotient space with respect to the kernel of $\| \cdot \|p$. Since for any measurable function f, we have that $\| f \|p = 0$ if and only if $f = 0$ almost everywhere, the kernel of $\| \cdot \|p$ does not depend upon p,

$$\mathcal{N} \equiv \ker(\|\cdot\|_p) = \{f : f = 0 \ \mu\text{everywhere -almost}\}$$

In the quotient space, two functions f and g are identified if $f = g$ almost everywhere. The resulting normed vector space is, by definition,

$$L^p(S, \mu) \equiv \mathcal{L}^p(S, \mu)/\mathcal{N}$$

For $p = \infty$, the space $L^\infty(S, \mu)$ is defined as follows. We start with the set of all measurable functions from S to **C** or **R** which are bounded. Again two such functions are identified if they are equal almost everywhere. Denote this set by $L^\infty(S, \mu)$. For a function f in this set, its essential supremum serves as an appropriate norm:

$$\|f\|_\infty \equiv \inf\{C \geq 0 : |f(x)| \leq C \text{ for almost every } x\}.$$

As before, if there exists $q < \infty$ such that $f \in L^\infty(S, \mu) \cap L^q(S, \mu)$, then

$$\|f\|_\infty = \lim_{p \to \infty} \|f\|_p.$$

For $1 \leq p \leq \infty$, $L^p(S, \mu)$ is a Banach space. The fact that L^p is complete is often referred to as the Riesz-Fischer theorem. Completeness can be checked using the convergence theorems for Lebesgue integrals.

When the underlying measure space S is understood, $L^p(S, \mu)$ is often abbreviated $L^p(\mu)$, or just L^p. The above definitions generalize to Bochner spaces.

3.3.1 Special cases

Similar to the ℓ^p spaces, L^2 is the only Hilbert space among L^p spaces. In the complex case, the inner product on L^2 is defined by

$$\langle f, g \rangle = \int_S f(x)\overline{g(x)} \, d\mu(x)$$

The additional inner product structure allows for a richer theory, with applications to, for instance, Fourier series and quantum mechanics. Functions in L^2 are sometimes called **quadratically integrable functions**, **square-integrable functions** or **square-summable functions**, but sometimes these terms are reserved for functions that are square-integrable in some other sense, such as in the sense of a Riemann integral (Titchmarsh 1976).

If we use complex-valued functions, the space L^∞ is a commutative C*-algebra with pointwise multiplication and conjugation. For many measure spaces, including all sigma-finite ones, it is in fact a commutative von Neumann algebra. An element of L^∞ defines a bounded operator on any L^p space by multiplication.

For $1 \leq p \leq \infty$ the ℓ^p spaces are a special case of L^p spaces, when $S = \mathbf{N}$, and μ is the counting measure on **N**. More generally, if one considers any set S with the counting measure, the resulting L^p space is denoted $\ell^p(S)$. For example, the space $\ell^p(\mathbf{Z})$ is the space of all sequences indexed by the integers, and when defining the p-norm on such a space, one sums over all the integers. The space $\ell^p(n)$, where n is the set with n elements, is $\mathbf{R}^n$ with its p-norm as defined above. As any Hilbert space, every space L^2 is linearly isometric to a suitable $\ell^2(I)$, where the cardinality of the set I is the cardinality of an arbitrary Hilbertian basis for this particular L^2.

3.4   Properties of L^p spaces

3.4.1   Dual spaces

The dual space (the Banach space of all continuous linear functionals) of $L^p(\mu)$ for $1 < p < \infty$ has a natural isomorphism with $L^q(\mu)$, where q is such that $1/p + 1/q = 1$. This isomorphism associates $g \in L^q(\mu)$ with the functional $\kappa p(g) \in L^p(\mu)^*$ defined by

$$f \mapsto \kappa_p(g)(f) = \int fg \, d\mu \quad \text{for every } f \in L^p(\mu)$$

The fact that $\kappa p(g)$ is well defined and continuous follows from Hölder's inequality. $\kappa p : L^q(\mu) \to L^p(\mu)^*$ is a linear mapping which is an isometry by the extremal case of Hölder's inequality. It is also possible to show (for example with the Radon–Nikodym theorem, see[4]) that any $G \in L^p(\mu)^*$ can be expressed this way: i.e., that κp is *onto*. Since κp is onto and isometric, it is an isomorphism of Banach spaces. With this (isometric) isomorphism in mind, it is usual to say simply that L^q is the dual Banach space of L^p.

For $1 < p < \infty$, the space $L^p(\mu)$ is reflexive. Let κp be as above and let $\kappa q : L^p(\mu) \to L^q(\mu)^*$ be the corresponding linear isometry. Consider the map from $L^p(\mu)$ to $L^p(\mu)^{**}$, obtained by composing κq with the transpose (or adjoint) of the inverse of κp:

$$j_p : L^p(\mu) \xrightarrow{\kappa_q} L^q(\mu)^* \xrightarrow{\left(\kappa_p^{-1}\right)^*} L^p(\mu)^{**}$$

This map coincides with the canonical embedding J of $L^p(\mu)$ into its bidual. Moreover, the map jp is onto, as composition of two onto isometries, and this proves reflexivity.

If the measure μ on S is sigma-finite, then the dual of $L^1(\mu)$ is isometrically isomorphic to $L^\infty(\mu)$ (more precisely, the map κ_1 corresponding to $p = 1$ is an isometry from $L^\infty(\mu)$ onto $L^1(\mu)^*$).

The dual of L^∞ is subtler. Elements of $L^\infty(\mu)^*$ can be identified with bounded signed *finitely* additive measures on S that are absolutely continuous with respect to μ. See ba space for more details. If we assume the axiom of choice, this space is much bigger than $L^1(\mu)$ except in some trivial cases. However, Saharon Shelah proved that there are relatively consistent extensions of Zermelo-Fraenkel set theory (ZF + DC + "Every subset of the real numbers has the Baire property") in which the dual of ℓ^∞ is ℓ^1.[5]

3.4.2 Embeddings

Colloquially, if $1 \le p < q \le \infty$, then $L^p(S, \mu)$ contains functions that are more locally singular, while elements of $L^q(S, \mu)$ can be more spread out. Consider the Lebesgue measure on the half line $(0, \infty)$. A continuous function in L^1 might blow up near 0 but must decay sufficiently fast toward infinity. On the other hand, continuous functions in L^∞ need not decay at all but no blow-up is allowed. The precise technical result is the following:[6]

1. Let $1 \le p < q \le \infty$. $L^q(S, \mu) \subset L^p(S, \mu)$ iff S does not contain sets of arbitrarily large measure, and

2. Let $1 \le p < q \le \infty$. $L^p(S, \mu) \subset L^q(S, \mu)$ iff S does not contain arbitrarily small sets of non-zero measure.

In both cases the embedding is continuous, in that the identity operator is a bounded linear map from L^q to L^p in the first case, and L^p to L^q in the second. (This is a consequence of the closed graph theorem and properties of L^p spaces.) Indeed, if the domain S has finite measure, one can make the following explicit calculation via Jensen's inequality:

$$\|f\|_p \le \mu(S)^{\frac{1}{p}-\frac{1}{q}} \|f\|_q$$

The constant appearing in the above inequality is optimal, in the sense that the operator norm of the identity $I : L^q(S, \mu) \to L^p(S, \mu)$ is precisely

$$\|I\|_{q,p} = \mu(S)^{\frac{1}{p}-\frac{1}{q}}$$

the case of equality being achieved exactly when $f = 1$ μ-a.e.

3.4.3 Dense subspaces

Throughout this section we assume that: $1 \le p < \infty$.

Let (S, Σ, μ) be a measure space. An *integrable simple function* f on S is one of the form

$$f = \sum_{j=1}^{n} a_j \mathbf{1}_{A_j}$$

where aj is scalar, $Aj \in \Sigma$ has finite measure and $\mathbf{1}_{A_j}$ is the indicator function of the set A_j, for $j = 1, ..., n$. By construction of the integral, the vector space of integrable simple functions is dense in $L^p(S, \Sigma, \mu)$.

More can be said when S is a metrizable topological space and Σ its Borel σ–algebra, i.e., the smallest σ–algebra of subsets of S containing the open sets.

Suppose $V \subset S$ is an open set with $\mu(V) < \infty$. It can be proved that for every Borel set $A \in \Sigma$ contained in V, and for every $\varepsilon > 0$, there exist a closed set F and an open set U such that

$$F \subset A \subset U \subset V \quad \text{and} \quad \mu(U) - \mu(F) = \mu(U \setminus F) < \varepsilon$$

It follows that there exists φ continuous on S such that

$$0 \le \varphi \le \mathbf{1}_V \quad \text{and} \quad \int_S |\mathbf{1}_A - \varphi| \mathrm{d}\mu < \varepsilon$$

If S can be covered by an increasing sequence (Vn) of open sets that have finite measure, then the space of p–integrable continuous functions is dense in $L^p(S, \Sigma, \mu)$. More precisely, one can use bounded continuous functions that vanish outside one of the open sets Vn.

This applies in particular when $S = \mathbf{R}^d$ and when μ is the Lebesgue measure. The space of continuous and compactly supported functions is dense in $L^p(\mathbf{R}^d)$. Similarly, the space of integrable *step functions* is dense in $L^p(\mathbf{R}^d)$; this space is the linear span of indicator functions of bounded intervals when $d = 1$, of bounded rectangles when $d = 2$ and more generally of products of bounded intervals.

Several properties of general functions in $L^p(\mathbf{R}^d)$ are first proved for continuous and compactly supported functions (sometimes for step functions), then extended by density to all functions. For example, it is proved this way that translations are continuous on $L^p(\mathbf{R}^d)$, in the following sense:

$$\forall f \in L^p(\mathbf{R}^d) : \qquad \|\tau_t f - f\|_p \to 0, \quad \text{as } \mathbf{R}^d \ni t \to 0,$$

where

$$(\tau_t f)(x) = f(x - t).$$

3.5   Applications

L^p spaces are widely used in mathematics and applications.

3.5.1   Hausdorff–Young inequality

The Fourier transform for the real line (resp. for periodic functions, see Fourier series), maps $L^p(\mathbf{R})$ to $L^q(\mathbf{R})$ (resp. $L^p(\mathbf{T})$ to ℓ^q), where $1 \leq p \leq 2$ and $1/p + 1/q = 1$. This is a consequence of the Riesz-Thorin interpolation theorem, and is made precise with the Hausdorff–Young inequality.

By contrast, if $p > 2$, the Fourier transform does not map into L^q.

3.5.2   Hilbert spaces

Hilbert spaces are central to many applications, from quantum mechanics to stochastic calculus. The spaces L^2 and ℓ^2 are both Hilbert spaces. In fact, by choosing a Hilbert basis (i.e., a maximal orthonormal subset of L^2 or any Hilbert space), one sees that all Hilbert spaces are isometric to $\ell^2(E)$, where E is a set with an appropriate cardinality.

3.5.3   Statistics

In statistics, measures of central tendency and statistical dispersion, such as the mean, median, and standard deviation, are defined in terms of L^p metrics, and measures of central tendency can be characterized as solutions to variational problems.

3.6 L^p $(0 < p < 1)$

Let (S, Σ, μ) be a measure space. If $0 < p < 1$, then $L^p(\mu)$ can be defined as above: it is the vector space of those measurable functions f such that

$$N_p(f) = \int_S |f|^p \, d\mu < \infty$$

As before, we may introduce the p-norm $\| f \|p = Np(f)^{1/p}$, but $\| \cdot \|p$ does not satisfy the triangle inequality in this case, and defines only a quasi-norm. The inequality $(a + b)^p \leq a^p + b^p$, valid for $a, b \geq 0$ implies that (Rudin 1991, §1.47)

$$N_p(f + g) \leq N_p(f) + N_p(g)$$

and so the function

$$d_p(f, g) = N_p(f - g) = \| f - g \|_p^p$$

is a metric on $L^p(\mu)$. The resulting metric space is complete; the verification is similar to the familiar case when $p \geq 1$.

In this setting L^p satisfies a *reverse Minkowski inequality*, that is for u, v in L^p

$$\| |u| + |v| \|_p \geq \|u\|_p + \|v\|_p$$

This result may be used to prove Clarkson's inequalities, which are in turn used to establish the uniform convexity of the spaces L^p for $1 < p < \infty$ (Adams & Fournier 2003).

The space L^p for $0 < p < 1$ is an F-space: it admits a complete translation-invariant metric with respect to which the vector space operations are continuous. It is also locally bounded, much like the case $p \geq 1$. It is the prototypical example of an F-space that, for most reasonable measure spaces, is not locally convex: in ℓ^p or $L^p([0, 1])$, every open convex set containing the 0 function is unbounded for the p-quasi-norm; therefore, the 0 vector does not possess a fundamental system of convex neighborhoods. Specifically, this is true if the measure space S contains an infinite family of disjoint measurable sets of finite positive measure.

The only nonempty convex open set in $L^p([0, 1])$ is the entire space (Rudin 1991, §1.47). As a particular consequence, there are no nonzero linear functionals on $L^p([0, 1])$: the dual space is the zero space. In the case of the counting measure on the natural numbers (producing the sequence space $L^p(\mu) = \ell^p$), the bounded linear functionals on ℓ^p are exactly those that are bounded on ℓ^1, namely those given by sequences in ℓ^∞. Although ℓ^p does contain non-trivial convex open sets, it fails to have enough of them to give a base for the topology.

The situation of having no linear functionals is highly undesirable for the purposes of doing analysis. In the case of the Lebesgue measure on $\mathbf{R}^n$, rather than work with L^p for $0 < p < 1$, it is common to work with the Hardy space H^p whenever possible, as this has quite a few linear functionals: enough to distinguish points from one another. However, the Hahn–Banach theorem still fails in H^p for $p < 1$ (Duren 1970, §7.5).

3.6.1 L^0, the space of measurable functions

The vector space of (equivalence classes of) measurable functions on (S, Σ, μ) is denoted $L^0(S, \Sigma, \mu)$ (Kalton, Peck & Roberts 1984). By definition, it contains all the L^p, and is equipped with the topology of *convergence in measure*.

When μ is a probability measure (i.e., $\mu(S) = 1$), this mode of convergence is named *convergence in probability*.

The description is easier when μ is finite. If μ is a finite measure on (S, Σ), the 0 function admits for the convergence in measure the following fundamental system of neighborhoods

$$V_\varepsilon = \left\{ f : \mu\big(\{x : |f(x)| > \varepsilon\}\big) < \varepsilon \right\}, \qquad \varepsilon > 0$$

The topology can be defined by any metric d of the form

$$d(f, g) = \int_S \varphi\big(|f(x) - g(x)|\big) \, d\mu(x)$$

where φ is bounded continuous concave and non-decreasing on $[0, \infty)$, with $\varphi(0) = 0$ and $\varphi(t) > 0$ when $t > 0$ (for example, $\varphi(t) = \min(t, 1)$). Such a metric is called Lévy-metric for L^0. Under this metric the space L^0 is complete (it is again an F-space). The space L^0 is in general not locally bounded, and not locally convex.

For the infinite Lebesgue measure λ on $\mathbf{R}^n$, the definition of the fundamental system of neighborhoods could be modified as follows

$$W_\varepsilon = \left\{ f : \lambda\left(\left\{x : |f(x)| > \varepsilon \text{ and } |x| < \frac{1}{\varepsilon}\right\}\right) < \varepsilon \right\}$$

The resulting space $L^0(\mathbf{R}^n, \lambda)$ coincides as topological vector space with $L^0(\mathbf{R}^n, g(x) \, d\lambda(x))$, for any positive λ–integrable density g.

3.7 Weak L^p

Let (S, Σ, μ) be a measure space, and f a measurable function with real or complex values on S. The distribution function of f is defined for $t > 0$ by

$$\lambda_f(t) = \mu \left\{ x \in S : |f(x)| > t \right\}$$

If f is in $L^p(S, \mu)$ for some p with $1 \le p < \infty$, then by Markov's inequality,

$$\lambda_f(t) \le \frac{\|f\|_p^p}{t^p}$$

A function f is said to be in the space **weak** $L^p(\boldsymbol{S}, \mu)$, or $L^{p,w}(S, \mu)$, if there is a constant $C > 0$ such that, for all $t > 0$,

$$\lambda_f(t) \le \frac{C^p}{t^p}$$

The best constant C for this inequality is the $L^{p,w}$-norm of f, and is denoted by

$$\|f\|_{p,w} = \sup_{t>0} \, t \lambda_f^{\frac{1}{p}}(t)$$

The weak L^p coincide with the Lorentz spaces $L^{p,\infty}$, so this notation is also used to denote them.

The $L^{p,w}$-norm is not a true norm, since the triangle inequality fails to hold. Nevertheless, for f in $L^p(S, \mu)$,

$$\|f\|_{p,w} \le \|f\|_p$$

and in particular $L^p(S, \mu) \subset L^{p,w}(S, \mu)$. Under the convention that two functions are equal if they are equal μ almost everywhere, then the spaces $L^{p,w}$ are complete (Grafakos 2004).

For any $0 < r < p$ the expression

$$|||f|||_{L^{p,\infty}} = \sup_{0<\mu(E)<\infty} \mu(E)^{-\frac{1}{r}+\frac{1}{p}} \left(\int_E |f|^r \, d\mu \right)^{\frac{1}{r}}$$

is comparable to the $L^{p,w}$-norm. Further in the case $p > 1$, this expression defines a norm if $r = 1$. Hence for $p > 1$ the weak L^p spaces are Banach spaces (Grafakos 2004).

A major result that uses the $L^{p,w}$-spaces is the Marcinkiewicz interpolation theorem, which has broad applications to harmonic analysis and the study of singular integrals.

3.8 Weighted L^p spaces

As before, consider a measure space (S, Σ, μ). Let $w : S \to [0, \infty)$ be a measurable function. The w-**weighted** L^p **space** is defined as $L^p(S, w \, d\mu)$, where $w \, d\mu$ means the measure v defined by

$$\nu(A) \equiv \int_A w(x) \, d\mu(x), \qquad A \in \Sigma,$$

or, in terms of the Radon–Nikodym derivative, $w = d\nu/d\mu$ the norm for $L^p(S, w \, d\mu)$ is explicitly

$$\|u\|_{L^p(S, w \, d\mu)} \equiv \left(\int_S w(x) |u(x)|^p \, d\mu(x) \right)^{\frac{1}{p}}$$

As L^p-spaces, the weighted spaces have nothing special, since $L^p(S, w\,d\mu)$ is equal to $L^p(S, d\nu)$. But they are the natural framework for several results in harmonic analysis (Grafakos 2004); they appear for example in the Muckenhoupt theorem: for $1 < p < \infty$, the classical Hilbert transform is defined on $L^p(\mathbf{T}, \lambda)$ where $\mathbf{T}$ denotes the unit circle and λ the Lebesgue measure; the (nonlinear) Hardy–Littlewood maximal operator is bounded on $L^p(\mathbf{R}^n, \lambda)$. Muckenhoupt's theorem describes weights w such that the Hilbert transform remains bounded on $L^p(\mathbf{T}, w\,d\lambda)$ and the maximal operator on $L^p(\mathbf{R}^n, w\,d\lambda)$.

3.9 L^p spaces on manifolds

One may also define spaces $L^p(M)$ on a manifold, called the **intrinsic L^p spaces** of the manifold, using densities.

3.10 See also

- Birnbaum–Orlicz space

- Hardy space

- Riesz–Thorin theorem

- Hölder mean

- Hölder space

- Root mean square

- Locally integrable function $\left(L^1_{\mathrm{loc}}\right)$

- $L^p(G)$ spaces over a locally compact group G

- Minkowski distance

- L-infinity

3.11 Notes

[1] Rolewicz, Stefan (1987), *Functional analysis and control theory: Linear systems*, Mathematics and its Applications (East European Series) **29** (Translated from the Polish by Ewa Bednarczuk ed.), Dordrecht; Warsaw: D. Reidel Publishing Co.; PWN—Polish Scientific Publishers, pp. xvi+524, ISBN 90-277-2186-6, MR 920371, OCLC 13064804

[2] Maddox, I.J. (1988), *Elements of Functional Analysis* (2nd ed.), Cambridge: CUP, page 16

[3] We could just say "integrable". Since the integrand is a non-negative real-valued function, there is no difference between having a finite Lebesgue integral and having a finite improper integral (as there is say for the function $\sin(x)/x$ when integrated over the entire real line).

[4] Rudin, Walter (1980), *Real and Complex Analysis* (2nd ed.), New Delhi: Tata McGraw-Hill, ISBN 9780070542341, Theorem 6.16

[5] Schechter, Eric (1997), *Handbook of Analysis and its Foundations*, London: Academic Press Inc. See Sections 14.77 and 27.44-—47

[6] Villani, Alfonso (1985), "Another note on the inclusion $L^p(\mu) \subset L^q(\mu)$", *Amer. Math. Monthly* **92** (7): 485–487, doi:10.2307/2322503, MR 801221

3.12 References

- Adams, Robert A.; Fournier, John F. (2003), *Sobolev Spaces* (Second ed.), Academic Press, ISBN 978-0-12-044143-3.

- Bourbaki, Nicolas (1987), *Topological vector spaces*, Elements of mathematics, Berlin: Springer-Verlag, ISBN 978-3-540-13627-9.

- DiBenedetto, Emmanuele (2002), *Real analysis*, Birkhäuser, ISBN 3-7643-4231-5.

- Dunford, Nelson; Schwartz, Jacob T. (1958), *Linear operators, volume I*, Wiley-Interscience.

- Duren, P. (1970), *Theory of H^p-Spaces*, New York: Academic Press

- Grafakos, Loukas (2004), *Classical and Modern Fourier Analysis*, Pearson Education, Inc., pp. 253–257, ISBN 0-13-035399-X.

- Hewitt, Edwin; Stromberg, Karl (1965), *Real and abstract analysis*, Springer-Verlag.

- Kalton, Nigel J.; Peck, N. Tenney; Roberts, James W. (1984), *An F-space sampler*, London Mathematical Society Lecture Note Series **89**, Cambridge: Cambridge University Press, ISBN 0-521-27585-7, MR 808777

- Riesz, Frigyes (1910), "Untersuchungen über Systeme integrierbarer Funktionen", *Mathematische Annalen* **69** (4): 449–497, doi:10.1007/BF01457637

- Rudin, Walter (1991), *Functional Analysis*, McGraw-Hill Science/Engineering/Math, ISBN 978-0-07-054236-5

- Rudin, Walter (1987), *Real and complex analysis* (3rd ed.), New York: McGraw-Hill, ISBN 978-0-07-054234-1, MR 924157

- Titchmarsh, EC (1976), *The theory of functions*, Oxford University Press, ISBN 978-0-19-853349-8

3.13 External links

- Hazewinkel, Michiel, ed. (2001), "Lebesgue space", *Encyclopedia of Mathematics*, Springer, ISBN 978-1-55608-010-4

- Proof that L^p spaces are complete at PlanetMath.org.

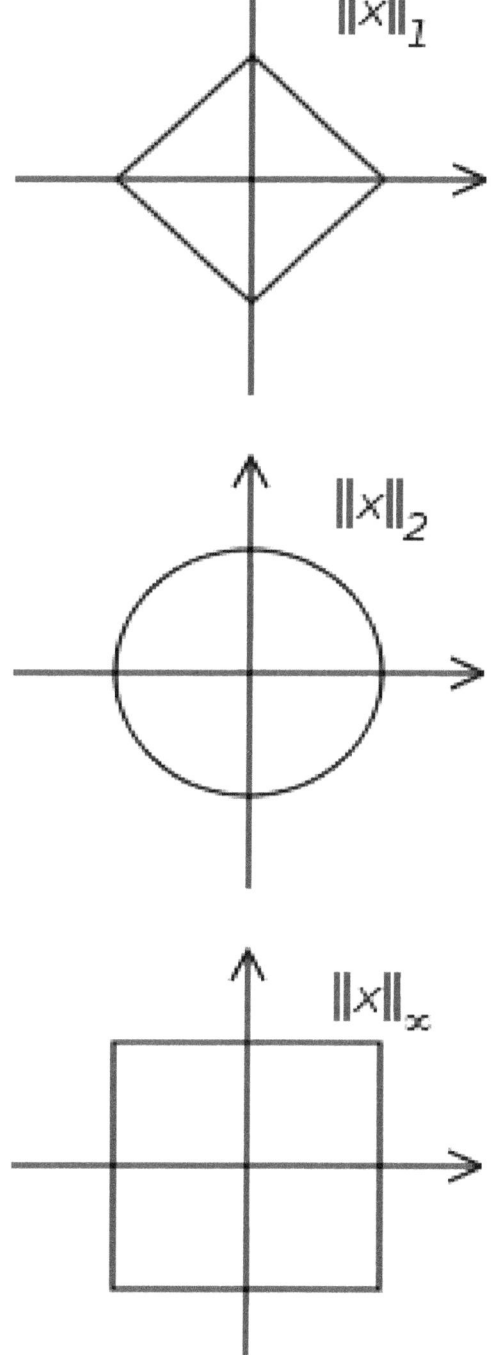

Illustrations of unit circles in different p-norms (every vector from the origin to the unit circle has a length of one, the length being calculated with length-formula of the corresponding p).

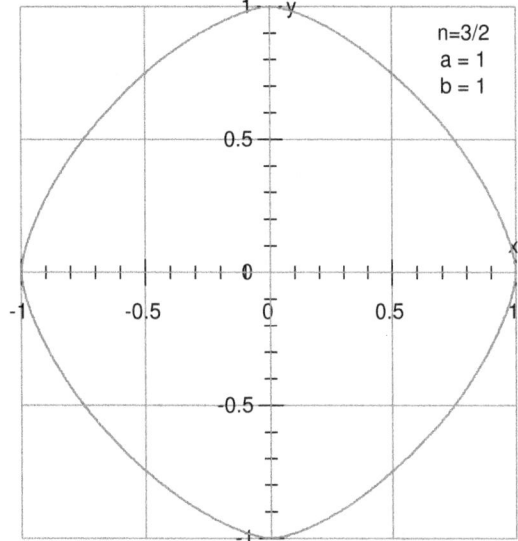

Unit circle (superellipse) in p = 3/2 *norm*

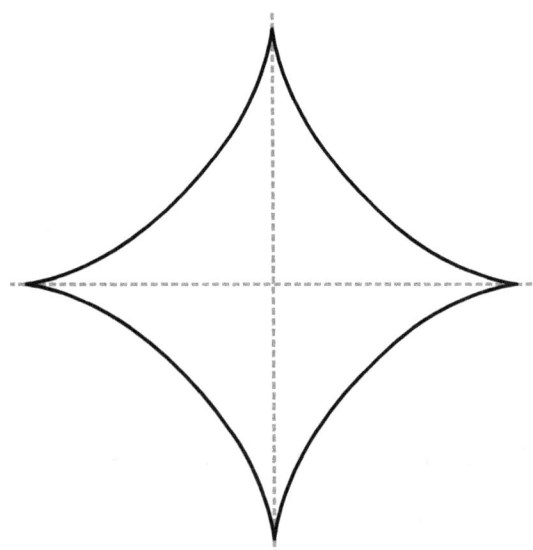

Astroid, unit circle in p = 2/3 *metric*

Chapter 4

Sequence space

For usage in evolutionary biology, see Sequence space (evolution).

In functional analysis and related areas of mathematics, a **sequence space** is a vector space whose elements are infinite sequences of real or complex numbers. Equivalently, it is a function space whose elements are functions from the natural numbers to the field **K** of real or complex numbers. The set of all such functions is naturally identified with the set of all possible infinite sequences with elements in **K**, and can be turned into a vector space under the operations of pointwise addition of functions and pointwise scalar multiplication. All sequence spaces are linear subspaces of this space. Sequence spaces are typically equipped with a norm, or at least the structure of a topological vector space.

The most important sequences spaces in analysis are the ℓ^p spaces, consisting of the p-power summable sequences, with the p-norm. These are special cases of L^p spaces for the counting measure on the set of natural numbers. Other important classes of sequences like convergent sequences or null sequences form sequence spaces, respectively denoted c and c_0, with the sup norm. Any sequence space can also be equipped with the topology of pointwise convergence, under which it becomes a special kind of Fréchet space called FK-space.

4.1 Definition

Let **K** denote the field either of real or complex numbers. Denote by $\mathbf{K}^\mathbf{N}$ the set of all sequences of scalars

$$(x_n)_{n\in\mathbf{N}}, \quad x_n \in \mathbf{K}.$$

This can be turned into a vector space by defining vector addition as

$$(x_n)_{n\in\mathbf{N}} + (y_n)_{n\in\mathbf{N}} \overset{\text{def}}{=} (x_n + y_n)_{n\in\mathbf{N}}$$

and the scalar multiplication as

$$\alpha(x_n)_{n\in\mathbf{N}} := (\alpha x_n)_{n\in\mathbf{N}}.$$

A **sequence space** is any linear subspace of $\mathbf{K}^\mathbf{N}$.

4.1.1 ℓ^p spaces

See also: L^p space and L-infinity

For $0 < p < \infty$, ℓ^p is the subspace of $\mathbf{K}^\mathbf{N}$ consisting of all sequences $x = (\mathbf{x}n)$ satisfying

$$\sum_n |x_n|^p < \infty.$$

If $p \geq 1$, then the real-valued operation $\|\cdot\|_p$ defined by

$$\|x\|_p = \left(\sum_n |x_n|^p \right)^{1/p}$$

defines a norm on ℓ^p. In fact, ℓ^p is a complete metric space with respect to this norm, and therefore is a Banach space.

If $0 < p < 1$, then ℓ^p does not carry a norm, but rather a metric defined by

$$d(x,y) = \sum_n |x_n - y_n|^p.$$

If $p = \infty$, then ℓ^∞ is defined to be the space of all bounded sequences. With respect to the norm

$$\|x\|_\infty = \sup_n |x_n|,$$

ℓ^∞ is also a Banach space.

37

4.1.2   *c* and *c*$_0$

The space of convergent sequences c is a sequence space. This consists of all $x \in \mathbf{K}^{\mathbf{N}}$ such that $\lim_{n \to \infty} x_n$ exists. Since every convergent sequence is bounded, c is a linear subspace of ℓ^∞. It is, moreover, a closed subspace with respect to the infinity norm, and so a Banach space in its own right.

The subspace of null sequences c_0 consists of all sequences whose limit is zero. This is a closed subspace of c, and so again a Banach space.

4.1.3   Other sequence spaces

The space of bounded series, denote by *bs*, is the space of sequences x for which

$$\sup_n \left| \sum_{i=0}^{n} x_i \right| < \infty.$$

This space, when equipped with the norm

$$\|x\|_{bs} = \sup_n \left| \sum_{i=0}^{n} x_i \right|,$$

is a Banach space isometrically isomorphic to ℓ^∞, via the linear mapping

$$(x_n)_{n \in \mathbf{N}} \mapsto \left(\sum_{i=0}^{n} x_i \right)_{n \in \mathbf{N}}.$$

The subspace *cs* consisting of all convergent series is a subspace that goes over to the space c under this isomorphism.

The space Φ or c_{00} is defined to be the space of all infinite sequences with only a finite number of non-zero terms (sequences with finite support). This set is dense in many sequence spaces.

4.2   Properties of ℓ^p spaces and the space c_0

See also: c space

The space ℓ^2 is the only ℓ^p space that is a Hilbert space, since any norm that is induced by an inner product should satisfy the parallelogram identity $\|x + y\|_p^2 + \|x - y\|_p^2 =$ $2\|x\|_p^2 + 2\|y\|_p^2$. Substituting two distinct unit vectors for x and y directly shows that the identity is not true unless $p = 2$.

Each ℓ^p is distinct, in that ℓ^p is a strict subset of ℓ^s whenever $p < s$; furthermore, ℓ^p is not linearly isomorphic to ℓ^s when $p \neq s$. In fact, by Pitt's theorem (Pitt 1936), every bounded linear operator from ℓ^s to ℓ^p is compact when $p < s$. No such operator can be an isomorphism; and further, it cannot be an isomorphism on any infinite-dimensional subspace of ℓ^s, and is thus said to be strictly singular.

If $1 < p < \infty$, then the (continuous) dual space of ℓ^p is isometrically isomorphic to ℓ^q, where q is the Hölder conjugate of p: $1/p + 1/q = 1$. The specific isomorphism associates to an element x of ℓ^q the functional

$$L_x(y) = \sum_n x_n y_n$$

for y in ℓ^p. Hölder's inequality implies that Lx is a bounded linear functional on ℓ^p, and in fact

$$|L_x(y)| \leq \|x\|_q \|y\|_p$$

so that the operator norm satisfies

$$\|L_x\|_{(\ell^p)^*} \overset{\text{def}}{=} \sup_{y \in \ell^p, y \neq 0} \frac{|L_x(y)|}{\|y\|_p} \leq \|x\|_q.$$

In fact, taking y to be the element of ℓ^p with

$$y_n = \begin{cases} 0 & \text{if } x_n = 0 \\ x_n^{-1}|x_n|^q & \text{if } x_n \neq 0 \end{cases}$$

gives $Lx(y) = \|x\|_q$, so that in fact

$$\|L_x\|_{(\ell^p)^*} = \|x\|_q.$$

Conversely, given a bounded linear functional L on ℓ^p, the sequence defined by $x_n = L(e_n)$ lies in ℓ^q. Thus the mapping $x \mapsto L_x$ gives an isometry

$$\kappa_q : \ell^q \to (\ell^p)^*.$$

The map

$$\ell^q \overset{\kappa_q}{\longrightarrow} (\ell^p)^* \overset{(\kappa_q^*)^{-1}}{\longrightarrow}$$

obtained by composing κp with the inverse of its transpose coincides with the canonical injection of ℓ^q into its double dual. As a consequence ℓ^q is a reflexive space. By abuse of notation, it is typical to identify ℓ^q with the dual of ℓ^p: $(\ell^p)^* = \ell^q$. Then reflexivity is understood by the sequence of identifications $(\ell^p)^{**} = (\ell^q)^* = \ell^p$.

The space c_0 is defined as the space of all sequences converging to zero, with norm identical to $\|x\|_\infty$. It is a closed subspace of ℓ^∞, hence a Banach space. The dual of c_0 is ℓ^1; the dual of ℓ^1 is ℓ^∞. For the case of natural numbers index set, the ℓ^p and c_0 are separable, with the sole exception of ℓ^∞. The dual of ℓ^∞ is the ba space.

The spaces c_0 and ℓ^p (for $1 \le p < \infty$) have a canonical unconditional Schauder basis $\{ei \mid i = 1, 2, \ldots\}$, where ei is the sequence which is zero but for a 1 in the ith entry.

The space ℓ^1 has the Schur property: In ℓ^1, any sequence that is weakly convergent is also strongly convergent (Schur 1921). However, since the weak topology on infinite-dimensional spaces is strictly weaker than the strong topology, there are nets in ℓ^1 that are weak convergent but not strong convergent.

The ℓ^p spaces can be embedded into many Banach spaces. The question of whether every infinite-dimensional Banach space contains an isomorph of some ℓ^p or of c_0, was answered negatively by B. S. Tsirelson's construction of Tsirelson space in 1974. The dual statement, that every separable Banach space is linearly isometric to a quotient space of ℓ^1, was answered in the affirmative by Banach & Mazur (1933). That is, for every separable Banach space X, there exists a quotient map $Q : \ell^1 \to X$, so that X is isomorphic to $\ell^1 / \ker Q$. In general, $\ker Q$ is not complemented in ℓ^1, that is, there does not exist a subspace Y of ℓ^1 such that $\ell^1 = Y \oplus \ker Q$. In fact, ℓ^1 has uncountably many uncomplemented subspaces that are not isomorphic to one another (for example, take $X = \ell^p$; since there are uncountably many such X's, and since no ℓ^p is isomorphic to any other, there are thus uncountably many $\ker Q$'s).

Except for the trivial finite-dimensional case, an unusual feature of ℓ^p is that it is not polynomially reflexive.

4.3 See also

- L^p space

- Tsirelson space

- beta-dual space

4.4 References

- Banach, S.; Mazur, S. (1933), "Zur Theorie der linearen Dimension", *Studia Mathematica* **4**: 100–112.

- Dunford, Nelson; Schwartz, Jacob T. (1958), *Linear operators, volume I*, Wiley-Interscience.

- Pitt, H.R. (1936), "A note on bilinear forms", *J. London Math. Soc.* **11** (3): 174–180, doi:10.1112/jlms/s1-11.3.174.

- Schur, J. (1921), "Über lineare Transformationen in der Theorie der unendlichen Reihen", *Journal für die reine und angewandte Mathematik* **151**: 79–111, doi:10.1515/crll.1921.151.79.

4.2.1 ℓ^p spaces are increasing in p

For $p \in [1, +\infty]$, the spaces ℓ^p are increasing in p, with $1 \le p < q \le +\infty$ implying $\|f\|_q \le \|f\|_p$.

This follows from defining $F := \frac{f}{\|f\|_p}$ for $f \in \ell^p$, and noting that $|F(m)| \le 1$ for all $m \in \mathbb{N}$, which can be shown to imply $\|F\|_q^q \le 1$.

Chapter 5

Sobolev space

In mathematics, a **Sobolev space** is a vector space of functions equipped with a norm that is a combination of L^p-norms of the function itself and its derivatives up to a given order. The derivatives are understood in a suitable weak sense to make the space complete, thus a Banach space. Intuitively, a Sobolev space is a space of functions with sufficiently many derivatives for some application domain, such as partial differential equations, and equipped with a norm that measures both the size and regularity of a function.

Sobolev spaces are named after the Russian mathematician Sergei Sobolev. Their importance comes from the fact that solutions of partial differential equations are naturally found in Sobolev spaces, rather than in spaces of continuous functions and with the derivatives understood in the classical sense.

5.1 Motivation

There are many criteria for smoothness of mathematical functions. The most basic criterion may be that of continuity. A stronger notion of smoothness is that of differentiability (because functions that are differentiable are also continuous) and a yet stronger notion of smoothness is that the derivative also be continuous (these functions are said to be of class C^1 — see Differentiability class). Differentiable functions are important in many areas, and in particular for differential equations. In the twentieth century, however, it was observed that the space C^1 (or C^2, etc.) was not exactly the right space to study solutions of differential equations. The Sobolev spaces are the modern replacement for these spaces in which to look for solutions of partial differential equations.

Quantities or properties of the underlying model of the differential equation are usually expressed in terms of integral norms, rather than the uniform norm. A typical example is measuring the energy of a temperature or velocity distribution by an L^2-norm. It is therefore important to develop a tool for differentiating Lebesgue space functions.

The integration by parts formula yields that for every $u \in C^k(\Omega)$, where k is a natural number and for all infinitely differentiable functions with compact support $\varphi \in C_c^\infty(\Omega)$,

$$\int_\Omega u D^\alpha \varphi \, dx = (-1)^{|\alpha|} \int_\Omega \varphi D^\alpha u \, dx,$$

where α a multi-index of order $|\alpha| = k$ and Ω is an open subset in $\mathbb{R}^n$. Here, the notation

$$D^\alpha f = \frac{\partial^{|\alpha|} f}{\partial x_1^{\alpha_1} \dots \partial x_n^{\alpha_n}},$$

is used.

The left-hand side of this equation still makes sense if we only assume u to be locally integrable. If there exists a locally integrable function v, such that

$$\int_\Omega u D^\alpha \varphi \, dx = (-1)^{|\alpha|} \int_\Omega \varphi v \, dx, \quad \varphi \in C_c^\infty(\Omega),$$

we call v the weak α-th partial derivative of u. If there exists a weak α-th partial derivative of u, then it is uniquely defined almost everywhere. On the other hand, if $u \in C^k(\Omega)$, then the classical and the weak derivative coincide. Thus, if v is a weak α-th partial derivative of u, we may denote it by $D^\alpha u := v$.

For example, the function

$$u(x) = \begin{cases} 1 + x & \text{if } -1 < x < 0 \\ 10 & \text{if } x = 0 \\ 1 - x & \text{if } 0 < x < 1 \\ 0 & \text{otherwise} \end{cases}$$

is not continuous at zero, and not differentiable at -1, 0, or 1. Yet the function

$$v(x) = \begin{cases} 1 & \text{if} -1 < x < 0 \\ -1 & \text{if} 0 < x < 1 \\ 0 & \text{otherwise} \end{cases}$$

satisfies the definition for being the weak derivative of $u(x)$, which then qualifies as being in the Sobolev space $W^{1,p}$ (for any allowed p, see definition below).

The Sobolev spaces $W^{k,p}(\Omega)$ combine the concepts of weak differentiability and Lebesgue norms.

5.2 Sobolev spaces with integer k

5.2.1 One-dimensional case

In the one-dimensional case (functions on **R**) the Sobolev space $W^{k,p}$ is defined to be the subset of functions f in $L^p(\mathbf{R})$ such that the function f and its weak derivatives up to some order k have a finite L^p norm, for given p ($1 \le p \le +\infty$). As mentioned above, some care must be taken to define derivatives in the proper sense. In the one-dimensional problem it is enough to assume that $f^{(k-1)}$, the $(k-1)$-th derivative of the function f, is differentiable almost everywhere and is equal almost everywhere to the Lebesgue integral of its derivative (this gets rid of examples such as Cantor's function which are irrelevant to what the definition is trying to accomplish).

With this definition, the Sobolev spaces admit a natural norm,

$$\|f\|_{k,p} = \left(\sum_{i=0}^{k} \left\| f^{(i)} \right\|_p^p \right)^{\frac{1}{p}} = \left(\sum_{i=0}^{k} \int \left| f^{(i)}(t) \right|^p dt \right)^{\frac{1}{p}}.$$

Equipped with the norm $\| \cdot \|_{k,p}$, $W^{k,p}$ becomes a Banach space. It turns out that it is enough to take only the first and last in the sequence, i.e., the norm defined by

$$\left\| f^{(k)} \right\|_p + \|f\|_p$$

is equivalent to the norm above (i.e. the induced topologies of the norms are the same).

The case $p = 2$

Sobolev spaces with $p = 2$ (at least on a one-dimensional finite interval) are especially important because of their connection with Fourier series and because they form a Hilbert space. A special notation has arisen to cover this case, since the space is a Hilbert space:

$$H^k = W^{k,2}.$$

The space H^k can be defined naturally in terms of Fourier series whose coefficients decay sufficiently rapidly, namely,

$$H^k(\mathbb{T}) = \left\{ f \in L^2(\mathbb{T}) : \sum_{n=-\infty}^{\infty} \left(1 + n^2 + n^4 \right. \right.$$
$$\left. \left. + \cdots + n^{2k} \right) \left| \widehat{f}(n) \right|^2 < \infty \right\}$$

where $\widehat{f}$ is the Fourier series of f. As above, one can use the equivalent norm

$$\|f\|_{k,2}^2 = \sum_{n=-\infty}^{\infty} \left(1 + |n|^2 \right)^k \left| \widehat{f}(n) \right|^2.$$

Both representations follow easily from Parseval's theorem and the fact that differentiation is equivalent to multiplying the Fourier coefficient by in.

Furthermore, the space H^k admits an inner product, like the space $H^0 = L^2$. In fact, the H^k inner product is defined in terms of the L^2 inner product:

$$\langle u, v \rangle_{H^k} = \sum_{i=0}^{k} \left\langle D^i u, D^i v \right\rangle_{L^2}.$$

The space H^k becomes a Hilbert space with this inner product.

Other examples

Some other Sobolev spaces permit a simpler description. For example, $W^{1,1}(0, 1)$ is the space of absolutely continuous functions on $(0, 1)$ (or rather, equivalence classes of functions that are equal almost everywhere to such), while $W^{1,\infty}(I)$ is the space of Lipschitz functions on I, for every interval I. All spaces $W^{k,\infty}$ are (normed) algebras, i.e. the product of two elements is once again a function of this Sobolev space, which is not the case for $p < +\infty$. (E.g., functions behaving like $|x|^{-1/3}$ at the origin are in L^2, but the product of two such functions is not in L^2).

5.2.2 Multidimensional case

The transition to multiple dimensions brings more difficulties, starting from the very definition. The requirement that $f^{(k-1)}$ be the integral of $f^{(k)}$ does not generalize, and the simplest solution is to consider derivatives in the sense of distribution theory.

A formal definition now follows. Let Ω be an open set in $\mathbf{R}^n$, let k be a natural number and let $1 \leq p \leq +\infty$. The Sobolev space $W^{k,p}(\Omega)$ is defined to be the set of all functions f defined on Ω such that for every multi-index α with $|\alpha| \leq k$, the mixed partial derivative

$$f^{(\alpha)} = \frac{\partial^{|\alpha|} f}{\partial x_1^{\alpha_1} \dots \partial x_n^{\alpha_n}}$$

is both locally integrable and in $L^p(\Omega)$, i.e.

$$\left\| f^{(\alpha)} \right\|_{L^p} < \infty.$$

That is, the Sobolev space $W^{k,p}(\Omega)$ is defined as

$$W^{k,p}(\Omega) = \{u \in L^p(\Omega) : D^\alpha u \in L^p(\Omega) \; \forall |\alpha| \leq k\}.$$

The natural number k is called the order of the Sobolev space $W^{k,p}(\Omega)$.

There are several choices for a norm for $W^{k,p}(\Omega)$. The following two are common and are equivalent in the sense of equivalence of norms:

$$\|u\|_{W^{k,p}(\Omega)} := \begin{cases} \left(\sum_{|\alpha| \leq k} \|D^\alpha u\|_{L^p(\Omega)}^p \right)^{\frac{1}{p}}, \leq p < +\infty; \\ \max_{|\alpha| \leq k} \|D^\alpha u\|_{L^\infty(\Omega)}, \; p = +\infty; \end{cases}$$

and

$$\|u\|'_{W^{k,p}(\Omega)} := \begin{cases} \sum_{|\alpha| \leq k} \|D^\alpha u\|_{L^p(\Omega)}, & 1 \leq p < +\infty; \\ \sum_{|\alpha| \leq k} \|D^\alpha u\|_{L^\infty(\Omega)}, & p = +\infty. \end{cases}$$

With respect to either of these norms, $W^{k,p}(\Omega)$ is a Banach space. For $p < +\infty$, $W^{k,p}(\Omega)$ is also a separable space. It is conventional to denote $W^{k,2}(\Omega)$ by $H^k(\Omega)$ for it is a Hilbert space with the norm $\| \cdot \|_{W^{k,2}(\Omega)}$.[1]

Approximation by smooth functions

Many of the properties of the Sobolev spaces cannot be seen directly from the definition. It is therefore interesting to investigate under which conditions a function $u \in W^{k,p}(\Omega)$ can be approximated by smooth functions. If p is finite and Ω is bounded with Lipschitz boundary, then for any $u \in W^{k,p}(\Omega)$ there exists an approximating sequence of functions $um \in C^\infty(\Omega)$, smooth up to the boundary such that:[2]

$$\|u_m - u\|_{W^{k,p}(\Omega)} \to 0.$$

Examples

In higher dimensions, it is no longer true that, for example, $W^{1,1}$ contains only continuous functions. For example, $1/|x|$ belongs to $W^{1,1}(\mathbf{B}^3)$ where $\mathbf{B}^3$ is the unit ball in three dimensions. For $k > n/p$ the space $W^{k,p}(\Omega)$ will contain only continuous functions, but for which k this is already true depends both on p and on the dimension. For example, as can be easily checked using spherical polar coordinates for the function $f : \mathbf{B}^n \to \mathbf{R} \cup \{+\infty\}$, defined on the n-dimensional ball we have:

$$f(x) = |x|^{-\alpha} \in W^{k,p}(\mathbf{B}^n) \; \Leftrightarrow \; \alpha < \frac{n}{p} - k.$$

Intuitively, the blow-up of f at 0 "counts for less" when n is large since the unit ball has "more outside and less inside" in higher dimensions.

Absolutely continuous on lines (ACL) characterization of Sobolev functions

Let Ω be an open set in $\mathbf{R}^n$ and $1 \leq p \leq +\infty$. If a function is in $W^{1,p}(\Omega)$, then, possibly after modifying the function on a set of measure zero, the restriction to almost every line parallel to the coordinate directions in $\mathbf{R}^n$ is absolutely continuous; what's more, the classical derivative along the lines that are parallel to the coordinate directions are in $L^p(\Omega)$. Conversely, if the restriction of f to almost every line parallel to the coordinate directions is absolutely continuous, then the pointwise gradient ∇f exists almost everywhere, and f is in $W^{1,p}(\Omega)$ provided f and $|\nabla f|$ are both in $L^p(\Omega)$. In particular, in this case the weak partial derivatives of f and pointwise partial derivatives of f agree almost everywhere. The ACL characterization of the Sobolev spaces was established by Otto M. Nikodym (1933); see (Maz'ya 1985, §1.1.3).

A stronger result holds in the case $p > n$. A function in $W^{1,p}(\Omega)$ is, after modifying on a set of measure zero, Hölder continuous of exponent $\gamma = 1 - n/p$, by Morrey's inequality. In particular, if $p = +\infty$, then the function is Lipschitz continuous.

Functions vanishing at the boundary

Let Ω be an open set in $\mathbf{R}^n$. The Sobolev space $W^{1,2}(\Omega)$ is also denoted by $H^1(\Omega)$. It is a Hilbert space, with an important subspace $H1$

$0(\Omega)$ defined to be the closure in $H^1(\Omega)$ of the infinitely differentiable functions compactly supported in Ω. The Sobolev norm defined above reduces here to

$$\|f\|_{H^1} = \left(\int_\Omega \left(|f|^2 + |\nabla f|^2 \right) \right)^{\frac{1}{2}}.$$

When Ω has a regular boundary, $H1$

$0(\Omega)$ can be described as the space of functions in $H^1(\Omega)$ that vanish at the boundary, in the sense of traces (see below). When $n = 1$, if $\Omega = (a, b)$ is a bounded interval, then $H1$

$0(a, b)$ consists of continuous functions on $[a, b]$ of the form

$$f(x) = \int_a^x f'(t)\,\mathrm{d}t, \qquad x \in [a, b]$$

where the generalized derivative f' is in $L^2(a, b)$ and has 0 integral, so that $f(b) = f(a) = 0$.

When Ω is bounded, the Poincaré inequality states that there is a constant $C = C(\Omega)$ such that

$$\int_\Omega |f|^2 \le C^2 \int_\Omega |\nabla f|^2, \quad f \in H^1_0(\Omega).$$

When Ω is bounded, the injection from $H1$

$0(\Omega)$ to $L^2(\Omega)$ is compact. This fact plays a role in the study of the Dirichlet problem, and in the fact that there exists an orthonormal basis of $L^2(\Omega)$ consisting of eigenvectors of the Laplace operator (with Dirichlet boundary condition).

5.3 Sobolev spaces with non-integer k

5.3.1 Bessel potential spaces

For a natural number k and $1 < p < \infty$ one can show (by using Fourier multipliers[3][4]) that the space $W^{k,p}(\mathbb{R}^n)$ can equivalently be defined as

$$W^{k,p}(\mathbb{R}^n) = H^{k,p}(\mathbb{R}^n) := \Big\{ f \in L^p(\mathbb{R}^n) :$$
$$F^{-1}[(1 + |\xi|^2)^{\frac{k}{2}} \mathcal{F} f] \in L^p(\mathbb{R}^n) \Big\}$$

with the norm

$$\|f\|_{H^{k,p}(\mathbb{R}^n)} := \left\| \mathcal{F}^{-1}[(1 + |\xi|^2)^{\frac{k}{2}} \mathcal{F} f] \right\|_{L^p(\mathbb{R}^n)}$$

This motivates Sobolev spaces with non-integer order since in the above definition we can replace k by any real number s. The resulting spaces

$$H^{s,p}(\mathbb{R}^n) := \Big\{$$
$$f \in L^p(\mathbb{R}^n) : \mathcal{F}^{-1}[(1 + |\xi|^2)^{\frac{s}{2}} \mathcal{F} f] \in L^p(\mathbb{R}^n) \Big\}$$

are called Bessel potential spaces[5] (named after Friedrich Bessel). They are Banach spaces in general and Hilbert spaces in the special case $p = 2$.

For an open set $\Omega \subseteq \mathbb{R}^n$, $H^{s,p}(\Omega)$ is the set of restrictions of functions from $H^{s,p}(\mathbb{R}^n)$ to Ω equipped with the norm

$$\|f\|_{H^{s,p}(\Omega)} := \inf \big\{ \|g\|_{H^{s,p}(\mathbb{R}^n)} : g \in H^{s,p}(\mathbb{R}^n), g|_\Omega = f \big\}$$

Again, $H^{s,p}(\Omega)$ is a Banach space and in the case $p = 2$ a Hilbert space.

Using extension theorems for Sobolev spaces, it can be shown that also $W^{k,p}(\Omega) = H^{k,p}(\Omega)$ holds in the sense of equivalent norms, if Ω is domain with uniform C^k-*boundary,* k *a natural number and $1 < p < \infty$. By the embeddings*

$$H^{k+1,p}(\mathbb{R}^n) \hookrightarrow H^{s',p}(\mathbb{R}^n) \hookrightarrow H^{s,p}(\mathbb{R}^n) \hookrightarrow H^{k,p}(\mathbb{R}^n),$$
$$, k \le s \le s' \le k+1$$

the Bessel potential spaces $H^{s,p}(\mathbb{R}^n)$ form a continuous scale between the Sobolev spaces $W^{k,p}(\mathbb{R}^n)$. From an abstract point of view, the Bessel potential spaces occur as complex interpolation spaces of Sobolev spaces, i.e. in the sense of equivalent norms it holds that

$$\left[W^{k,p}(\mathbb{R}^n), W^{k+1,p}(\mathbb{R}^n) \right]_\theta = H^{s,p}(\mathbb{R}^n),$$

where:

$$1 \le p \le \infty, \, 0 < \theta < 1, \, s = (1-\theta)k + \theta(k+1) = k+\theta.$$

5.3.2 Sobolev–Slobodeckij spaces

Another approach to define fractional order Sobolev spaces arises from the idea to generalize the Hölder condition to the L^p-setting.[6] For an open subset Ω of $\mathbb{R}^n$, $1 \le p < \infty$, $\theta \in (0,1)$ and $f \in L^p(\Omega)$, the **Slobodeckij seminorm** (roughly analogous to the Hölder seminorm) is defined by

$$[f]_{\theta,p,\Omega} := \left(\int_\Omega \int_\Omega \frac{|f(x) - f(y)|^p}{|x - y|^{\theta p + n}} \, dx \, dy \right)^{\frac{1}{p}}$$

Let $s > 0$ be not an integer and set $\theta = s - \lfloor s \rfloor \in (0,1)$. Using the same idea as for the Hölder spaces, the **Sobolev–Slobodeckij space**[7] $W^{s,p}(\Omega)$ is defined as

$$W^{s,p}(\Omega) := \left\{ f \in W^{\lfloor s \rfloor, p}(\Omega) : \sup_{|\alpha| = \lfloor s \rfloor} [D^\alpha f]_{\theta, p, \Omega} < \infty \right\}$$

It is a Banach space for the norm

$$\|f\|_{W^{s,p}(\Omega)} := \|f\|_{W^{\lfloor s \rfloor, p}(\Omega)} + \sup_{|\alpha| = \lfloor s \rfloor} [D^\alpha f]_{\theta, p, \Omega}$$

If the open subset Ω is suitably regular in the sense that there exist certain extension operators, then also the Sobolev–Slobodeckij spaces form a scale of Banach spaces, i.e. one has the continuous injections or embeddings

$$W^{k+1,p}(\Omega) \hookrightarrow W^{s',p}(\Omega) \hookrightarrow W^{s,p}(\Omega) \hookrightarrow W^{k,p}(\Omega), \quad k \leq s \leq s' \leq k+1.$$

There are examples of irregular Ω such that $W^{1,p}(\Omega)$ is not even a vector subspace of $W^{s,p}(\Omega)$ for $0 < s < 1$.

From an abstract point of view, the spaces $W^{s,p}(\Omega)$ coincide with the real interpolation spaces of Sobolev spaces, i.e. in the sense of equivalent norms the following holds:

$$W^{s,p}(\Omega) = \left(W^{k,p}(\Omega), W^{k+1,p}(\Omega) \right)_{\theta, p}$$

$$, \quad k \in \mathbb{N}, s \in (k, k+1), \theta = s - \lfloor s \rfloor$$

Sobolev–Slobodeckij spaces play an important role in the study of traces of Sobolev functions. They are special cases of Besov spaces.[4]

5.4   Traces

Sobolev spaces are often considered when investigating partial differential equations. It is essential to consider boundary values of Sobolev functions. If $u \in C(\Omega)$, those boundary values are described by the restriction $u|_{\partial\Omega}$. However, it is not clear how to describe values at the boundary for $u \in W^{k,p}(\Omega)$, as the n-dimensional measure of the boundary is zero. The following theorem[2] resolves the problem:

> **Trace Theorem.** Assume Ω is bounded with Lipschitz boundary. Then there exists a bounded linear operator $T : W^{1,p}(\Omega) \to L^p(\partial\Omega)$ such that
>
> $$Tu = u|_{\partial\Omega} \qquad u \in W^{1,p}(\Omega) \cap C(\overline{\Omega})$$
>
> $$\|Tu\|_{L^p(\partial\Omega)} \leq c(p, \Omega)\|u\|_{W^{1,p}(\Omega)} \quad u \in W^{1,p}(\Omega).$$

Tu is called the trace of u. Roughly speaking, this theorem extends the restriction operator to the Sobolev space

$W^{1,p}(\Omega)$ for well-behaved Ω. Note that the trace operator T is in general not surjective, but for $1 < p < \infty$ it maps onto the Sobolev-Slobodeckij space $W^{1 - \frac{1}{p}, p}(\partial\Omega)$.

Intuitively, taking the trace costs $1/p$ of a derivative. The functions u in $W^{1,p}(\Omega)$ with zero trace, i.e. $Tu = 0$, can be characterized by the equality

$$W_0^{1,p}(\Omega) = \left\{ u \in W^{1,p}(\Omega) : Tu = 0 \right\},$$

where

$$W_0^{1,p}(\Omega) := \left\{ u \in W^{1,p}(\Omega) : \exists \{u_m\}_{m=1}^\infty \subset C^\infty{}_c \right.$$

$$(\Omega), \text{ such that } u_m \to u \text{ in } W$$

In other words, for Ω bounded with Lipschitz boundary, trace-zero functions in $W^{1,p}(\Omega)$ can be approximated by smooth functions with compact support.

5.5   Extension operators

If X is an open domain whose boundary is not too poorly behaved (e.g., if its boundary is a manifold, or satisfies the more permissive "cone condition") then there is an operator A mapping functions of X to functions of $\mathbf{R}^n$ such that:

1. $Au(x) = u(x)$ for almost every x in X and

2. A is continuous from $W^{k,p}(X)$ to $W^{k,p}(\mathbb{R}^n)$, for any $1 \leq p \leq \infty$ and integer k.

We will call such an operator A an extension operator for X.

5.5.1   Case of $p = 2$

Extension operators are the most natural way to define $H^s(X)$ for non-integer s (we cannot work directly on X since taking Fourier transform is a global operation). We define $H^s(X)$ by saying that u is in $H^s(X)$ if and only if Au is in $H^s(\mathbb{R}^n)$. Equivalently, complex interpolation yields the same $H^s(X)$ spaces so long as X has an extension operator. If X does not have an extension operator, complex interpolation is the only way to obtain the $H^s(X)$ spaces.

As a result, the interpolation inequality still holds.

5.5.2   Extension by zero

As in the section #Functions vanishing at the boundary, we define $H_0^s(X)$ to be the closure in $H^s(X)$ of the space

$C_c^\infty(X)$ of infinitely differentiable compactly supported functions. Given the definition of a trace, above, we may state the following

Theorem *Let X be uniformly C^m regular, $m \geq s$ and let P be the linear map sending u in $H^s(X)$ to*

$$\left(u, \frac{du}{dn}, \cdots, \frac{d^k u}{dn^k}\right)\Bigg|_G$$

where d/dn is the derivative normal to G, and k is the largest integer less than s. Then H_0^s is precisely the kernel of P.

If $u \in H_0^s(X)$ we may define its **extension by zero** $\tilde{u} \in L^2(\mathbb{R}^n)$ in the natural way, namely

$$\tilde{u}(x) = u(x) \text{ if } x \in X, 0 \text{ otherwise.}$$

Theorem *Let $s > \frac{1}{2}$. The map taking u to $\tilde{u}$ is continuous into $H^s(\mathbb{R}^n)$ if and only if s is not of the form $n + \frac{1}{2}$ for n an integer.*

For a function $f \in L^p(\Omega)$ on an open subset Ω of $\mathbb{R}^n$, its extension by zero

$$Ef := \begin{cases} f & \text{on } \Omega, \\ 0 & \text{otherwise} \end{cases}$$

is an element of $L^p(\mathbb{R}^n)$. Furthermore,

$$\|Ef\|_{L^p(\mathbb{R}^n)} = \|f\|_{L^p(\Omega)}.$$

In the case of the Sobolev space $W^{1,p}(\Omega)$ for $1 \leq p \leq \infty$, extending a function u by zero will not necessarily yield an element of $W^{1,p}(\mathbb{R}^n)$. But if Ω is bounded with Lipschitz boundary (e.g. $\partial\Omega$ is C^1), then for any bounded open set O such that $\Omega \subset\subset O$ (i.e. Ω is compactly contained in O), there exists a bounded linear operator[2]

$$E : W^{1,p}(\Omega) \to W^{1,p}(\mathbb{R}^n),$$

such that for each $u \in W^{1,p}(\Omega)$: $Eu = u$ a.e. on Ω, Eu has compact support within O, and there exists a constant C depending only on p, Ω, O and the dimension n, such that

$$\|Eu\|_{W^{1,p}(\mathbb{R}^n)} \leq C \|u\|_{W^{1,p}(\Omega)}.$$

We call Eu an extension of u to $\mathbb{R}^n$.

5.6 Sobolev embeddings

Main article: Sobolev inequality

It is a natural question to ask if a Sobolev function is continuous or even continuously differentiable. Roughly speaking, sufficiently many weak derivatives or large p result in a classical derivative. This idea is generalized and made precise in the Sobolev embedding theorem.

Write $W^{k,p}$ for the Sobolev space of some compact Riemannian manifold of dimension n. Here k can be any real number, and $1 \leq p \leq \infty$. (For $p = \infty$ the Sobolev space $W^{k,\infty}$ is defined to be the Hölder space $C^{n,\alpha}$ where $k = n + \alpha$ and $0 < \alpha \leq 1$.) The Sobolev embedding theorem states that if $k \geq m$ and $k - n/p \geq m - n/q$ then

$$W^{k,p} \subseteq W^{m,q}$$

and the embedding is continuous. Moreover if $k > m$ and $k - n/p > m - n/q$ then the embedding is completely continuous (this is sometimes called **Kondrachov's theorem** or the **Rellich-Kondrachov theorem**). Functions in $W^{m,\infty}$ have all derivatives of order less than m are continuous, so in particular this gives conditions on Sobolev spaces for various derivatives to be continuous. Informally these embeddings say that to convert an L^p estimate to a boundedness estimate costs $1/p$ derivatives per dimension.

There are similar variations of the embedding theorem for non-compact manifolds such as $\mathbf{R}^n$ (Stein 1970).

5.7 Notes

[1] Evans 1998, Chapter 5.2

[2] Adams 1975

[3] Bergh & Löfström 1976

[4] Triebel 1995

[5] Bessel potential spaces with variable integrability have been independently introduced by Almeida & Samko (A. Almeida and S. Samko, "Characterization of Riesz and Bessel potentials on variable Lebesgue spaces", J. Function Spaces Appl. 4 (2006), no. 2, 113–144) and Gurka, Harjulehto & Nekvinda (P. Gurka, P. Harjulehto and A. Nekvinda: "Bessel potential spaces with variable exponent", Math. Inequal. Appl. 10 (2007), no. 3, 661–676).

[6] Lunardi 1995

[7] In the literature, fractional Sobolev-type spaces are also called *Aronszajn spaces*, *Gagliardo spaces* or *Slobodeckij*

spaces, after the names of the mathematicians who introduced them in the 1950s: N. Aronszajn ("Boundary values of functions with finite Dirichlet integral", Techn. Report of Univ. of Kansas 14 (1955), 77–94), E. Gagliardo ("Proprietà di alcune classi di funzioni in più variabili", *Ricerche Mat.* 7 (1958), 102–137), and L. N. Slobodeckij ("Generalized Sobolev spaces and their applications to boundary value problems of partial differential equations", Leningrad. *Gos. Ped. Inst. Učep. Zap.* 197 (1958), 54–112).

5.8   References

- Adams, Robert A. (1975), *Sobolev Spaces*, Boston, MA: Academic Press, ISBN 978-0-12-044150-1.

- Aubin, Thierry (1982), *Nonlinear analysis on manifolds. Monge-Ampère equations*, Grundlehren der Mathematischen Wissenschaften [Fundamental Principles of Mathematical Sciences] **252**, Berlin, New York: Springer-Verlag, ISBN 978-0-387-90704-8, MR 681859.

- Bergh, Jöran; Löfström, Jörgen (1976), *Interpolation Spaces, An Introduction*, Grundlehren der Mathematischen Wissenschaften **223**, Springer-Verlag, pp. X + 207, ISBN 978-7-5062-6011-4, MR 0482275, Zbl 0344.46071

- Evans, L.C. (1998), *Partial Differential Equations*, AMS_Chelsea.

- Maz'ja, Vladimir G. (1985), *Sobolev Spaces*, Springer Series in Soviet Mathematics, Berlin–Heidelberg–New York: Springer-Verlag, pp. xix+486, ISBN 0-387-13589-8, MR 817985, Zbl 0692.46023.

- Maz'ya, Vladimir G.; Poborchi, Sergei V. (1997), *Differentiable Functions on Bad Domains*, Singapore–New Jersey–London–Hong Kong: World Scientific, pp. xx+481, ISBN 981-02-2767-1, MR 1643072, Zbl 0918.46033.

- Maz'ya, Vladimir G. (2011) [1985], *Sobolev Spaces. With Applications to Elliptic Partial Differential Equations.*, Grundlehren der Mathematischen Wissenschaften **342** (2nd revised and augmented ed.), Berlin–Heidelberg–New York: Springer Verlag, pp. xxviii+866, ISBN 978-3-642-15563-5, MR 2777530, Zbl 1217.46002.

- Lunardi, Alessandra (1995), *Analytic semigroups and optimal regularity in parabolic problems*, Basel: Birkhäuser Verlag.

- Nikodym, Otto (1933), "Sur une classe de fonctions considérée dans l'étude du problème de Dirichlet", *Fund. Math.* **21**: 129–150.

- Nikol'skii, S.M. (2001), "Imbedding theorems", in Hazewinkel, Michiel, *Encyclopedia of Mathematics*, Springer, ISBN 978-1-55608-010-4.

- Nikol'skii, S.M. (2001), "Sobolev space", in Hazewinkel, Michiel, *Encyclopedia of Mathematics*, Springer, ISBN 978-1-55608-010-4.

- Sobolev, S.L. (1963), "On a theorem of functional analysis", *Transl. Amer. Math. Soc.* **34** (2): 39–68; translation of Mat. Sb., 4 (1938) pp. 471–497.

- Sobolev, S.L. (1963), *Some applications of functional analysis in mathematical physics*, Amer. Math. Soc..

- Stein, E (1970), *Singular Integrals and Differentiability Properties of Functions*, Princeton Univ. Press, ISBN 0-691-08079-8.

- Triebel, H. (1995), *Interpolation Theory, Function Spaces, Differential Operators*, Heidelberg: Johann Ambrosius Barth.

- Ziemer, William P. (1989), *Weakly differentiable functions*, Graduate Texts in Mathematics **120**, Berlin, New York: Springer-Verlag, ISBN 978-0-387-97017-2, MR 1014685.

5.9   External links

- Eleonora Di Nezza, Giampiero Palatucci, Enrico Valdinoci (2011). "Hitchhiker's guide to the fractional Sobolev spaces".

Chapter 6

Generalized function

In mathematics, **generalized functions**, or **distributions**, are objects extending the notion of functions. There is more than one recognized theory. Generalized functions are especially useful in making discontinuous functions more like smooth functions, and describing discrete physical phenomena such as point charges. They are applied extensively, especially in physics and engineering.

A common feature of some of the approaches is that they build on operator aspects of everyday, numerical functions. The early history is connected with some ideas on operational calculus, and more contemporary developments in certain directions are closely related to ideas of Mikio Sato, on what he calls algebraic analysis. Important influences on the subject have been the technical requirements of theories of partial differential equations, and group representation theory.

6.1   Some early history

In the mathematics of the nineteenth century, aspects of generalized function theory appeared, for example in the definition of the Green's function, in the Laplace transform, and in Riemann's theory of trigonometric series, which were not necessarily the Fourier series of an integrable function. These were disconnected aspects of mathematical analysis at the time.

The intensive use of the Laplace transform in engineering led to the heuristic use of symbolic methods, called operational calculus. Since justifications were given that used divergent series, these methods had a bad reputation from the point of view of pure mathematics. They are typical of later application of generalized function methods. An influential book on operational calculus was Oliver Heaviside's *Electromagnetic Theory* of 1899.

When the Lebesgue integral was introduced, there was for the first time a notion of generalized function central to mathematics. An integrable function, in Lebesgue's theory, is equivalent to any other which is the same almost every-

where. That means its value at a given point is (in a sense) not its most important feature. In functional analysis a clear formulation is given of the *essential* feature of an integrable function, namely the way it defines a linear functional on other functions. This allows a definition of weak derivative.

During the late 1920s and 1930s further steps were taken, basic to future work. The Dirac delta function was boldly defined by Paul Dirac (an aspect of his scientific formalism); this was to treat measures, thought of as densities (such as charge density) like honest functions. Sergei Sobolev, working in partial differential equation theory, defined the first adequate theory of generalized functions, from the mathematical point of view, in order to work with weak solutions of PDEs.[1] Others proposing related theories at the time were Salomon Bochner and Kurt Friedrichs. Sobolev's work was further developed in an extended form by L. Schwartz.[2]

6.2   Schwartz distributions

The realization of such a concept that was to become accepted as definitive, for many purposes, was the theory of distributions, developed by Laurent Schwartz. It can be called a principled theory, based on duality theory for topological vector spaces. Its main rival, in applied mathematics, is to use sequences of smooth approximations (the 'James Lighthill' explanation), which is more *ad hoc*. This now enters the theory as mollifier theory.[3]

This theory was very successful and is still widely used, but suffers from the main drawback that it allows only linear operations. In other words, distributions cannot be multiplied (except for very special cases): unlike most classical function spaces, they are not an algebra. For example it is not meaningful to square the Dirac delta function. Work of Schwartz from around 1954 showed that this was an intrinsic difficulty.

Some solutions to the multiplication problem have been

proposed. One is based on a very simple and intuitive definition a generalized function given by Yu. V. Egorov[4] (see also his article in Demidov's book in the book list below) that allows arbitrary operations on, and between, generalized functions.

Another solution of the multiplication problem is dictated by the path integral formulation of quantum mechanics. Since this is required to be equivalent to the Schrödinger theory of quantum mechanics which is invariant under coordinate transformations, this property must be shared by path integrals. This fixes all products of generalized functions as shown by H. Kleinert and A. Chervyakov.[5] The result is equivalent to what can be derived from dimensional regularization.[6]

6.3 Algebras of generalized functions

Several constructions of algebras of generalized functions have been proposed, among others those by Yu. M. Shirokov [7] and those by E. Rosinger, Y. Egorov, and R. Robinson.[8] In the first case, the multiplication is determined with some regularization of generalized function. In the second case, the algebra is constructed as *multiplication of distributions*. Both cases are discussed below.

6.3.1 Non-commutative algebra of generalized functions

The algebra of generalized functions can be built-up with an appropriate procedure of projection of a function $F = F(x)$ to its smooth F_{smooth} and its singular $F_{singular}$ parts. The product of generalized functions F and G appears as

(1) $FG = F_{smooth} G_{smooth} + F_{smooth} G_{singular} + F_{singular} G_{smooth}.$

Such a rule applies to both the space of main functions and the space of operators which act on the space of the main functions. The associativity of multiplication is achieved; and the function signum is defined in such a way, that its square is unity everywhere (including the origin of coordinates). Note that the product of singular parts does not appear in the right-hand side of (1); in particular, $\delta(x)^2 = 0$. Such a formalism includes the conventional theory of generalized functions (without their product) as a special case. However, the resulting algebra is non-commutative: generalized functions signum and delta anticommute.[7] Few applications of the algebra were suggested.[9][10]

6.3.2 Multiplication of distributions

The problem of *multiplication of distributions*, a limitation of the Schwartz distribution theory, becomes serious for non-linear problems.

Various approaches are used today. The simplest one is based on the definition of generalized function given by Yu. V. Egorov.[4] Another approach to construct associative differential algebras is based on J.-F. Colombeau's construction: see Colombeau algebra. These are factor spaces

$$G = M/N$$

of "moderate" modulo "negligible" nets of functions, where "moderateness" and "negligibility" refers to growth with respect to the index of the family.

6.3.3 Example: Colombeau algebra

A simple example is obtained by using the polynomial scale on **N**, $s = \{a_m : \mathbb{N} \to \mathbb{R}, n \mapsto n^m; \ m \in \mathbb{Z}\}$. Then for any semi normed algebra (E,P), the factor space will be

$$G_s(E, P) = \frac{\{f \in E^{\mathbb{N}} \mid \forall p \in P, \exists m \in \mathbb{Z} : p(f_n) = o(n^m)\}}{\{f \in E^{\mathbb{N}} \mid \forall p \in P, \forall m \in \mathbb{Z} : p(f_n) = o(n^m)\}}$$

In particular, for $(E, P) = (\mathbf{C}, |.|)$ one gets (Colombeau's) generalized complex numbers (which can be "infinitely large" and "infinitesimally small" and still allow for rigorous arithmetics, very similar to nonstandard numbers). For $(E, P) = (C^\infty(\mathbf{R}), \{pk\})$ (where *pk* is the supremum of all derivatives of order less than or equal to *k* on the ball of radius *k*) one gets Colombeau's simplified algebra.

6.3.4 Injection of Schwartz distributions

This algebra "contains" all distributions T of D' via the injection

$$j(T) = (\varphi n * T)n + N,$$

where $*$ is the convolution operation, and

$$\varphi n(x) = n \, \varphi(nx).$$

This injection is *non-canonical* in the sense that it depends on the choice of the mollifier φ, which should be C^∞, of integral one and have all its derivatives at 0 vanishing. To obtain a canonical injection, the indexing set can be modified to be $\mathbf{N} \times D(\mathbf{R})$, with a convenient filter base on $D(\mathbf{R})$ (functions of vanishing moments up to order q).

6.3.5 Sheaf structure

If (E,P) is a (pre-)sheaf of semi normed algebras on some topological space X, then $Gs(E, P)$ will also have this property. This means that the notion of restriction will be defined, which allows to define the support of a generalized function w.r.t. a subsheaf, in particular:

- For the subsheaf $\{0\}$, one gets the usual support (complement of the largest open subset where the function is zero).

- For the subsheaf E (embedded using the canonical (constant) injection), one gets what is called the singular support, i.e., roughly speaking, the closure of the set where the generalized function is not a smooth function (for $E = C^\infty$).

6.3.6 Microlocal analysis

The Fourier transformation being (well-)defined for compactly supported generalized functions (component-wise), one can apply the same construction as for distributions, and define Lars Hörmander's *wave front set* also for generalized functions.

This has an especially important application in the analysis of propagation of singularities.

6.4 Other theories

These include: the *convolution quotient* theory of Jan Mikusinski, based on the field of fractions of convolution algebras that are integral domains; and the theories of hyperfunctions, based (in their initial conception) on boundary values of analytic functions, and now making use of sheaf theory.

6.5 Topological groups

Bruhat introduced a class of test functions, the Schwartz–Bruhat functions as they are now known, on a class of locally compact groups that goes beyond the manifolds that are the typical function domains. The applications are mostly in number theory, particularly to adelic algebraic groups. André Weil rewrote Tate's thesis in this language, characterizing the zeta distribution on the idele group; and has also applied it to the explicit formula of an L-function.

6.6 Generalized section

A further way in which the theory has been extended is as **generalized sections** of a smooth vector bundle. This is on the Schwartz pattern, constructing objects dual to the test objects, smooth sections of a bundle that have compact support. The most developed theory is that of De Rham currents, dual to differential forms. These are homological in nature, in the way that differential forms give rise to De Rham cohomology. They can be used to formulate a very general Stokes' theorem.

6.7 See also

- Beppo-Levi space
- Dirac delta function
- Generalized eigenfunction
- Distribution (mathematics)
- Hyperfunction
- Laplacian of the indicator
- Rigged Hilbert space

6.8 Books

- L. Schwartz: Théorie des distributions
- L. Schwartz: Sur l'impossibilité de la multiplication des distributions. Comptes Rendus de L'Academie des Sciences, Paris, 239 (1954) 847-848.
- I.M. Gel'fand et al.: Generalized Functions, vols I–VI, Academic Press, 1964. (Translated from Russian.)
- L. Hörmander: The Analysis of Linear Partial Differential Operators, Springer Verlag, 1983.
- A. S. Demidov: Generalized Functions in Mathematical Physics: Main Ideas and Concepts (Nova Science Publishers, Huntington, 2001). With an addition by Yu. V. Egorov.
- M. Oberguggenberger: Multiplication of distributions and applications to partial differential equations (Longman, Harlow, 1992).
- Oberguggenberger, M. (2001). "Generalized functions in nonlinear models - a survey". *Nonlinear Analysis* **47** (8): 5029–5040. doi:10.1016/s0362-546x(01)00614-9.

- J.-F. Colombeau: New Generalized Functions and Multiplication of Distributions, North Holland, 1983.

- M. Grosser et al.: Geometric theory of generalized functions with applications to general relativity, Kluwer Academic Publishers, 2001.

- H. Kleinert, *Path Integrals in Quantum Mechanics, Statistics, Polymer Physics, and Financial Markets,* 4th edition, World Scientific (Singapore, 2006)(online here). See Chapter 11 for products of generalized functions.

6.9 References

[1] Kolmogorov, A. N., Fomin, S. V., & Fomin, S. V. (1999). Elements of the theory of functions and functional analysis (Vol. 1). Courier Dover Publications.

[2] Schwartz, L (1952). "Théorie des distributions". *Bull. Amer. Math. Soc.* **58**: 78–85. doi:10.1090/S0002-9904-1952-09555-0.

[3] Halperin, I., & Schwartz, L. (1952). Introduction to the Theory of Distributions. Toronto: University of Toronto Press. (Short lecture by Halpering on Schwartz's theory)

[4] Yu. V. Egorov (1990). "A contribution to the theory of generalized functions". *Russ. Math. Surveys (Uspekhi Mat. Nauk)* **45** (5): 1–49. Bibcode:1990RuMaS..45....1E. doi:10.1070/rm1990v045n05abeh002683.

[5] H. Kleinert and A. Chervyakov (2001). "Rules for integrals over products of distributions from coordinate independence of path integrals" (PDF). *Europ. Phys. J.* **C 19** (4): 743–747. arXiv:quant-ph/0002067.   Bibcode:2001EPJC...19..743K. doi:10.1007/s100520100600.

[6] H. Kleinert and A. Chervyakov (2000). "Coordinate Independence of Quantum-Mechanical Path Integrals" (PDF). *Phys. Lett.* A 269: 63. Bibcode:2000PhLA..273....1K. doi:10.1016/S0375-9601(00)00475-8.

[7] Yu. M. Shirokov (1979). "Algebra of one-dimensional generalized functions". *Theoretical and Mathematical Physics* **39**: 291–301.

[8] cite wanted

[9] O. G. Goryaga; Yu. M. Shirokov (1981). "Energy levels of an oscillator with singular concentrated potential". *Theoretical and Mathematical Physics* **46** (3): 321–324. Bibcode:1981TMP....46..210G. doi:10.1007/BF01032729.

[10] G. K. Tolokonnikov (1982). "Differential rings used in Shirokov algebras". *Theoretical and Mathematical Physics* **53** (1): 952–954. Bibcode:1982TMP....53..952T. doi:10.1007/BF01014789.

Chapter 7

Hardy space

"Hardy class" redirects here. For the warships, see Hardy class destroyer.

In complex analysis, the **Hardy spaces** (or **Hardy classes**) H^p are certain spaces of holomorphic functions on the unit disk or upper half plane. They were introduced by Frigyes Riesz (Riesz 1923), who named them after G. H. Hardy, because of the paper (Hardy 1915). In real analysis **Hardy spaces** are certain spaces of distributions on the real line, which are (in the sense of distributions) boundary values of the holomorphic functions of the complex Hardy spaces, and are related to the L^p spaces of functional analysis. For $1 \leq p \leq \infty$ these real Hardy spaces H^p are certain subsets of L^p, while for $p < 1$ the L^p spaces have some undesirable properties, and the Hardy spaces are much better behaved.

There are also higher-dimensional generalizations, consisting of certain holomorphic functions on tube domains in the complex case, or certain spaces of distributions on $\mathbf{R}^n$ in the real case.

Hardy spaces have a number of applications in mathematical analysis itself, as well as in control theory (such as H^∞ methods) and in scattering theory.

7.1   Hardy spaces for the unit disk

For spaces of holomorphic functions on the open unit disk, the Hardy space H^2 consists of the functions f whose mean square value on the circle of radius r remains bounded as $r \to 1$ from below.

More generally, the Hardy space H^p for $0 < p < \infty$ is the class of holomorphic functions f on the open unit disk satisfying

$$\sup_{0<r<1} \left(\frac{1}{2\pi} \int_0^{2\pi} \left| f\left(re^{i\theta}\right) \right|^p \, d\theta \right)^{\frac{1}{p}} < \infty.$$

This class H^p is a vector space. The number on the left side of the above inequality is the Hardy space p-norm for f,

denoted by $\|f\|_{H^p}$. It is a norm when $p \geq 1$, but not when $0 < p < 1$.

The space H^∞ is defined as the vector space of bounded holomorphic functions on the disk, with the norm

$$\|f\|_{H^\infty} = \sup_{|z|<1} |f(z)|.$$

For $0 < p \leq q \leq \infty$, the class H^q is a subset of H^p, and the H^p-norm is increasing with p (it is a consequence of Hölder's inequality that the L^p-norm is increasing for probability measures, i.e. measures with total mass 1).

7.2   Hardy spaces on the unit circle

The Hardy spaces defined in the preceding section can also be viewed as certain closed vector subspaces of the complex L^p spaces on the unit circle. This connection is provided by the following theorem (Katznelson 1976, Thm 3.8): Given $f \in H^p$, with $p \geq 0$, the radial limit

$$\tilde{f}\left(e^{i\theta}\right) = \lim_{r\to 1} f\left(re^{i\theta}\right)$$

exists for almost every θ. The function $\tilde{f}$ belongs to the L^p space for the unit circle, and one has that

$$\|\tilde{f}\|_{L^p} = \|f\|_{H^p}.$$

Denoting the unit circle by $\mathbf{T}$, and by $H^p(\mathbf{T})$ the vector subspace of $L^p(\mathbf{T})$ consisting of all limit functions $\tilde{f}$, when f varies in H^p, one then has that for $p \geq 1$,(Katznelson 1976)

$$g \in II^p\left(\mathbf{T}\right) \text{ if only and if } g \in L^p\left(\mathbf{T}\right) \text{ and } \hat{g}($$
$$n) = 0 \text{ all for } n < 0,$$

where the $\hat{g}(n)$ are the Fourier coefficients of a function g integrable on the unit circle,

$$\forall n \in \mathbf{Z}, \quad \hat{g}(n) = \frac{1}{2\pi} \int_0^{2\pi} g\left(e^{i\phi}\right) e^{-in\phi} \, d\phi.$$

The space $H^p(\mathbf{T})$ is a closed subspace of $L^p(\mathbf{T})$. Since $L^p(\mathbf{T})$ is a Banach space (for $1 \le p \le \infty$), so is $H^p(\mathbf{T})$.

The above can be turned around. Given a function $\tilde{f} \in L^p(\mathbf{T})$, with $p \ge 1$, one can regain a (harmonic) function f on the unit disk by means of the Poisson kernel P_r:

$$f\left(re^{i\theta}\right) = \frac{1}{2\pi} \int_0^{2\pi} P_r(\theta - \phi)\tilde{f}\left(e^{i\phi}\right) \, d\phi, \quad r < 1,$$

and f belongs to H^p exactly when $\tilde{f}$ is in $H^p(\mathbf{T})$. Supposing that $\tilde{f}$ is in $H^p(\mathbf{T})$. *i.e.* that $\tilde{f}$ has Fourier coefficients $(an)n{\in}\mathbf{Z}$ with $an = 0$ for every $n < 0$,then the element f of the Hardy space H^p associated to $\tilde{f}$ is the holomorphic function

$$f(z) = \sum_{n=0}^{\infty} a_n z^n, \quad |z| < 1.$$

In applications, those functions with vanishing negative Fourier coefficients are commonly interpreted as the causal solutions. Thus, the space H^2 is seen to sit naturally inside L^2 space, and is represented by infinite sequences indexed by $\mathbf{N}$; whereas L^2 consists of bi-infinite sequences indexed by $\mathbf{Z}$.

7.2.1   Connection to real Hardy spaces on the circle

When $1 \le p < \infty$, the *real Hardy spaces* H^p discussed further down in this article are easy to describe in the present context. A real function f on the unit circle belongs to the real Hardy space $H^p(\mathbf{T})$ if it is the real part of a function in $H^p(\mathbf{T})$, and a complex function f belongs to the real Hardy space iff $\mathrm{Re}(f)$ and $\mathrm{Im}(f)$ belong to the space (see the section on real Hardy spaces below).

For $p < 1$, such tools as Fourier coefficients, Poisson integral, conjugate function, are no longer valid. For example, consider

$$F(z) = \frac{1+z}{1-z}, \quad |z| < 1$$

for which

$$f(e^{i\theta}) := \tilde{F}(e^{i\theta}) = i \cot(\tfrac{\theta}{2}).$$

The function F is in H^p for every $p < 1$, the radial limit f is in $H^p(\mathbf{T})$ but $\mathrm{Re}(f)$ is 0 almost everywhere. It is no longer possible to recover F from $\mathrm{Re}(f)$, and one cannot define real-$H^p(\mathbf{T})$ in the simple way above.

For the same function F, let $fr(e^{i\theta}) = F(re^{i\theta})$. The limit when $r \to 1$ of $\mathrm{Re}(fr)$, *in the sense of distributions* on the circle, is a non-zero multiple of the Dirac distribution at $z = 1$. The Dirac distribution at any point of the unit circle belongs to real-$H^p(\mathbf{T})$ for every $p < 1$ (see below).

7.3   Factorization into inner and outer functions (Beurling)

For $0 < p \le \infty$, every non-zero function f in H^p can be written as the product $f = Gh$ where G is an *outer function* and h is an *inner function*, as defined below (Rudin 1987, Thm 17.17). This "Beurling factorization" allows the Hardy space to be completely characterized by the spaces of inner and outer functions.

One says that $G(z)$ is an **outer (exterior) function** if it takes the form

$$G(z) = c \exp\left(\frac{1}{2\pi} \int_{-\pi}^{\pi} \frac{e^{i\theta} + z}{e^{i\theta} - z} \log\left(\varphi(e^{i\theta})\right) \, d\theta \right)$$

for some complex number c with $|c| = 1$, and some positive measurable function φ on the unit circle such that $\log(\varphi)$ is integrable on the circle. In particular, when φ is integrable on the circle, G is in H^1 because the above takes the form of the Poisson kernel (Rudin 1987, Thm 17.16). This implies that

$$\lim_{r \to 1^-} \left| G\left(re^{i\theta}\right) \right| = \varphi\left(e^{i\theta}\right)$$

for almost every θ.

One says that h is an **inner (interior) function** if and only if $|h| \le 1$ on the unit disk and the limit

$$\lim_{r \to 1^-} h(re^{i\theta})$$

exists for almost all θ and its modulus is equal to 1. In particular, h is in H^∞. The inner function can be further factored into a form involving a Blaschke product.

The function f, decomposed as $f = Gh$, is in H^p if and only if the positive function φ belongs to $L^p(\mathbf{T})$, where φ is the function in the representation of the outer function G.

Let G be an outer function represented as above from a function φ on the circle. Replacing φ by φ^{α}, $\alpha > 0$, a family $(G\alpha)$ of outer functions is obtained, with the properties:

$G_1 = G$, $G\alpha_{+\beta} = G\alpha\,G_{\beta}$ and $|G\alpha| = |G|^{\alpha}$ almost everywhere on the circle.

It follows that whenever $0 < p, q, r < \infty$ and $1/r = 1/p + 1/q$, every function f in H^r can be expressed as the product of a function in H^p and a function in H^q. For example: every function in H^1 is the product of two functions in H^2; every function in H^p, $p < 1$, can be expressed as product of several functions in some H^q, $q > 1$.

7.4 Real-variable techniques on the unit circle

Real-variable techniques, mainly associated to the study of *real Hardy spaces* defined on $\mathbf{R}^n$ (see below), are also used in the simpler framework of the circle. It is a common practice to allow for complex functions (or distributions) in these "real" spaces. The definition that follows does not distinguish between real or complex case.

Let Pr denote the Poisson kernel on the unit circle $\mathbf{T}$. For a distribution f on the unit circle, set

$$(Mf)(e^{i\theta}) = \sup_{0<r<1} \left|(f * P_r)\left(e^{i\theta}\right)\right|,$$

where the *star* indicates convolution between the distribution f and the function $e^{i\theta} \to Pr(\theta)$ on the circle. Namely, $(f * Pr)(e^{i\theta})$ is the result of the action of f on the C^{∞}-function defined on the unit circle by

$$e^{i\varphi} \to P_r(\theta - \varphi).$$

For $0 < p < \infty$, the *real Hardy space* $H^p(\mathbf{T})$ consists of distributions f such that Mf is in $L^p(\mathbf{T})$.

The function F defined on the unit disk by $F(re^{i\theta}) = (f * Pr)(e^{i\theta})$ is harmonic, and Mf is the *radial maximal function* of F. When Mf belongs to $L^p(\mathbf{T})$ and $p \geq 1$, the distribution f "is" a function in $L^p(\mathbf{T})$, namely the boundary value of F. For $p \geq 1$, the *real Hardy space* $H^p(\mathbf{T})$ is a subset of $L^p(\mathbf{T})$.

7.4.1 Conjugate function

To every real trigonometric polynomial u on the unit circle, one associates the real *conjugate polynomial* v such that $u + iv$ extends to a holomorphic function in the unit disk,

$$u(e^{i\theta}) = \frac{a_0}{2} + \sum_{k \geq 1} a_k \cos(k\theta) + b_k \sin(k\theta) \longrightarrow v(e^{i\theta}$$
$$) = \sum_{k \geq 1} a_k \sin(k\theta) - b_k \cos(k\theta).$$

This mapping $u \to v$ extends to a bounded linear operator H on $L^p(\mathbf{T})$, when $1 < p < \infty$ (up to a scalar multiple, it is the Hilbert transform on the unit circle), and H also maps $L^1(\mathbf{T})$ to weak-$L^1(\mathbf{T})$. When $1 \leq p < \infty$, the following are equivalent for a *real valued* integrable function f on the unit circle:

- the function f is the real part of some function $g \in H^p(\mathbf{T})$

- the function f and its conjugate $H(f)$ belong to $L^p(\mathbf{T})$

- the radial maximal function Mf belongs to $L^p(\mathbf{T})$.

When $1 < p < \infty$, $H(f)$ belongs to $L^p(\mathbf{T})$ when $f \in L^p(\mathbf{T})$, hence the real Hardy space $H^p(\mathbf{T})$ coincides with $L^p(\mathbf{T})$ in this case. For $p = 1$, the real Hardy space $H^1(\mathbf{T})$ is a proper subspace of $L^1(\mathbf{T})$.

The case of $p = \infty$ was excluded from the definition of real Hardy spaces, because the maximal function Mf of an L^{∞} function is always bounded, and because it is not desirable that real-H^{∞} be equal to L^{∞}. However, the two following properties are equivalent for a real valued function f

- the function f is the real part of some function $g \in H^{\infty}(\mathbf{T})$

- the function f and its conjugate $H(f)$ belong to $L^{\infty}(\mathbf{T})$.

7.4.2 Real Hardy spaces for $0 < p < 1$

When $0 < p < 1$, a function F in H^p cannot be reconstructed from the real part of its boundary limit *function* on the circle, because of the lack of convexity of L^p in this case. Convexity fails but a kind of "*complex convexity*" remains, namely the fact that $z \to |z|^q$ is subharmonic for every $q > 0$. As a consequence, if

$$F(z) = \sum_{n=0}^{+\infty} c_n z^n, \quad |z| < 1$$

is in H^p, it can be shown that $cn = O(n^{1/p-1})$. It follows that the Fourier series

$$\sum_{n=0}^{+\infty} c_n e^{in\theta}$$

converges in the sense of distributions to a distribution f on the unit circle, and $F(re^{i\theta}) = (f * Pr)(\theta)$. The function $F \in H^p$ can be reconstructed from the real distribution $\mathrm{Re}(f)$ on the circle, because the Taylor coefficients cn of F can be computed from the Fourier coefficients of $\mathrm{Re}(f)$: distributions on the circle are general enough for handling Hardy spaces when $p < 1$. Distributions do appear, as it is seen with functions $F(z) = (1-z)^{-N}$ (for $|z| < 1$), that belong to H^p when $0 < Np < 1$ (and N an integer ≥ 1).

A real distribution on the circle belongs to real-$H^p(\mathbf{T})$ iff it is the boundary value of the real part of some $F \in H^p$. A Dirac distribution δx, at any point x of the unit circle, belongs to real-$H^p(\mathbf{T})$ for every $p < 1$; derivatives $\delta' x$ belong when $p < 1/2$, second derivatives $\delta'' x$ when $p < 1/3$, and so on.

7.5 Hardy spaces for the upper half plane

It is possible to define Hardy spaces on other domains than the disc, and in many applications Hardy spaces on a complex half-plane (usually the right half-plane or upper half-plane) are used.

The Hardy space $H^p(\mathbf{H})$ on the upper half-plane $\mathbf{H}$ is defined to be the space of holomorphic functions f on $\mathbf{H}$ with bounded (quasi-)norm, the norm being given by

$$\|f\|_{H^p} = \sup_{y>0} \left(\int |f(x+iy)|^p \, dx \right)^{\frac{1}{p}}.$$

The corresponding $H^\infty(\mathbf{H})$ is defined as functions of bounded norm, with the norm given by

$$\|f\|_{H^\infty} = \sup_{z \in \mathbf{H}} |f(z)|.$$

Although the unit disk $\mathbf{D}$ and the upper half-plane $\mathbf{H}$ can be mapped to one another by means of Möbius transformations, they are not interchangeable as domains for Hardy spaces. Contributing to this difference is the fact that the unit circle has finite (one-dimensional) Lebesgue measure while the real line does not. However, for H^2, one may still state the following theorem: Given the Möbius transformation $m : \mathbf{D} \to \mathbf{H}$ with

$$m(z) = i \cdot \frac{1+z}{1-z}$$

then there is an isometric isomorphism $M : H^2(\mathbf{H}) \to H^2(\mathbf{D})$ with

$$(Mf)(z) = \frac{\sqrt{\pi}}{1-z} f(m(z)).$$

7.6 Real Hardy spaces for $\mathbf{R}^n$

In analysis on the real vector space $\mathbf{R}^n$, the Hardy space H^p (for $0 < p \leq \infty$) consists of tempered distributions f such that for some Schwartz function Φ with $\int \Phi = 1$, the maximal function

$$(M_\Phi f)(x) = \sup_{t>0} |(f * \Phi_t)(x)|$$

is in $L^p(\mathbf{R}^n)$, where $*$ is convolution and $\Phi t\,(x) = t^{-n} \Phi(x/t)$. The H^p-quasinorm $\|f\|_{Hp}$ of a distribution f of H^p is defined to be the L^p norm of $M\Phi f$ (this depends on the choice of Φ, but different choices of Schwartz functions Φ give equivalent norms). The H^p-quasinorm is a norm when $p \geq 1$, but not when $p < 1$.

If $1 < p < \infty$, the Hardy space H^p is the same vector space as L^p, with equivalent norm. When $p = 1$, the Hardy space H^1 is a proper subspace of L^1. One can find sequences in H^1 that are bounded in L^1 but unbounded in H^1, for example on the line

$$f_k(x) = \mathbf{1}_{[0,1]}(x-k) - \mathbf{1}_{[0,1]}(x+k), \quad k > 0.$$

The L^1 and H^1 norms are not equivalent on H^1, and H^1 is not closed in L^1. The dual of H^1 is the space BMO of functions of bounded mean oscillation. The space BMO contains unbounded functions (proving again that H^1 is not closed in L^1).

If $p < 1$ then the Hardy space H^p has elements that are not functions, and its dual is the homogeneous Lipschitz space of order $n(1/p - 1)$. When $p < 1$, the H^p-quasinorm is not a norm, as it is not subadditive. The pth power $\|f\|_{Hp}^p$ is subadditive for $p < 1$ and so defines a metric on the Hardy space H^p, which defines the topology and makes H^p into a complete metric space.

7.6.1 Atomic decomposition

When $0 < p \leq 1$, a bounded measurable function f of compact support is in the Hardy space H^p if and only if all its moments

$$\int_{\mathbf{R}^n} f(x) x_1^{i_1} \dots x_n^{i_n} \, dx,$$

whose order $i_1 + \ldots + i_n$ is at most $n(1/p - 1)$, vanish. For example, the integral of f must vanish in order that $f \in H^p$, $0 < p \leq 1$, and as long as $p > n / (n+1)$ this is also sufficient.

If in addition f has support in some ball B and is bounded by $|B|^{-1/p}$ then f is called an **H^p-atom** (here $|B|$ denotes the Euclidean volume of B in $\mathbf{R}^n$). The H^p-quasinorm of an arbitrary H^p-atom is bounded by a constant depending only on p and on the Schwartz function Φ.

When $0 < p \leq 1$, any element f of H^p has an **atomic decomposition** as a convergent infinite combination of H^p-atoms,

$$f = \sum c_j a_j, \quad \sum |c_j|^p < \infty$$

where the a_j are H^p-atoms and the c_j are scalars.

On the line for example, the difference of Dirac distributions $f = \delta_1 - \delta_0$ can be represented as a series of Haar functions, convergent in H^p-quasinorm when $1/2 < p < 1$ (on the circle, the corresponding representation is valid for $0 < p < 1$, but on the line, Haar functions do not belong to H^p when $p \leq 1/2$ because their maximal function is equivalent at infinity to $a\, x^{-2}$ for some $a \neq 0$).

7.7 Martingale H^p

Let $(M_n)_{n\geq 0}$ be a martingale on some probability space (Ω, Σ, P), with respect to an increasing sequence of σ-fields $(\Sigma_n)_{n\geq 0}$. *Assume for simplicity that Σ is equal to the σ-field generated by the sequence $(\Sigma n)_{n\geq 0}$. The* maximal function *of the martingale is defined by*

$$M^* = \sup_{n\geq 0} |M_n|.$$

Let $1 \leq p < \infty$. The martingale $(M_n)_{n\geq 0}$ belongs to *martingale-H^p* when $M^* \in L^p$.

If $M^* \in L^p$, the martingale $(M_n)_{n\geq 0}$ is bounded in L^p, hence it converges almost surely to some function f by the martingale convergence theorem. Moreover, M_n converges to f in L^p-norm by the dominated convergence theorem, hence M_n can be expressed as conditional expectation of f on Σ_n. It is thus possible to identify martingale-H^p with the subspace of $L^p(\Omega, \Sigma, P)$ consisting of those f such that the martingale

$$M_n = E\big(f|\Sigma_n\big)$$

belongs to martingale-H^p.

Doob's maximal inequality implies that martingale-H^p coincides with $L^p(\Omega, \Sigma, P)$ when $1 < p < \infty$. The interesting space is martingale-H^1, whose dual is martingale-BMO (Garsia 1973).

The Burkholder–Gundy inequalities (when $p > 1$) and the Burgess Davis inequality (when $p = 1$) relate the L^p-norm of the maximal function to that of the *square function* of the martingale

$$S(f) = \left(|M_0|^2 + \sum_{n=0}^{\infty} |M_{n+1} - M_n|^2 \right)^{\frac{1}{2}}.$$

Martingale-H^p can be defined by saying that $S(f) \in L^p$.(Garsia 1973).

Martingales with continuous time parameter can also be considered. A direct link with the classical theory is obtained via the complex Brownian motion (B_t) in the complex plane, starting from the point $z = 0$ at time $t = 0$. Let τ denote the hitting time of the unit circle. For every holomorphic function F in the unit disk,

$$M_t = F(B_{t\wedge\tau})$$

is a martingale, that belongs to martingale-H^p iff $F \in H^p$ (Burkholder, Gundy & Silverstein 1971).

7.7.1 Example: dyadic martingale-H^1

In this example, $\Omega = [0, 1]$ and Σ_n is the finite field generated by the dyadic partition of $[0, 1]$ into 2^n intervals of length 2^{-n}, for every $n \geq 0$. If a function f on $[0, 1]$ is represented by its expansion on the Haar system (h_k)

$$f = \sum c_k h_k,$$

then the martingale-H^1 norm of f can be defined by the L^1 norm of the square function

$$\int_0^1 \left(\sum |c_k h_k(x)|^2 \right)^{\frac{1}{2}} dx.$$

This space, sometimes denoted by $H^1(\delta)$, is isomorphic to the classical real H^1 space on the circle (Müller 2005). The Haar system is an unconditional basis for $H^1(\delta)$.

7.8 References

- Burkholder, Donald L.; Gundy, Richard F.; Silverstein, Martin L. (1971), "A maximal function characterization of the class H^p", *Transactions of the American Mathematical Society* **157**: 137–153, doi:10.2307/1995838, JSTOR 1995838, MR 0274767.

- Cima, Joseph A.; Ross, William T. (2000), *The Backward Shift on the Hardy Space*, American Mathematical Society, ISBN 0-8218-2083-4

- Colwell, Peter (1985), *Blaschke Products - Bounded Analytic Functions*, Ann Arbor: University of Michigan Press, ISBN 0-472-10065-3

- Duren, P. (1970), *Theory of H^p-Spaces*, New York: Academic Press

- Fefferman, Charles; Stein, Elias M. (1972), "H^p spaces of several variables", *Acta Math.* **129** (3–4): 137–193, doi:10.1007/BF02392215, MR 0447953.

- Folland, G.B. (2001), "Hardy spaces", in Hazewinkel, Michiel, *Encyclopedia of Mathematics*, Springer, ISBN 978-1-55608-010-4

- Garsia, Adriano M. (1973), *Martingale inequalities: Seminar notes on recent progress*, Mathematics Lecture Notes Series, Reading, Mass.-London-Amsterdam: W. A. Benjamin, Inc., pp. viii+184 MR 0448538

- Hardy, G. H. (1915), "On the mean value of the modulus of an analytic function", *Proceedings of the London Mathematical Society* **14**: 269–277, doi:10.1112/plms/s2_14.1.269, JFM 45.1331.03

- Hoffman, Kenneth (1988), *Banach spaces of analytic functions*, New York: Dover Publications, ISBN 0-486-65785-X

- Katznelson, Yitzhak (1976), *An introduction to Harmonic Analysis*, Dover, ISBN 0-486-63331-4

- Müller, Paul F. X. (2005), *Isomorphisms between H^1 spaces*, Basel: Mathematics Institute of the Polish Academy of Sciences. Mathematical Monographs (New Series), Birkhäuser Verlag, pp. xiv+453, ISBN 978-3-7643-2431-5, MR 2157745

- Riesz, F. (1923), "Über die Randwerte einer analytischen Funktion", *Math. Z.* **18**: 87–95, doi:10.1007/BF01192397

- Mashreghi, J. (2009), *Representation Theorems in Hardy Spaces*, Cambridge University Press

- Rudin, Walter (1987), *Real and Complex Analysis*, McGraw-Hill, ISBN 0-07-100276-6

- Shvedenko, S.V. (2001), "Hardy classes", in Hazewinkel, Michiel, *Encyclopedia of Mathematics*, Springer, ISBN 978-1-55608-010-4

Chapter 8

Holomorphic function

This article is about a set of mathematical functions. For the life cycles of fungi, see Teleomorph, anamorph and holomorph.

In mathematics, **holomorphic functions** are the central objects of study in complex analysis. A holomorphic function is a complex-valued function of one or more complex variables that is complex differentiable in a neighborhood of every point in its domain. The existence of a complex derivative in a neighborhood is a very strong condition, for it implies that any holomorphic function is actually infinitely differentiable and equal to its own Taylor series.

The term *analytic function* is often used interchangeably with "holomorphic function", although the word "analytic" is also used in a broader sense to describe any function (real, complex, or of more general type) that can be written as a convergent power series in a neighborhood of each point in its domain. The fact that all holomorphic functions are complex analytic functions, and vice versa, is a major theorem in complex analysis.[1]

Holomorphic functions are also sometimes referred to as *regular functions*[2] or as *conformal maps*. A holomorphic function whose domain is the whole complex plane is called an entire function. The phrase "holomorphic at a point z_0" means not just differentiable at z_0, but differentiable everywhere within some neighborhood of z_0 in the complex plane.

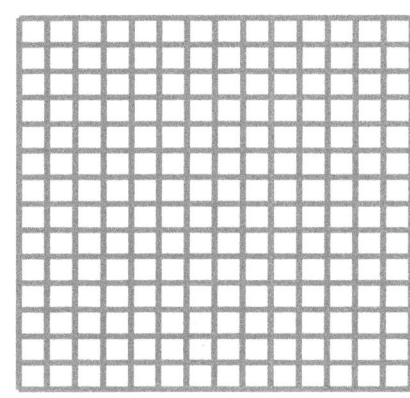

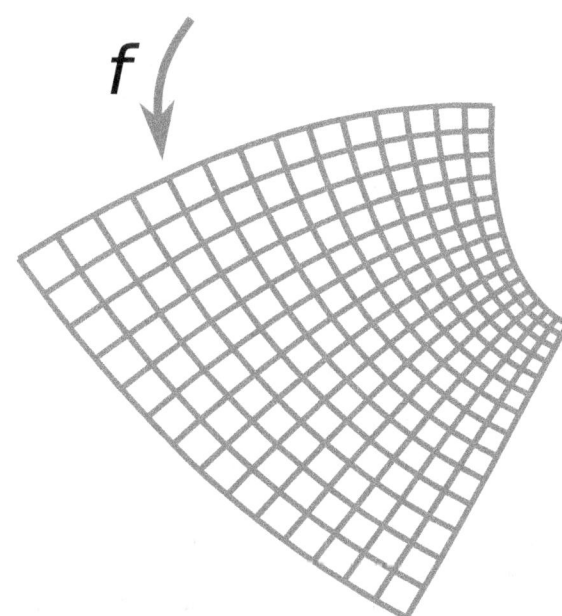

A rectangular grid (top) and its image under a conformal map f (bottom).

8.1 Definition

Given a complex-valued function f of a single complex variable, the **derivative** of f at a point z_0 in its domain is defined by the limit[3]

$$f'(z_0) = \lim_{z \to z_0} \frac{f(z) - f(z_0)}{z - z_0}.$$

This is the same as the definition of the derivative for real functions, except that all of the quantities are complex. In particular, the limit is taken as the complex number z approaches z_0, and must have the same value for any sequence of complex values for z that approach z_0 on the complex plane. If the limit exists, we say that f is **complex-**

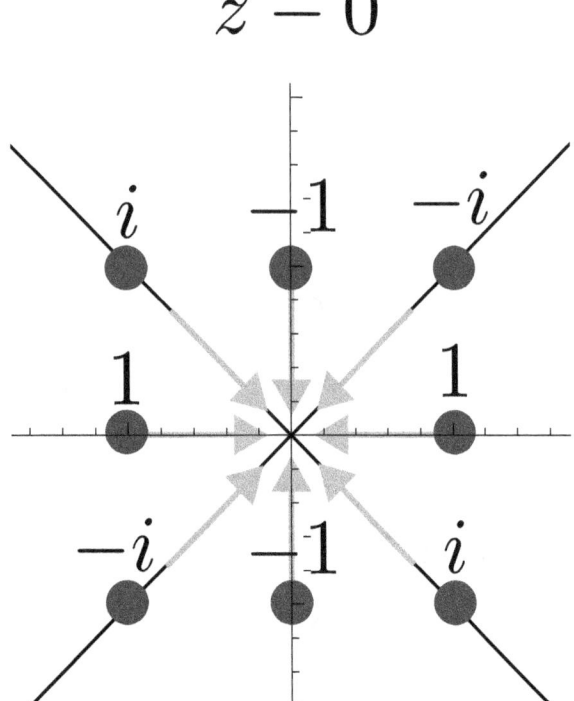

$$\frac{\bar{z} - \bar{0}}{z - 0}$$

The function $f(z) = \bar{z}$ is not complex-differentiable at zero, because as shown above, the value of $\frac{f(z) - f(0)}{z - 0}$ varies depending on the direction from which zero is approached. Along the real axis, f equals the function g(z) = z and the limit is 1, while along the imaginary axis, f equals h(z) = −z and the limit is −1. Other directions yield yet other limits.

differentiable at the point z_0. This concept of complex differentiability shares several properties with real differentiability: it is linear and obeys the product rule, quotient rule, and chain rule.[4]

If f is *complex differentiable* at *every* point z_0 in an open set U, we say that f is **holomorphic on U**. We say that f is **holomorphic at the point** z_0 if it is holomorphic on some neighborhood of z_0.[5] We say that f is holomorphic on some non-open set A if it is holomorphic in an open set containing A.

The relationship between real differentiability and complex differentiability is the following. If a complex function $f(x + \mathrm{i}\, y) = u(x, y) + \mathrm{i}\, v(x, y)$ is holomorphic, then u and v have first partial derivatives with respect to x and y, and satisfy the Cauchy–Riemann equations:[6]

$$\frac{\partial u}{\partial x} = \frac{\partial v}{\partial y} \quad \text{and} \quad \frac{\partial u}{\partial y} = -\frac{\partial v}{\partial x}$$

or, equivalently, the Wirtinger derivative of f with respect to the complex conjugate of z is zero:[7]

$$\frac{\partial f}{\partial \bar{z}} = 0,$$

which is to say that, roughly, f is functionally independent from the complex conjugate of z.

If continuity is not a given, the converse is not necessarily true. A simple converse is that if u and v have *continuous* first partial derivatives and satisfy the Cauchy–Riemann equations, then f is holomorphic. A more satisfying converse, which is much harder to prove, is the Looman–Menchoff theorem: if f is continuous, u and v have first partial derivatives (but not necessarily continuous), and they satisfy the Cauchy–Riemann equations, then f is holomorphic.[8]

8.2 Terminology

The word "holomorphic" was introduced by two of Cauchy's students, Briot (1817–1882) and Bouquet (1819–1895), and derives from the Greek ὅλος (*holos*) meaning "entire", and μορφή (*morphē*) meaning "form" or "appearance".[9]

Today, the term "holomorphic function" is sometimes preferred to "analytic function", as the latter is a more general concept. This is also because an important result in complex analysis is that every holomorphic function is complex analytic, a fact that does not follow directly from the definitions. The term "analytic" is however also in wide use.

8.3 Properties

Because complex differentiation is linear and obeys the product, quotient, and chain rules; the sums, products and compositions of holomorphic functions are holomorphic, and the quotient of two holomorphic functions is holomorphic wherever the denominator is not zero.[10]

If one identifies **C** with **R**^{2}, then the holomorphic functions coincide with those functions of two real variables with continuous first derivatives which solve the Cauchy–Riemann equations, a set of two partial differential equations.[6]

Every holomorphic function can be separated into its real and imaginary parts, and each of these is a solution of

Laplace's equation on $\mathbf{R}^2$. In other words, if we express a holomorphic function $f(z)$ as $u(x, y) + i\, v(x, y)$ both u and v are harmonic functions, where v is the harmonic conjugate of u and vice versa.[11]

Cauchy's integral theorem implies that the line integral of every holomorphic function along a loop vanishes:[12]

$$\oint_\gamma f(z)\, dz = 0.$$

Here γ is a rectifiable path in a simply connected open subset U of the complex plane $\mathbf{C}$ whose start point is equal to its end point, and $f : U \to \mathbf{C}$ is a holomorphic function.

Cauchy's integral formula states that every function holomorphic inside a disk is completely determined by its values on the disk's boundary.[12] Furthermore: Suppose U is an open subset of $\mathbf{C}$, $f : U \to \mathbf{C}$ is a holomorphic function and the closed disk $D = \{z : |z - z_0| \le r\}$ is completely contained in U. Let γ be the circle forming the boundary of D. Then for every a in the interior of D:

$$f(a) = \frac{1}{2\pi i} \oint_\gamma \frac{f(z)}{z - a}\, dz$$

where the contour integral is taken counter-clockwise.

The derivative $f'(a)$ can be written as a contour integral[12] using **Cauchy's differentiation formula**:

$$f'(a) = \frac{1}{2\pi i} \oint_\gamma \frac{f(z)}{(z - a)^2}\, dz,$$

for any simple loop positively winding once around a, and

$$f'(a) = \lim_{\gamma \to a} \frac{i}{2\mathcal{A}(\gamma)} \oint_\gamma f(z)d\bar{z},$$

for infinitesimal positive loops γ around a.

In regions where the first derivative is not zero, holomorphic functions are conformal in the sense that they preserve angles and the shape (but not size) of small figures.[13]

Every holomorphic function is analytic. That is, a holomorphic function f has derivatives of every order at each point a in its domain, and it coincides with its own Taylor series at a in a neighborhood of a. In fact, f coincides with its Taylor series at a in any disk centered at that point and lying within the domain of the function.

From an algebraic point of view, the set of holomorphic functions on an open set is a commutative ring and a complex vector space.[7] In fact, it is a locally convex topological vector space, with the seminorms being the suprema on compact subsets.

From a geometric perspective, a function f is holomorphic at z_0 if and only if its exterior derivative df in a neighborhood U of z_0 is equal to $f'(z)\, dz$ for some continuous function f'. It follows from

$$0 = d^2 f = d(f' dz) = df' \wedge dz$$

that df' is also proportional to dz, implying that the derivative f' is itself holomorphic and thus that f is infinitely differentiable. Similarly, the fact that $d(f\, dz) = f'\, dz \wedge dz = 0$ implies that any function f that is holomorphic on the simply connected region U is also integrable on U. (For a path γ from z_0 to z lying entirely in U, define

$$F_\gamma(z) = F_0 + \int_\gamma f dz$$

in light of the Jordan curve theorem and the generalized Stokes' theorem, $F_\gamma(z)$ is independent of the particular choice of path γ, and thus $F(z)$ is a well-defined function on U having $F(z_0) = F_0$ and $dF = f\, dz$.)

8.4 Examples

All polynomial functions in z with complex coefficients are holomorphic on $\mathbf{C}$, and so are sine, cosine and the exponential function. (The trigonometric functions are in fact closely related to and can be defined via the exponential function using Euler's formula). The principal branch of the complex logarithm function is holomorphic on the set $\mathbf{C} \setminus \{z \in \mathbf{R} : z \le 0\}$. The square root function can be defined as

$$\sqrt{z} = e^{\frac{1}{2} \log z}$$

and is therefore holomorphic wherever the logarithm $\log(z)$ is. The function $1/z$ is holomorphic on $\{z : z \ne 0\}$.

As a consequence of the Cauchy–Riemann equations, a real-valued holomorphic function must be constant. Therefore, the absolute value of z, the argument of z, the real part of z and the imaginary part of z are not holomorphic. Another typical example of a continuous function which is not holomorphic is the complex conjugate z formed by complex conjugation.

8.5 Several variables

The definition of a holomorphic function generalizes to several complex variables in a straightforward way. Let D denote an open subset of $\mathbf{C}^n$, and let $f : D \to \mathbf{C}$. The function f is **analytic** at a point p in D if there exists an open neighborhood of p in which f is equal to a convergent power series in n complex variables.[14] Define f to be **holomorphic** if it is analytic at each point in its domain. Osgood's lemma shows (using the multivariate Cauchy integral formula) that, for a continuous function f, this is equivalent to f being holomorphic in each variable separately (meaning that if any n − 1 coordinates are fixed, then the restriction of f is a holomorphic function of the remaining coordinate). The much deeper Hartogs' theorem proves that the continuity hypothesis is unnecessary: f is holomorphic if and only if it is holomorphic in each variable separately.

More generally, a function of several complex variables that is square integrable over every compact subset of its domain is analytic if and only if it satisfies the Cauchy–Riemann equations in the sense of distributions.

Functions of several complex variables are in some basic ways more complicated than functions of a single complex variable. For example, the region of convergence of a power series is not necessarily an open ball; these regions are Reinhardt domains, the simplest example of which is a polydisk. However, they also come with some fundamental restrictions. Unlike functions of a single complex variable, the possible domains on which there are holomorphic functions that cannot be extended to larger domains are highly limited. Such a set is called a domain of holomorphy.

8.6 Extension to functional analysis

Main article: infinite-dimensional holomorphy

The concept of a holomorphic function can be extended to the infinite-dimensional spaces of functional analysis. For instance, the Fréchet or Gâteaux derivative can be used to define a notion of a holomorphic function on a Banach space over the field of complex numbers.

8.7 See also

- Antiderivative (complex analysis)
- Antiholomorphic function
- Biholomorphy
- Meromorphic function
- Quadrature domains
- Harmonic maps
- Harmonic morphisms

8.8 References

[1] *Analytic functions of one complex variable*, Encyclopedia of Mathematics. (European Mathematical Society ft. Springer, 2015)

[2] Springer Online Reference Books, Wolfram MathWorld

[3] Ahlfors, L., *Complex Analysis, 3 ed.* (McGraw-Hill, 1979).

[4] Henrici, P., *Applied and Computational Complex Analysis* (Wiley). [Three volumes: 1974, 1977, 1986.]

[5] Peter Ebenfelt, Norbert Hungerbühler, Joseph J. Kohn, Ngaiming Mok, Emil J. Straube (2011) *Complex Analysis* Springer Science & Business Media

[6] Markushevich, A.I., *Theory of Functions of a Complex Variable* (Prentice-Hall, 1965). [Three volumes.]

[7] Gunning, Robert C.; Rossi, Hugo (1965), *Analytic Functions of Several Complex Variables*, Prentice-Hall series in Modern Analysis, Englewood Cliffs, N.J.: Prentice-Hall, pp. xiv+317, MR 0180696, Zbl 0141.08601

[8] Gray, J. D.; Morris, S. A. (1978), "When is a Function that Satisfies the Cauchy-Riemann Equations Analytic?", *The American Mathematical Monthly* (April 1978) **85** (4): 246–256, doi:10.2307/2321164, JSTOR 2321164.

[9] Markushevich, A. I. (2005) [1977]. Silverman, Richard A., ed. *Theory of functions of a Complex Variable* (2nd ed.). New York: American Mathematical Society. p. 112. ISBN 0-8218-3780-X.

[10] Henrici, Peter (1993) [1986], *Applied and Computational Complex Analysis Volume 3*, Wiley Classics Library (Reprint ed.), New York - Chichester - Brisbane - Toronto - Singapore: John Wiley & Sons, pp. X+637, ISBN 0-471-58986-1, MR 0822470, Zbl 1107.30300.

[11] Evans, Lawrence C. (1998), *Partial Differential Equations*, American Mathematical Society.

[12] Lang, Serge (2003), *Complex Analysis*, Springer Verlag GTM, Springer Verlag

[13] Rudin, Walter (1987), *Real and complex analysis* (3rd ed.), New York: McGraw–Hill Book Co., ISBN 978-0-07-054234-1, MR 924157

[14] Gunning and Rossi, *Analytic Functions of Several Complex Variables*, p. 2.

University mathematics, Blakey,J,PhD. Blackwell ad Sons, 2nd. Edtn, 1958.

8.9 External links

- Hazewinkel, Michiel, ed. (2001), "Analytic function", *Encyclopedia of Mathematics*, Springer, ISBN 978-1-55608-010-4

Chapter 9

Pythagorean theorem

See also: Pythagorean trigonometric identity

In mathematics, the **Pythagorean theorem**, also known

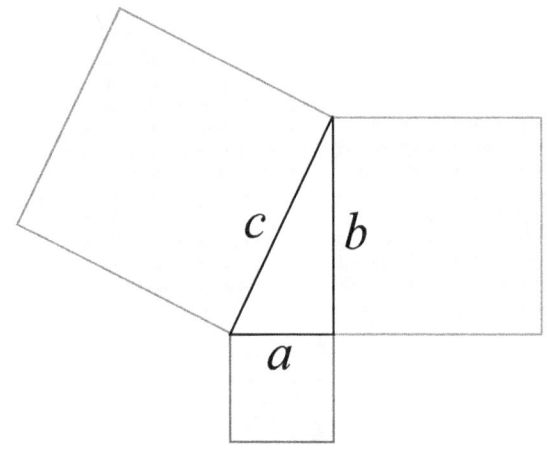

Pythagorean theorem
The sum of the areas of the two squares on the legs (a and b) equals the area of the square on the hypotenuse (c).

as **Pythagoras' theorem**, is a relation in Euclidean geometry among the three sides of a right triangle. It states that the square of the hypotenuse (the side opposite the right angle) is equal to the sum of the squares of the other two sides. The theorem can be written as an equation relating the lengths of the sides a, b and c, often called the "Pythagorean equation":[1]

$$a^2 + b^2 = c^2,$$

where c represents the length of the hypotenuse and a and b the lengths of the triangle's other two sides.

Although it is often argued that knowledge of the theorem predates him,[2] the theorem is named after the ancient Greek mathematician Pythagoras (c. 570 – c. 495 BC) as it is he who, by tradition, is credited with its first recorded proof.[3][4][5] There is some evidence that Babylonian mathematicians understood the formula, although little of it indicates an application within a mathematical framework.[6][7] Mesopotamian, Indian and Chinese mathematicians all discovered the theorem independently and, in some cases, provided proofs for special cases.

The theorem has been given numerous proofs – possibly the most for any mathematical theorem. They are very diverse, including both geometric proofs and algebraic proofs, with some dating back thousands of years. The theorem can be generalized in various ways, including higher-dimensional spaces, to spaces that are not Euclidean, to objects that are not right triangles, and indeed, to objects that are not triangles at all, but n-dimensional solids. The Pythagorean theorem has attracted interest outside mathematics as a symbol of mathematical abstruseness, mystique, or intellectual power; popular references in literature, plays, musicals, songs, stamps and cartoons abound.

9.1   Pythagorean proof

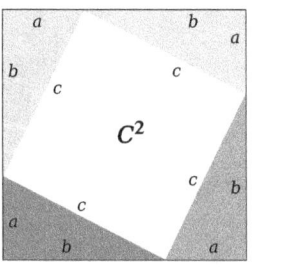 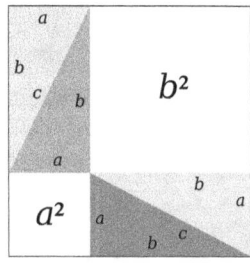

$$c^2 = a^2 + b^2$$

The Pythagorean proof (click to view animation)

The Pythagorean Theorem was known long before Pythagoras, but he may well have been the first to prove it.[2] In any event, the proof attributed to him is very simple, and is called a proof by rearrangement.

The two large squares shown in the figure each contain four

identical triangles, and the only difference between the two large squares is that the triangles are arranged differently. Therefore, the white space within each of the two large squares must have equal area. Equating the area of the white space yields the Pythagorean Theorem, Q.E.D.[8]

That Pythagoras originated this very simple proof is sometimes inferred from the writings of the later Greek philosopher and mathematician Proclus.[9] Several other proofs of this theorem are described below, but this is known as the Pythagorean one.

9.2 Other forms of the theorem

As pointed out in the introduction, if c denotes the length of the hypotenuse and a and b denote the lengths of the other two sides, the Pythagorean theorem can be expressed as the Pythagorean equation:

$$a^2 + b^2 = c^2.$$

If the length of both a and b are known, then c can be calculated as

$$c = \sqrt{a^2 + b^2}.$$

If the length of the hypotenuse c and of one side (a or b) are known, then the length of the other side can be calculated as

$$a = \sqrt{c^2 - b^2}$$

or

$$b = \sqrt{c^2 - a^2}.$$

The Pythagorean equation relates the sides of a right triangle in a simple way, so that if the lengths of any two sides are known the length of the third side can be found. Another corollary of the theorem is that in any right triangle, the hypotenuse is greater than any one of the other sides, but less than their sum.

A generalization of this theorem is the law of cosines, which allows the computation of the length of any side of any triangle, given the lengths of the other two sides and the angle between them. If the angle between the other sides is a right angle, the law of cosines reduces to the Pythagorean equation.

9.3 Other proofs of the theorem

This theorem may have more known proofs than any other (the law of quadratic reciprocity being another contender for that distinction); the book *The Pythagorean Proposition* contains 370 proofs.[10]

9.3.1 Proof using similar triangles

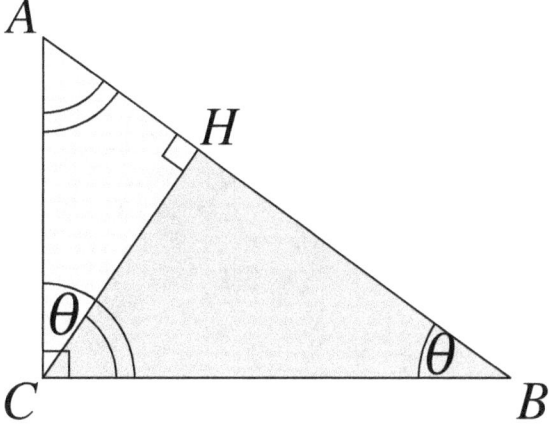

Proof using similar triangles

This proof is based on the proportionality of the sides of two similar triangles, that is, upon the fact that the ratio of any two corresponding sides of similar triangles is the same regardless of the size of the triangles.

Let *ABC* represent a right triangle, with the right angle located at *C*, as shown on the figure. Draw the altitude from point *C*, and call *H* its intersection with the side *AB*. Point *H* divides the length of the hypotenuse c into parts d and e. The new triangle *ACH* is similar to triangle *ABC*, because they both have a right angle (by definition of the altitude), and they share the angle at *A*, meaning that the third angle will be the same in both triangles as well, marked as θ in the figure. By a similar reasoning, the triangle *CBH* is also similar to *ABC*. The proof of similarity of the triangles requires the Triangle postulate: the sum of the angles in a triangle is two right angles, and is equivalent to the parallel postulate. Similarity of the triangles leads to the equality of ratios of corresponding sides:

$$\frac{BC}{AB} = \frac{BH}{BC} \text{ and } \frac{AC}{AB} = \frac{AH}{AC}.$$

The first result equates the cosines of the angles θ, whereas the second result equates their sines.

These ratios can be written as

$BC^2 = AB \times BH$ and $AC^2 = AB \times AH$.

Summing these two equalities results in

$BC^2 + AC^2$

$= AB \times BH + AB \times AH = AB \times (AH + BH) = AB^2,$

which, after simplification, expresses the Pythagorean theorem:

$BC^2 + AC^2 = AB^2$.

The role of this proof in history is the subject of much speculation. The underlying question is why Euclid did not use this proof, but invented another. One conjecture is that the proof by similar triangles involved a theory of proportions, a topic not discussed until later in the *Elements*, and that the theory of proportions needed further development at that time.[11][12]

9.3.2 Euclid's proof

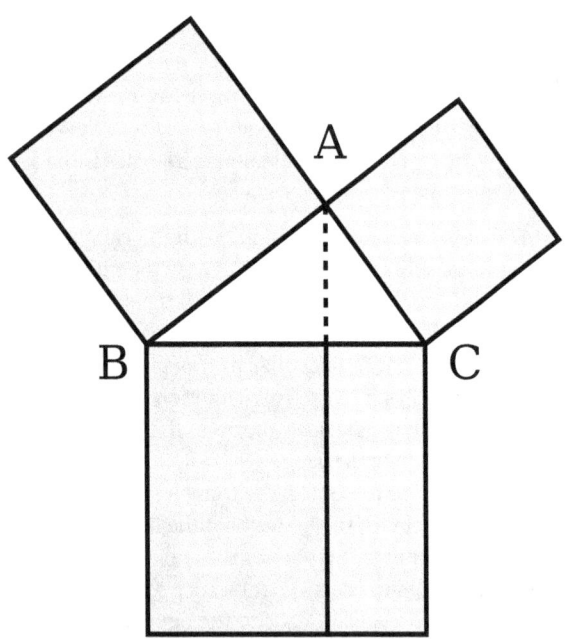

Proof in Euclid's Elements

In outline, here is how the proof in Euclid's *Elements* proceeds. The large square is divided into a left and right rectangle. A triangle is constructed that has half the area of the left rectangle. Then another triangle is constructed that has half the area of the square on the left-most side. These two triangles are shown to be congruent, proving this square

has the same area as the left rectangle. This argument is followed by a similar version for the right rectangle and the remaining square. Putting the two rectangles together to reform the square on the hypotenuse, its area is the same as the sum of the area of the other two squares. The details follow.

Let *A, B, C* be the vertices of a right triangle, with a right angle at *A*. Drop a perpendicular from *A* to the side opposite the hypotenuse in the square on the hypotenuse. That line divides the square on the hypotenuse into two rectangles, each having the same area as one of the two squares on the legs.

For the formal proof, we require four elementary lemmata:

1. If two triangles have two sides of the one equal to two sides of the other, each to each, and the angles included by those sides equal, then the triangles are congruent (side-angle-side).

2. The area of a triangle is half the area of any parallelogram on the same base and having the same altitude.

3. The area of a rectangle is equal to the product of two adjacent sides.

4. The area of a square is equal to the product of two of its sides (follows from 3).

Next, each top square is related to a triangle congruent with another triangle related in turn to one of two rectangles making up the lower square.[13]

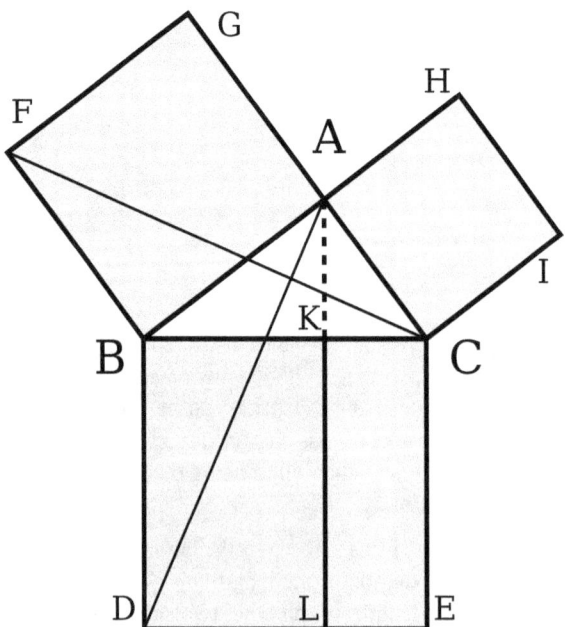

Illustration including the new lines

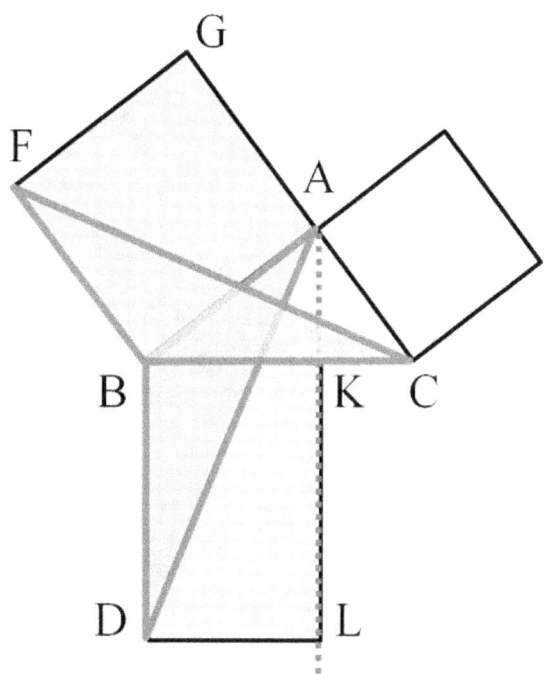

Showing the two congruent triangles of half the area of rectangle BDLK and square BAGF

The proof is as follows:

1. Let ACB be a right-angled triangle with right angle CAB.

2. On each of the sides BC, AB, and CA, squares are drawn, CBDE, BAGF, and ACIH, in that order. The construction of squares requires the immediately preceding theorems in Euclid, and depends upon the parallel postulate.[14]

3. From A, draw a line parallel to BD and CE. It will perpendicularly intersect BC and DE at K and L, respectively.

4. Join CF and AD, to form the triangles BCF and BDA.

5. Angles CAB and BAG are both right angles; therefore C, A, and G are collinear. Similarly for B, A, and H.

6. Angles CBD and FBA are both right angles; therefore angle ABD equals angle FBC, since both are the sum of a right angle and angle ABC.

7. Since AB is equal to FB and BD is equal to BC, triangle ABD must be congruent to triangle FBC.

8. Since A-K-L is a straight line, parallel to BD, then rectangle BDLK has twice the area of triangle ABD

because they share the base BD and have the same altitude BK, i.e., a line normal to their common base, connecting the parallel lines BD and AL. (lemma 2)

9. Since C is collinear with A and G, square BAGF must be twice in area to triangle FBC.

10. Therefore, rectangle BDLK must have the same area as square BAGF = AB^2.

11. Similarly, it can be shown that rectangle CKLE must have the same area as square ACIH = AC^2.

12. Adding these two results, $AB^2 + AC^2 = BD \times BK + KL \times KC$

13. Since BD = KL, $BD \times BK + KL \times KC = BD(BK + KC) = BD \times BC$

14. Therefore, $AB^2 + AC^2 = BC^2$, since CBDE is a square.

This proof, which appears in Euclid's *Elements* as that of Proposition 47 in Book 1,[15] demonstrates that the area of the square on the hypotenuse is the sum of the areas of the other two squares.[16] This is quite distinct from the proof by similarity of triangles, which is conjectured to be the proof that Pythagoras used.[12][17]

9.3.3 Proof by rearrangement

We have already discussed the Pythagorean proof, which was a proof by rearrangement. The same idea is conveyed by the leftmost animation below, which consists of a large square, side $a + b$, containing four identical right triangles. The triangles are shown in two arrangements, the first of which leaves two squares a^2 and b^2 uncovered, the second of which leaves square c^2 uncovered. The area encompassed by the outer square never changes, and the area of the four triangles is the same at the beginning and the end, so the black square areas must be equal, therefore $a^2 + b^2 = c^2$.

A second proof by rearrangement is given by the middle animation. A large square is formed with area c^2, from four identical right triangles with sides a, b and c, fitted around a small central square. Then two rectangles are formed with sides a and b by moving the triangles. Combining the smaller square with these rectangles produces two squares of areas a^2 and b^2, which must have the same area as the initial large square.[18]

The third, rightmost image also gives a proof. The upper two squares are divided as shown by the blue and green shading, into pieces that when rearranged can be made to fit in the lower square on the hypotenuse – or conversely the large square can be divided as shown into pieces that fill the other two. This shows the area of the large square equals that of the two smaller ones.[19]

9.3.4 Algebraic proofs

The theorem can be proved algebraically using four copies of a right triangle with sides a, b and c, arranged inside a square with side c as in the top half of the diagram.[20] The triangles are similar with area $\frac{1}{2}ab$, while the small square has side $b - a$ and area $(b - a)^2$. The area of the large square is therefore

$$(b - a)^2 + 4\frac{ab}{2} = (b - a)^2 + 2ab = a^2 + b^2.$$

But this is a square with side c and area c^2, so

$$c^2 = a^2 + b^2.$$

A similar proof uses four copies of the same triangle arranged symmetrically around a square with side c, as shown in the lower part of the diagram.[21] This results in a larger square, with side $a + b$ and area $(a + b)^2$. The four triangles and the square side c must have the same area as the larger square,

$$(b + a)^2 = c^2 + 4\frac{ab}{2} = c^2 + 2ab,$$

giving

$$c^2 = (b + a)^2 - 2ab = a^2 + b^2.$$

A related proof was published by future U.S. President James A. Garfield (then a U.S. Representative).[22][23] Instead of a square it uses a trapezoid, which can be constructed from the square in the second of the above proofs by bisecting along a diagonal of the inner square, to give the trapezoid as shown in the diagram. The area of the trapezoid can be calculated to be half the area of the square, that is

$$\frac{1}{2}(b + a)^2.$$

The inner square is similarly halved, and there are only two triangles so the proof proceeds as above except for a factor of $\frac{1}{2}$, which is removed by multiplying by two to give the result.

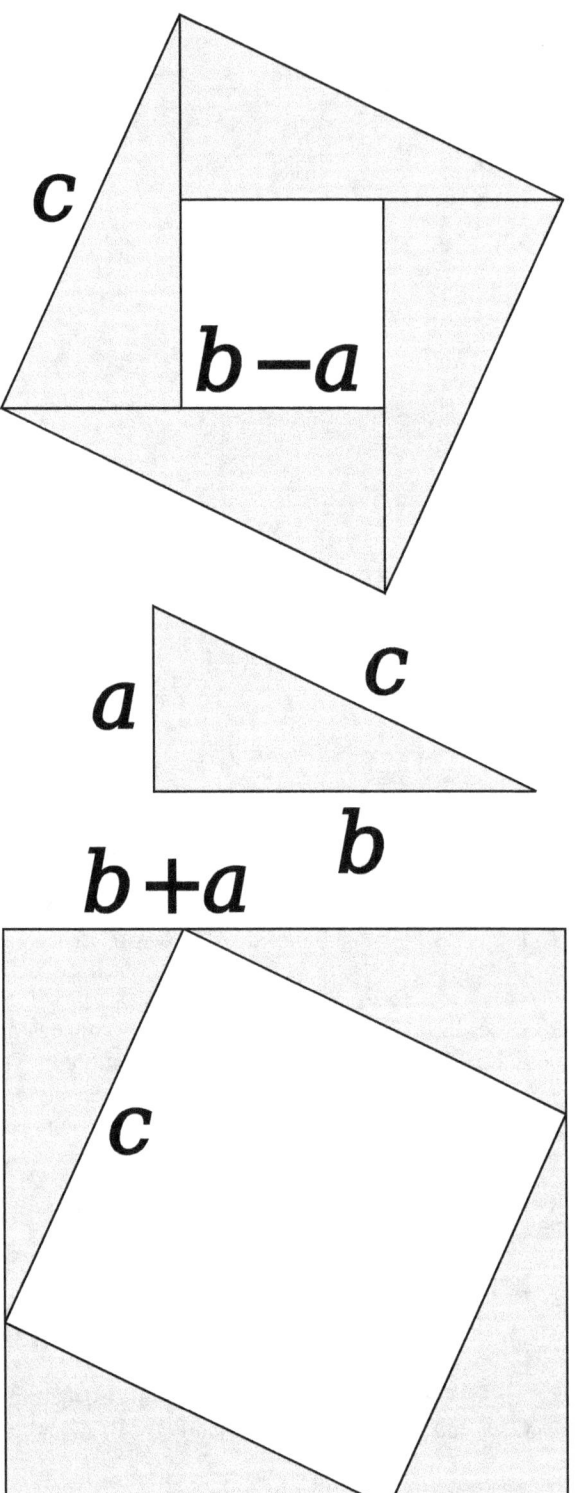

Diagram of the two algebraic proofs

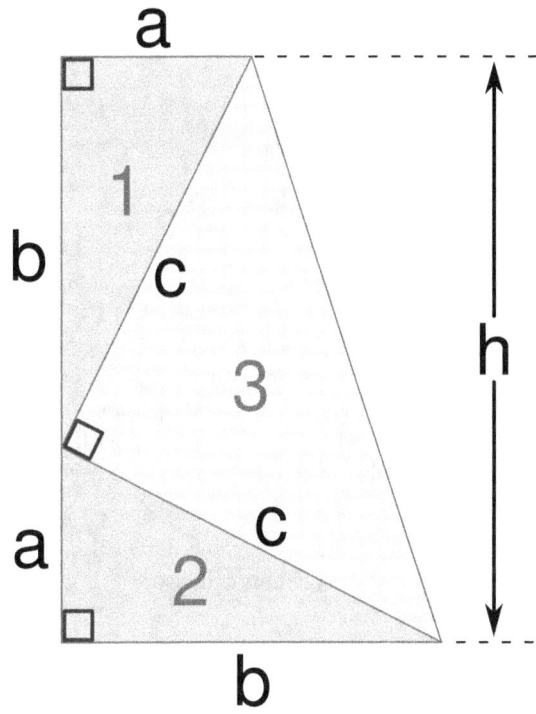

Diagram of Garfield's proof

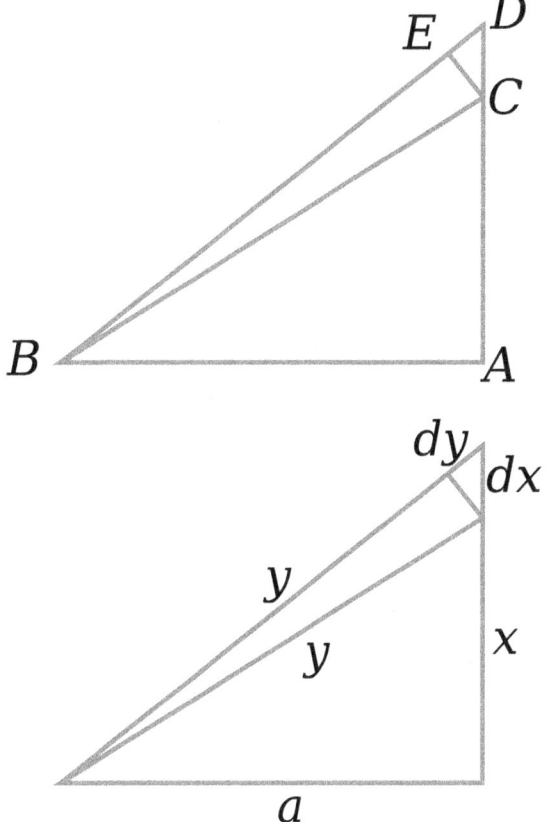

Diagram for differential proof

9.3.5 Proof using differentials

One can arrive at the Pythagorean theorem by studying how changes in a side produce a change in the hypotenuse and employing calculus.[24][25][26]

The triangle *ABC* is a right triangle, as shown in the upper part of the diagram, with *BC* the hypotenuse. At the same time the triangle lengths are measured as shown, with the hypotenuse of length y, the side *AC* of length x and the side *AB* of length a, as seen in the lower diagram part.

If x is increased by a small amount dx by extending the side *AC* slightly to *D*, then y also increases by dy. These form two sides of a triangle, *CDE*, which (with *E* chosen so *CE* is perpendicular to the hypotenuse) is a right triangle approximately similar to *ABC*. Therefore, the ratios of their sides must be the same, that is:

$$\frac{dy}{dx} = \frac{x}{y}.$$

This can be rewritten as follows:

$$y \cdot dy - x \cdot dx = 0.$$

This is a differential equation which is solved to give

$$y^2 - x^2 = C,$$

And the constant can be deduced from $x = 0$, $y = a$ to give the equation

$$y^2 = x^2 + a^2.$$

This is more of an intuitive proof than a formal one: it can be made more rigorous if proper limits are used in place of dx and dy.

9.4 Converse

The converse of the theorem is also true:[27]

> For any three positive numbers a, b, and c such that $a^2 + b^2 = c^2$, there exists a triangle with sides a, b and c, and every such triangle has a right angle between the sides of lengths a and b.

An alternative statement is:

> For any triangle with sides a, b, c, if $a^2 + b^2 = c^2$, then the angle between a and b measures $90°$.

This converse also appears in Euclid's *Elements* (Book I, Proposition 48):[28]

> "If in a triangle the square on one of the sides equals the sum of the squares on the remaining two sides of the triangle, then the angle contained by the remaining two sides of the triangle is right."

It can be proven using the law of cosines or as follows:

Let *ABC* be a triangle with side lengths a, b, and c, with $a^2 + b^2 = c^2$. Construct a second triangle with sides of length a and b containing a right angle. By the Pythagorean theorem, it follows that the hypotenuse of this triangle has length $c = \sqrt{a^2 + b^2}$, the same as the hypotenuse of the first triangle. Since both triangles' sides are the same lengths a, b and c, the triangles are congruent and must have the same angles. Therefore, the angle between the side of lengths a and b in the original triangle is a right angle.

The above proof of the converse makes use of the Pythagorean Theorem itself. The converse can also be proven without assuming the Pythagorean Theorem.[29][30]

A corollary of the Pythagorean theorem's converse is a simple means of determining whether a triangle is right, obtuse, or acute, as follows. Let c be chosen to be the longest of the three sides and $a + b > c$ (otherwise there is no triangle according to the triangle inequality). The following statements apply:[31]

- If $a^2 + b^2 = c^2$, then the triangle is right.

- If $a^2 + b^2 > c^2$, then the triangle is acute.

- If $a^2 + b^2 < c^2$, then the triangle is obtuse.

Edsger Dijkstra has stated this proposition about acute, right, and obtuse triangles in this language:

$$\operatorname{sgn}(\alpha + \beta - \gamma) = \operatorname{sgn}(a^2 + b^2 - c^2),$$

where α is the angle opposite to side a, β is the angle opposite to side b, γ is the angle opposite to side c, and sgn is the sign function.[32]

9.5 Consequences and uses of the theorem

9.5.1 Pythagorean triples

Main article: Pythagorean triple

A Pythagorean triple has three positive integers a, b, and c, such that $a^2 + b^2 = c^2$. In other words, a Pythagorean triple represents the lengths of the sides of a right triangle where all three sides have integer lengths.[1] Evidence from megalithic monuments in Northern Europe shows that such triples were known before the discovery of writing. Such a triple is commonly written (a, b, c). Some well-known examples are $(3, 4, 5)$ and $(5, 12, 13)$.

A primitive Pythagorean triple is one in which a, b and c are coprime (the greatest common divisor of a, b and c is 1).

The following is a list of primitive Pythagorean triples with values less than 100:

(3, 4, 5), (5, 12, 13), (7, 24, 25), (8, 15, 17), (9, 40, 41), (11, 60, 61), (12, 35, 37), (13, 84, 85), (16, 63, 65), (20, 21, 29), (28, 45, 53), (33, 56, 65), (36, 77, 85), (39, 80, 89), (48, 55, 73), (65, 72, 97)

9.5.2 Incommensurable lengths

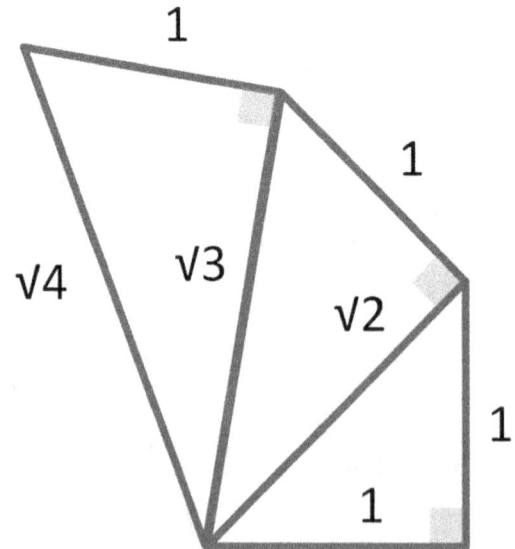

The spiral of Theodorus: A construction for line segments with lengths whose ratios are the square root of a positive integer

One of the consequences of the Pythagorean theorem is that line segments whose lengths are incommensurable (so the ratio of which is not a rational number) can be constructed using a straightedge and compass. Pythagoras's theorem enables construction of incommensurable lengths because the hypotenuse of a triangle is related to the sides by the square root operation.

The figure on the right shows how to construct line segments whose lengths are in the ratio of the square root of any positive integer.[33] Each triangle has a side (labeled "1") that is the chosen unit for measurement. In each right triangle, Pythagoras's theorem establishes the length of the hypotenuse in terms of this unit. If a hypotenuse is related to the unit by the square root of a positive integer that is not a perfect square, it is a realization of a length incommensurable with the unit, such as √2, √3, √5 . For more detail, see Quadratic irrational.

Incommensurable lengths conflicted with the Pythagorean school's concept of numbers as only whole numbers. The Pythagorean school dealt with proportions by comparison of integer multiples of a common subunit.[34] According to one legend, Hippasus of Metapontum (*ca.* 470 B.C.) was drowned at sea for making known the existence of the irrational or incommensurable.[35][36]

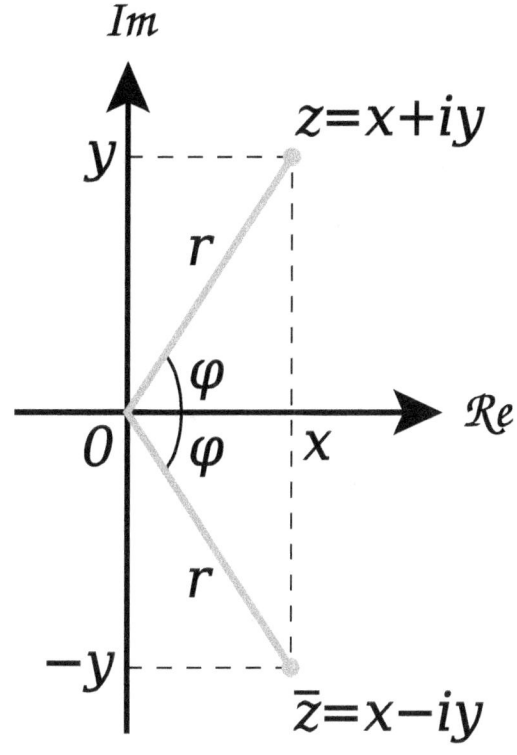

The absolute value of a complex number z is the distance r from z to the origin

9.5.3 Complex numbers

For any complex number

$$z = x + iy,$$

the absolute value or modulus is given by

$$r = |z| = \sqrt{x^2 + y^2}.$$

So the three quantities, r, x and y are related by the Pythagorean equation,

$$r^2 = x^2 + y^2.$$

Note that r is defined to be a positive number or zero but x and y can be negative as well as positive. Geometrically r is the distance of the z from zero or the origin O in the complex plane.

This can be generalised to find the distance between two points, z_1 and z_2 say. The required distance is given by

$$|z_1 - z_2| = \sqrt{(x_1 - x_2)^2 + (y_1 - y_2)^2},$$

so again they are related by a version of the Pythagorean equation,

$$|z_1 - z_2|^2 = (x_1 - x_2)^2 + (y_1 - y_2)^2.$$

9.5.4 Euclidean distance in various coordinate systems

The distance formula in Cartesian coordinates is derived from the Pythagorean theorem.[37] If (x_1, y_1) and (x_2, y_2) are points in the plane, then the distance between them, also called the Euclidean distance, is given by

$$\sqrt{(x_1 - x_2)^2 + (y_1 - y_2)^2}.$$

More generally, in Euclidean n-space, the Euclidean distance between two points, $A = (a_1, a_2, \ldots, a_n)$ and $B = (b_1, b_2, \ldots, b_n)$, is defined, by generalization of the Pythagorean theorem, as:

$$\sqrt{(a_1 - b_1)^2 + (a_2 - b_2)^2 + \cdots + (a_n - b_n)^2} =$$

$$\sqrt{\sum_{i=1}^{n} (a_i - b_i)^2}.$$

If Cartesian coordinates are not used, for example, if polar coordinates are used in two dimensions or, in more general terms, if curvilinear coordinates are used, the formulas expressing the Euclidean distance are more complicated than the Pythagorean theorem, but can be derived from it. A typical example where the straight-line distance between two points is converted to curvilinear coordinates can be found in the applications of Legendre polynomials in physics. The formulas can be discovered by using Pythagoras's theorem with the equations relating the curvilinear coordinates to Cartesian coordinates. For example, the polar coordinates (r, θ) can be introduced as:

$$x = r \cos\theta, \ y = r \sin\theta.$$

Then two points with locations (r_1, θ_1) and (r_2, θ_2) are separated by a distance s:

$$s^2 = (x_1 - x_2)^2 + (y_1 - y_2)^2 = (r_1 \cos\theta_1 - r_2 \cos\theta_2)^2$$

$$+ (r_1 \sin\theta_1 - r_2 \sin\theta_2)^2.$$

Performing the squares and combining terms, the Pythagorean formula for distance in Cartesian coordinates produces the separation in polar coordinates as:

$$s^2 = r_1^2 + r_2^2 - 2r_1 r_2 (\cos\theta_1 \cos\theta_2 + \sin\theta_1 \sin\theta_2)$$
$$= r_1^2 + r_2^2 - 2r_1 r_2 \cos(\theta_1 - \theta_2)$$
$$= r_1^2 + r_2^2 - 2r_1 r_2 \cos\Delta\theta,$$

using the trigonometric product-to-sum formulas. This formula is the law of cosines, sometimes called the Generalized Pythagorean Theorem.[38] From this result, for the case where the radii to the two locations are at right angles, the enclosed angle $\Delta\theta = \pi/2$, and the form corresponding to Pythagoras's theorem is regained: $s^2 = r_1^2 + r_2^2$. The Pythagorean theorem, valid for right triangles, therefore is a special case of the more general law of cosines, valid for arbitrary triangles.

9.5.5   Pythagorean trigonometric identity

Main article: Pythagorean trigonometric identity

In a right triangle with sides a, b and hypotenuse c, trigonometry determines the sine and cosine of the angle θ between side a and the hypotenuse as:

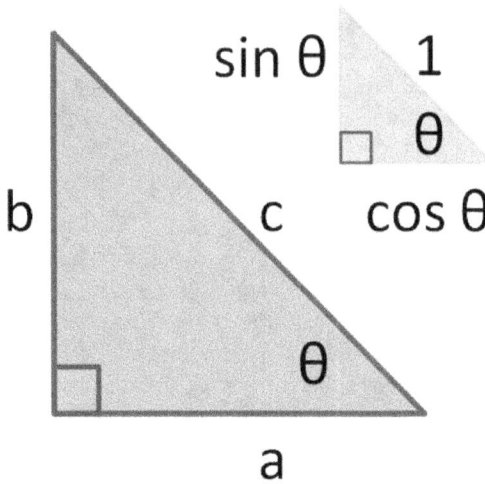

Similar right triangles showing sine and cosine of angle θ

$$\sin\theta = \frac{b}{c}, \quad \cos\theta = \frac{a}{c}.$$

From that it follows:

$$\cos^2\theta + \sin^2\theta = \frac{a^2 + b^2}{c^2} = 1,$$

where the last step applies Pythagoras's theorem. This relation between sine and cosine is sometimes called the fundamental Pythagorean trigonometric identity.[39] In similar triangles, the ratios of the sides are the same regardless of the size of the triangles, and depend upon the angles. Consequently, in the figure, the triangle with hypotenuse of unit size has opposite side of size $\sin\theta$ and adjacent side of size $\cos\theta$ in units of the hypotenuse.

9.5.6   Relation to the cross product

The Pythagorean theorem relates the cross product and dot product in a similar way:[40]

$$\|\mathbf{a} \times \mathbf{b}\|^2 + (\mathbf{a} \cdot \mathbf{b})^2 = \|\mathbf{a}\|^2 \|\mathbf{b}\|^2.$$

This can be seen from the definitions of the cross product and dot product, as

$$\mathbf{a} \times \mathbf{b} = ab\mathbf{n} \sin\theta$$
$$\mathbf{a} \cdot \mathbf{b} = ab \cos\theta,$$

with $\mathbf{n}$ a unit vector normal to both $\mathbf{a}$ and $\mathbf{b}$. The relationship follows from these definitions and the Pythagorean trigonometric identity.

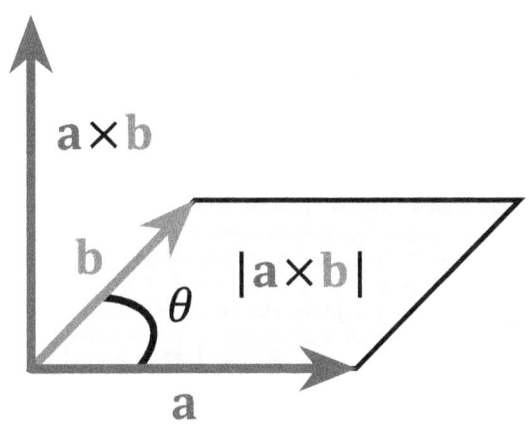

*The area of a parallelogram as a cross product; vectors **a** and **b** identify a plane and **a** × **b** is normal to this plane.*

This can also be used to define the cross product. By rearranging the following equation is obtained

$$\|\mathbf{a} \times \mathbf{b}\|^2 = \|\mathbf{a}\|^2 \|\mathbf{b}\|^2 - (\mathbf{a} \cdot \mathbf{b})^2.$$

This can be considered as a condition on the cross product and so part of its definition, for example in seven dimensions.[41][42]

9.6 Generalizations

9.6.1 Similar figures on the three sides

A generalization of the Pythagorean theorem extending beyond the areas of squares on the three sides to similar figures was known by Hippocrates of Chios in the 5th century BC,[43] and was included by Euclid in his *Elements*:[44]

> If one erects similar figures (see Euclidean geometry) with corresponding sides on the sides of a right triangle, then the sum of the areas of the ones on the two smaller sides equals the area of the one on the larger side.

This extension assumes that the sides of the original triangle are the corresponding sides of the three congruent figures (so the common ratios of sides between the similar figures are *a:b:c*).[45] While Euclid's proof only applied to convex polygons, the theorem also applies to concave polygons and even to similar figures that have curved boundaries (but still with part of a figure's boundary being the side of the original triangle).[45]

The basic idea behind this generalization is that the area of a plane figure is proportional to the square of any linear dimension, and in particular is proportional to the square of the length of any side. Thus, if similar figures with areas *A*, *B* and *C* are erected on sides with corresponding lengths *a*, *b* and *c* then:

$$\frac{A}{a^2} = \frac{B}{b^2} = \frac{C}{c^2},$$

$$\Rightarrow A + B = \frac{a^2}{c^2}C + \frac{b^2}{c^2}C.$$

But, by the Pythagorean theorem, $a^2 + b^2 = c^2$, so $A + B = C$.

Conversely, if we can prove that $A + B = C$ for three similar figures without using the Pythagorean theorem, then we can work backwards to construct a proof of the theorem. For example, the starting center triangle can be replicated and used as a triangle *C* on its hypotenuse, and two similar right triangles (*A* and *B*) constructed on the other two sides, formed by dividing the central triangle by its altitude. The sum of the areas of the two smaller triangles therefore is that of the third, thus $A + B = C$ and reversing the above logic leads to the Pythagorean theorem $a^2 + b^2 = c^2$.

9.6.2 Law of cosines

Main article: Law of cosines

The Pythagorean theorem is a special case of the more general theorem relating the lengths of sides in any triangle, the law of cosines:[46]

$$a^2 + b^2 - 2ab\cos\theta = c^2,$$

where θ is the angle between sides *a* and *b*.

When θ is 90 degrees, then cosθ = 0, and the formula reduces to the usual Pythagorean theorem.

9.6.3 Arbitrary triangle

At any selected angle of a general triangle of sides *a, b, c*, inscribe an isosceles triangle such that the equal angles at its base θ are the same as the selected angle. Suppose the selected angle θ is opposite the side labeled *c*. Inscribing the isosceles triangle forms triangle *ABD* with angle θ opposite side *a* and with side *r* along *c*. A second triangle is formed with angle θ opposite side *b* and a side with length *s* along

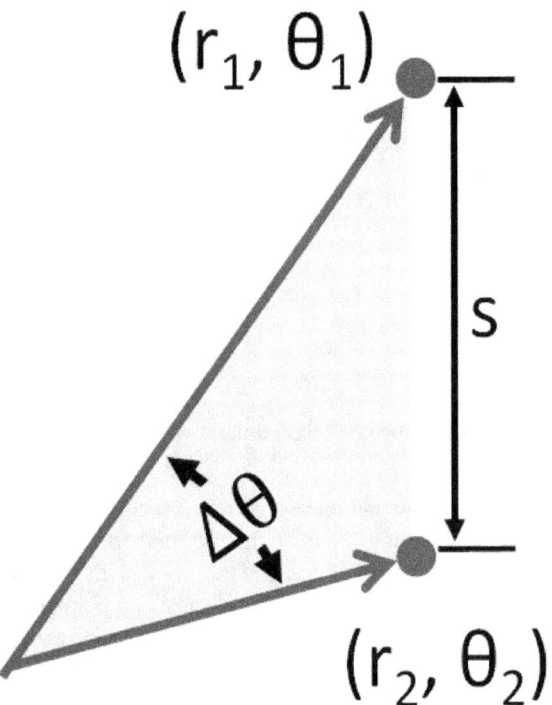

The separation s *of two points* (r_1, θ_1) *and* (r_2, θ_2) *in polar coordinates is given by the law of cosines. Interior angle* $\Delta\theta = \theta_1 - \theta_2$.

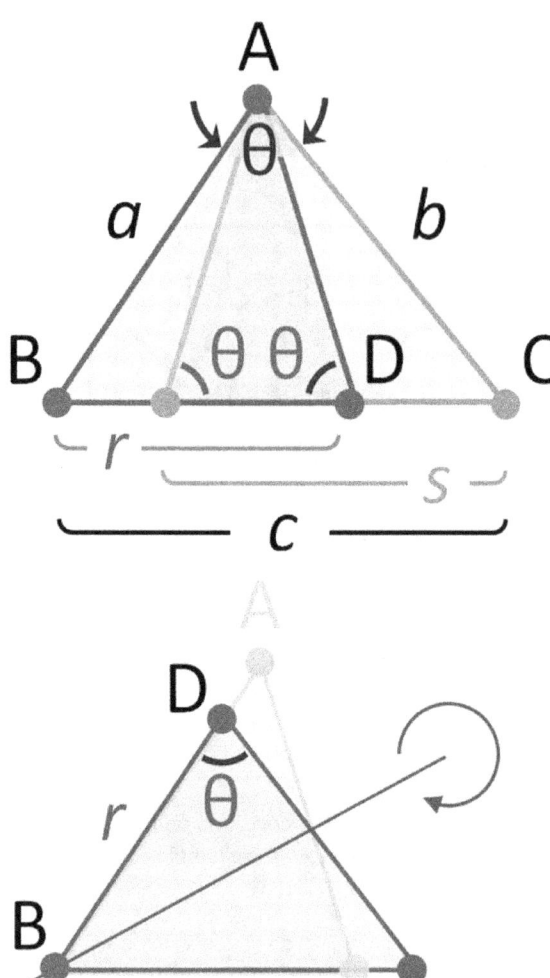

Generalization of Pythagoras's theorem by Tâbit ibn Qorra.[47] Lower panel: reflection of triangle ABD (top) to form triangle DBA, similar to triangle ABC (top).

c, as shown in the figure. Tâbit ibn Qorra[48] stated that the sides of the three triangles were related as:[49][50]

$$a^2 + b^2 = c(r + s) .$$

As the angle θ approaches $\pi/2$, the base of the isosceles triangle narrows, and lengths r and s overlap less and less. When $\theta = \pi/2$, *ADB* becomes a right triangle, $r + s = c$, and the original Pythagorean theorem is regained.

One proof observes that triangle *ABC* has the same angles as triangle *ABD*, but in opposite order. (The two triangles share the angle at vertex B, both contain the angle θ, and so also have the same third angle by the triangle postulate.) Consequently, *ABC* is similar to the reflection of *ABD*, the triangle *DBA* in the lower panel. Taking the ratio of sides opposite and adjacent to θ,

$$\frac{c}{a} = \frac{a}{r} .$$

Likewise, for the reflection of the other triangle,

$$\frac{c}{b} = \frac{b}{s} .$$

Clearing fractions and adding these two relations:

$$cr + cs = a^2 + b^2 ,$$

the required result.

The theorem remains valid if the angle θ is obtuse so the lengths r and s are non-overlapping.

9.6.4   General triangles using parallelograms

Pappus' area theorem is a further generalization, that applies to triangles that are not right triangles, using parallelograms on the three sides in place of squares (squares are a special

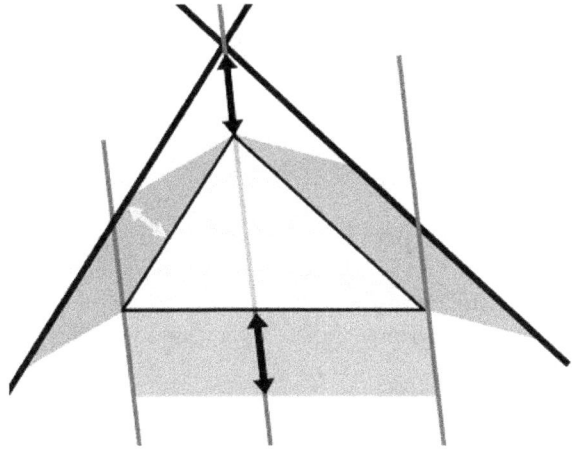

Generalization for arbitrary triangles, green area = blue area

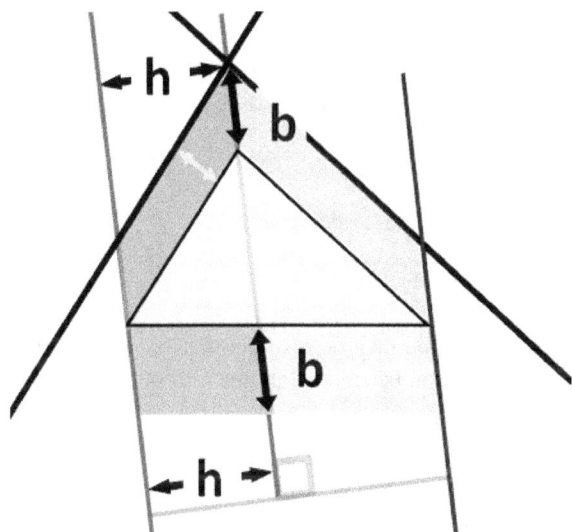

Construction for proof of parallelogram generalization

case, of course). The upper figure shows that for a scalene triangle, the area of the parallelogram on the longest side is the sum of the areas of the parallelograms on the other two sides, provided the parallelogram on the long side is constructed as indicated (the dimensions labeled with arrows are the same, and determine the sides of the bottom parallelogram). This replacement of squares with parallelograms bears a clear resemblance to the original Pythagoras's theorem, and was considered a generalization by Pappus of Alexandria in 4 A.D.[51][52]

The lower figure shows the elements of the proof. Focus on the left side of the figure. The left green parallelogram has the same area as the left, blue portion of the bottom parallelogram because both have the same base *b* and height *h*. However, the left green parallelogram also has the same area as the left green parallelogram of the upper figure, be-

cause they have the same base (the upper left side of the triangle) and the same height normal to that side of the triangle. Repeating the argument for the right side of the figure, the bottom parallelogram has the same area as the sum of the two green parallelograms.

9.6.5 Solid geometry

Main article: Solid geometry

In terms of solid geometry, Pythagoras's theorem can be

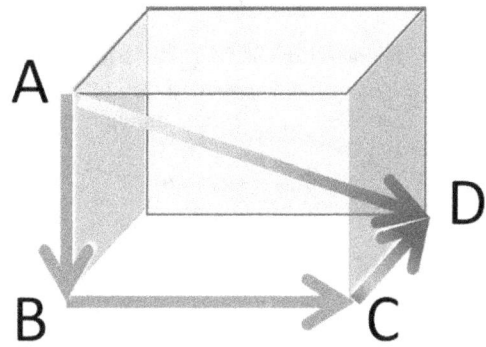

Pythagoras's theorem in three dimensions relates the diagonal AD to the three sides.

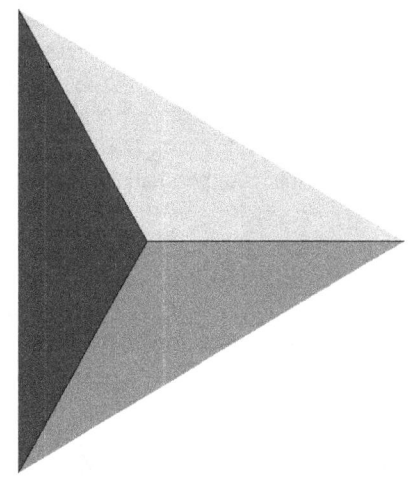

A tetrahedron with outward facing right-angle corner

applied to three dimensions as follows. Consider a rectangular solid as shown in the figure. The length of diagonal *BD* is found from Pythagoras's theorem as:

$$\overline{BD}^2 = \overline{BC}^2 + \overline{CD}^2 \,,$$

where these three sides form a right triangle. Using horizontal diagonal *BD* and the vertical edge *AB*, the length of diagonal *AD* then is found by a second application of Pythagoras's theorem as:

$$\overline{AD}^2 = \overline{AB}^2 + \overline{BD}^2 \, ,$$

or, doing it all in one step:

$$\overline{AD}^2 = \overline{AB}^2 + \overline{BC}^2 + \overline{CD}^2 \, .$$

This result is the three-dimensional expression for the magnitude of a vector **v** (the diagonal AD) in terms of its orthogonal components $\{v_k\}$ (the three mutually perpendicular sides):

$$\|\mathbf{v}\|^2 = \sum_{k=1}^{3} \|\mathbf{v}_k\|^2.$$

This one-step formulation may be viewed as a generalization of Pythagoras's theorem to higher dimensions. However, this result is really just the repeated application of the original Pythagoras's theorem to a succession of right triangles in a sequence of orthogonal planes.

A substantial generalization of the Pythagorean theorem to three dimensions is de Gua's theorem, named for Jean Paul de Gua de Malves: If a tetrahedron has a right angle corner (like a corner of a cube), then the square of the area of the face opposite the right angle corner is the sum of the squares of the areas of the other three faces. This result can be generalized as in the "*n*-dimensional Pythagorean theorem":[53]

> Let $x_1, x_2, \ldots, x_n$ be orthogonal vectors in $\mathbb{R}^n$. Consider the *n*-dimensional simplex *S* with vertices $0, x_1, \ldots, x_n$. (Think of the $(n-1)$-dimensional simplex with vertices $x_1, \ldots, x_n$ not including the origin as the "hypotenuse" of *S* and the remaining $(n-1)$-dimensional faces of *S* as its "legs".) Then the square of the volume of the hypotenuse of *S* is the sum of the squares of the volumes of the *n* legs.

This statement is illustrated in three dimensions by the tetrahedron in the figure. The "hypotenuse" is the base of the tetrahedron at the back of the figure, and the "legs" are the three sides emanating from the vertex in the foreground. As the depth of the base from the vertex increases, the area of the "legs" increases, while that of the base is fixed. The theorem suggests that when this depth is at the value creating a right vertex, the generalization of Pythagoras's theorem applies. In a different wording:[54]

Given an *n*-rectangular *n*-dimensional simplex, the square of the $(n-1)$-content of the facet opposing the right vertex will equal the sum of the squares of the $(n-1)$-contents of the remaining facets.

9.6.6 Inner product spaces

See also: Hilbert space

The Pythagorean theorem can be generalized to inner prod-

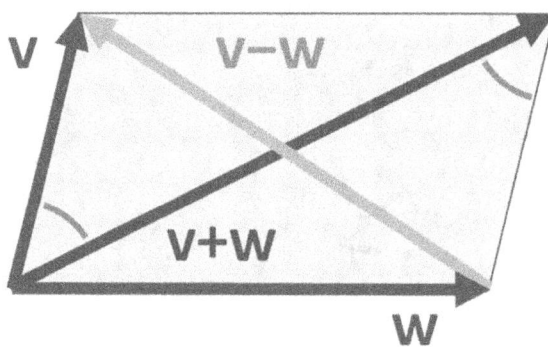

Vectors involved in the parallelogram law

uct spaces,[55] which are generalizations of the familiar 2-dimensional and 3-dimensional Euclidean spaces. For example, a function may be considered as a vector with infinitely many components in an inner product space, as in functional analysis.[56]

In an inner product space, the concept of perpendicularity is replaced by the concept of orthogonality: two vectors **v** and **w** are orthogonal if their inner product $\langle \mathbf{v}, \mathbf{w} \rangle$ is zero. The inner product is a generalization of the dot product of vectors. The dot product is called the *standard* inner product or the *Euclidean* inner product. However, other inner products are possible.[57]

The concept of length is replaced by the concept of the norm $\|\mathbf{v}\|$ of a vector **v**, defined as:[58]

$$\|\mathbf{v}\| \equiv \sqrt{\langle \mathbf{v}, \mathbf{v} \rangle} \, .$$

In an inner-product space, the **Pythagorean theorem** states that for any two orthogonal vectors **v** and **w** we have

$$\|\mathbf{v} + \mathbf{w}\|^2 = \|\mathbf{v}\|^2 + \|\mathbf{w}\|^2 \, .$$

Here the vectors **v** and **w** are akin to the sides of a right triangle with hypotenuse given by the vector sum **v** + **w**. This form of the Pythagorean theorem is a consequence of the properties of the inner product:

$$\|\mathbf{v} + \mathbf{w}\|^2 = \langle \mathbf{v} + \mathbf{w}, \mathbf{v} + \mathbf{w}\rangle = \langle \mathbf{v}, \mathbf{v}\rangle + \langle \mathbf{w}, \mathbf{w}\rangle + \langle$$
$$\mathbf{v}, \mathbf{w}\rangle + \langle \mathbf{w}, \mathbf{v}\rangle = \|\mathbf{v}\|^2 + \|\mathbf{w}\|^2,$$

where the inner products of the cross terms are zero, because of orthogonality.

A further generalization of the Pythagorean theorem in an inner product space to non-orthogonal vectors is the *parallelogram law* :[58]

$$2\|\mathbf{v}\|^2 + 2\|\mathbf{w}\|^2 = \|\mathbf{v} + \mathbf{w}\|^2 + \|\mathbf{v} - \mathbf{w}\|^2,$$

which says that twice the sum of the squares of the lengths of the sides of a parallelogram is the sum of the squares of the lengths of the diagonals. Any norm that satisfies this equality is *ipso facto* a norm corresponding to an inner product.[58]

The Pythagorean identity can be extended to sums of more than two orthogonal vectors. If $v1$, $v2$, ..., vn are pairwise-orthogonal vectors in an inner-product space, then application of the Pythagorean theorem to successive pairs of these vectors (as described for 3-dimensions in the section on solid geometry) results in the equation[59]

$$\|\sum_{k=1}^{n} \mathbf{v}_k\|^2 = \sum_{k=1}^{n} \|\mathbf{v}_k\|^2.$$

9.6.7 Non-Euclidean geometry

Main article: Non-Euclidean geometry
See also: Hilbert's axioms

The Pythagorean theorem is derived from the axioms of Euclidean geometry, and in fact, the Pythagorean theorem given above does not hold in a non-Euclidean geometry.[60] (The Pythagorean theorem has been shown, in fact, to be equivalent to Euclid's Parallel (Fifth) Postulate.[61][62]) In other words, in non-Euclidean geometry, the relation between the sides of a triangle must necessarily take a non-Pythagorean form. For example, in spherical geometry, all three sides of the right triangle (say a, b, and c) bounding an octant of the unit sphere have length equal to $\pi/2$, and all its angles are right angles, which violates the Pythagorean theorem because $a^2 + b^2 \neq c^2$.

Here two cases of non-Euclidean geometry are considered—spherical geometry and hyperbolic plane geometry; in each case, as in the Euclidean case for non-right triangles, the result replacing the Pythagorean theorem follows from the appropriate law of cosines.

However, the Pythagorean theorem remains true in hyperbolic geometry and elliptic geometry if the condition that the triangle be right is replaced with the condition that two of the angles sum to the third, say $A+B = C$. The sides are then related as follows: the sum of the areas of the circles with diameters a and b equals the area of the circle with diameter c.[63]

Spherical geometry

Main article: Spherical geometry
For any right triangle on a sphere of radius R (for example,

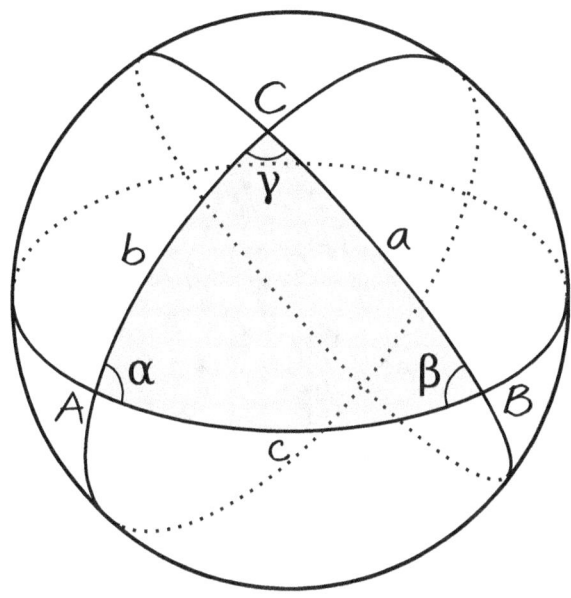

Spherical triangle

if γ in the figure is a right angle), with sides a, b, c, the relation between the sides takes the form:[64]

$$\cos\left(\frac{c}{R}\right) = \cos\left(\frac{a}{R}\right)\cos\left(\frac{b}{R}\right).$$

This equation can be derived as a special case of the spherical law of cosines that applies to all spherical triangles:

$$\cos\left(\frac{c}{R}\right) = \cos\left(\frac{a}{R}\right)\cos\left(\frac{b}{R}\right) + \sin\left(\frac{a}{R}\right)\sin\left(\frac{b}{R}\right)\cos\gamma.$$

By expressing the Maclaurin series for the cosine function as an asymptotic expansion with the remainder term in big O notation,

$$\cos x = 1 - \frac{x^2}{2} + O(x^4) \text{ as } x \to 0,$$

it can be shown that as the radius R approaches infinity and the arguments a/R, b/R, and c/R tend to zero, the spherical relation between the sides of a right triangle approaches the Euclidean form of the Pythagorean theorem. Substituting the asymptotic expansion for each of the cosines into the spherical relation for a right triangle yields

$$1-\frac{1}{2}\left(\frac{c}{R}\right)^2+O\left(\frac{1}{R^4}\right)=\left[1-\frac{1}{2}\left(\frac{a}{R}\right)^2+O\left(\frac{1}{R^4}\right)\right]\left[$$
$$1-\frac{1}{2}\left(\frac{b}{R}\right)^2+O\left(\frac{1}{R^4}\right)\right] \text{ as } R\to\infty.$$

The constants a^4, b^4, and c^4 have been absorbed into the big O remainder terms since they are independent of the radius R. This asymptotic relationship can be further simplified by multiplying out the bracketed quantities, cancelling the ones, multiplying through by -2, and collecting all the error terms together:

$$\left(\frac{c}{R}\right)^2=\left(\frac{a}{R}\right)^2+\left(\frac{b}{R}\right)^2+O\left(\frac{1}{R^4}\right) \text{ as } R\to\infty.$$

After multiplying through by R^2, the Euclidean Pythagorean relationship $c^2 = a^2 + b^2$ is recovered in the limit as the radius R approaches infinity (since the remainder term tends to zero):

$$c^2=a^2+b^2+O\left(\frac{1}{R^2}\right) \text{ as } R\to\infty.$$

Hyperbolic geometry

Main article: Hyperbolic geometry
See also: Gaussian curvature
For a right triangle in hyperbolic geometry with sides a, b,

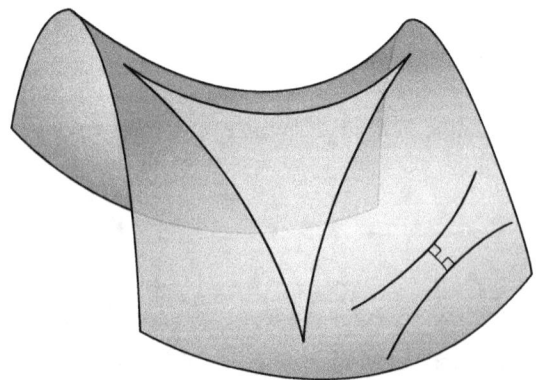

Hyperbolic triangle

c and with side c opposite a right angle, the relation between the sides takes the form:[65]

$$\cosh c = \cosh a \ \cosh b - \sinh a \ \sinh b \ \cos\gamma,$$

$$\cosh c = \cosh a \ \cosh b$$

where cosh is the hyperbolic cosine. This formula is a special form of the hyperbolic law of cosines that applies to all hyperbolic triangles:[66]

with γ the angle at the vertex opposite the side c.

By using the Maclaurin series for the hyperbolic cosine, $\cosh x \approx 1 + x^2/2$, it can be shown that as a hyperbolic triangle becomes very small (that is, as a, b, and c all approach zero), the hyperbolic relation for a right triangle approaches the form of Pythagoras's theorem.

9.6.8 Differential geometry

Main article: Differential geometry
On an infinitesimal level, in three dimensional space,

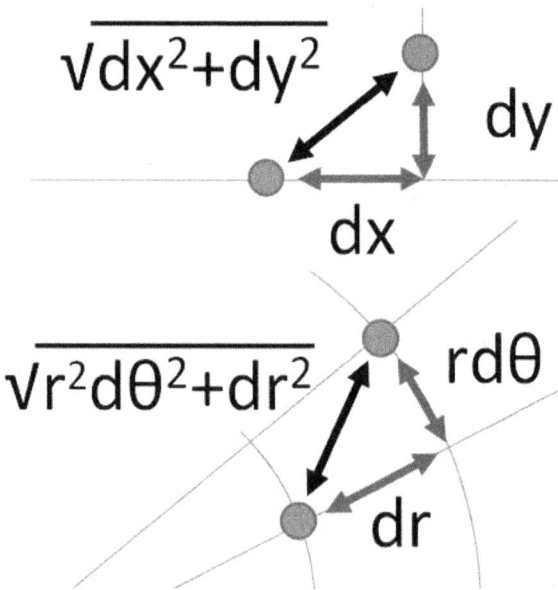

Distance between infinitesimally separated points in Cartesian coordinates (top) and polar coordinates (bottom), as given by Pythagoras's theorem

Pythagoras's theorem describes the distance between two infinitesimally separated points as:

$$ds^2 = dx^2 + dy^2 + dz^2,$$

with ds the element of distance and (dx, dy, dz) the components of the vector separating the two points. Such a space is called a Euclidean space. However, in Riemannian geometry, a generalization of this expression useful for general coordinates (not just Cartesian) and general spaces (not just Euclidean) takes the form:[67]

$$ds^2 = \sum_{i,j}^{n} g_{ij}\,dx_i\,dx_j$$

which is called the metric tensor. (Sometimes, by abuse of language, the same term is applied to the set of coefficients g_{ij}.) It may be a function of position, and often describes curved space. A simple example is Euclidean (flat) space expressed in curvilinear coordinates. For example, in polar coordinates:

$$ds^2 = dr^2 + r^2 d\theta^2 \ .$$

9.7 History

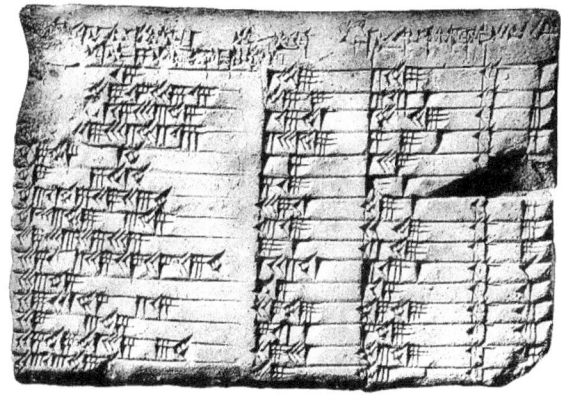

The Plimpton 322 tablet records Pythagorean triples from Babylonian times.[6]

There is debate whether the Pythagorean theorem was discovered once, or many times in many places, and the date of first discovery is uncertain, as is the date of the first proof. According to Joran Friberg, a historian of mathematics, evidence indicates that the Pythagorean Theorem was well-known to the mathematicians of the First Babylonian Dynasty (20th to 16th centuries BC), which would have been over a thousand years before Pythagoras was born.[68] Other sources, such as a book by Leon Lederman and Dick Teresi, mention that Pythagoras discovered the theorem,[69] although Teresi subsequently stated that the Babylonians developed the theorem "at least fifteen hundred years before Pythagoras was born."[70] The history of the theorem can be divided into four parts: knowledge of

Pythagorean triples, knowledge of the relationship among the sides of a right triangle, knowledge of the relationships among adjacent angles, and proofs of the theorem within some deductive system.

Bartel Leendert van der Waerden (1903–1996) conjectured that Pythagorean triples were discovered algebraically by the Babylonians.[71] Written between 2000 and 1786 BC, the Middle Kingdom Egyptian *Berlin Papyrus 6619* includes a problem whose solution is the Pythagorean triple 6:8:10, but the problem does not mention a triangle. The Mesopotamian tablet *Plimpton 322*, written between 1790 and 1750 BC during the reign of Hammurabi the Great, contains many entries closely related to Pythagorean triples.

In India, the *Baudhayana Sulba Sutra*, the dates of which are given variously as between the 8th and 2nd century BC, contains a list of Pythagorean triples discovered algebraically, a statement of the Pythagorean theorem, and a geometrical proof of the Pythagorean theorem for an isosceles right triangle. The *Apastamba Sulba Sutra* (c. 600 BC) contains a numerical proof of the general Pythagorean theorem, using an area computation. Van der Waerden believed that "it was certainly based on earlier traditions". Boyer (1991) thinks the elements found in the *Śulba-sūtram* may be of Mesopotamian derivation.[72]

 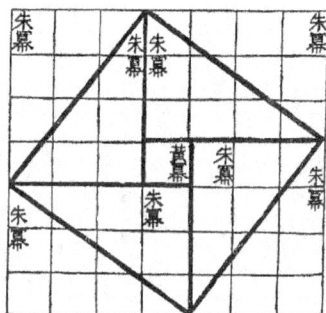

Geometric proof of the Pythagorean theorem from the Zhou Bi Suan Jing.

With contents known much earlier, but in surviving texts dating from roughly the 1st century BC, the Chinese text *Zhou Bi Suan Jing* (周髀算经), (*The Arithmetical Classic of the Gnomon and the Circular Paths of Heaven*) gives a reasoning for the Pythagorean theorem for the (3, 4, 5) triangle—in China it is called the "Gougu Theorem" (勾股定理).[73][74] During the Han Dynasty (202 BC to 220 AD), Pythagorean triples appear in *The Nine Chapters on the Mathematical Art*,[75] together with a mention of right triangles.[76] Some believe the theorem arose first in China,[77] where it is alternatively known as the "Shang Gao Theorem" (商高定理),[78] named after the Duke of Zhou's astronomer and mathematician, whose reasoning composed most of what was in the *Zhou Bi Suan Jing*.[79]

Pythagoras, whose dates are commonly given as 569–475 BC, used algebraic methods to construct Pythagorean triples, according to Proclus's commentary on Euclid. Proclus, however, wrote between 410 and 485 AD. According to Thomas L. Heath (1861–1940), no specific attribution of the theorem to Pythagoras exists in the surviving Greek literature from the five centuries after Pythagoras lived.[80] However, when authors such as Plutarch and Cicero attributed the theorem to Pythagoras, they did so in a way which suggests that the attribution was widely known and undoubted.[4][81] "Whether this formula is rightly attributed to Pythagoras personally, [...] one can safely assume that it belongs to the very oldest period of Pythagorean mathematics."[36]

Around 400 BC, according to Proclus, Plato gave a method for finding Pythagorean triples that combined algebra and geometry. Around 300 BC, in Euclid's *Elements*, the oldest extant axiomatic proof of the theorem is presented.[82]

9.8   In popular culture

Exhibit on the Pythagorean theorem at the Universum museum in Mexico City

The Pythagorean theorem has arisen in popular culture in a variety of ways.

- Hans Christian Andersen wrote in 1831 a poem about the Pythagorean theorem: *Formens Evige Magie (Et poetisk Spilfægteri)*.[83]

- A verse of the Major-General's Song in the Gilbert and Sullivan comic opera *The Pirates of Penzance*, "About binomial theorem I'm teeming with a lot o' news, With many cheerful facts about the square of the hypotenuse", makes an oblique reference to the theorem.[84]

- The Scarecrow in the film *The Wizard of Oz* makes a more specific reference to the theorem. Upon re-

ceiving his diploma from the Wizard, he immediately exhibits his "knowledge" by reciting a mangled and incorrect version of the theorem: "The sum of the square roots of any two sides of an isosceles triangle is equal to the square root of the remaining side. Oh, joy! Oh, rapture! I've got a brain!"[85][86]

- In 2000, Uganda released a coin with the shape of an isosceles right triangle. The coin's tail has an image of Pythagoras and the equation $\alpha^2 + \beta^2 = \gamma^2$, accompanied with the mention "PYTHAGORAS MILLENNIUM".[87][88]

- Greece, Japan, San Marino, Sierra Leone, and Suriname have issued postage stamps depicting Pythagoras and the Pythagorean theorem.[89]

- In Neal Stephenson's speculative fiction *Anathem*, the Pythagorean theorem is referred to as 'the Adrakhonic theorem'. A geometric proof of the theorem is displayed on the side of an alien ship to demonstrate the aliens' understanding of mathematics.

9.9   See also

- British flag theorem

- Dulcarnon

- Fermat's Last Theorem

- Linear algebra

- List of triangle topics

- L^p space

- Nonhypotenuse number

- Parallelogram law

- Ptolemy's theorem

- Pythagorean expectation

- Pythagorean tiling

- Rational trigonometry in Pythagoras's theorem

9.10   Notes

[1] Judith D. Sally, Paul Sally (2007). "Chapter 3: Pythagorean triples". *Roots to research: a vertical development of mathematical problems*. American Mathematical Society Bookstore. p. 63. ISBN 0-8218-4403-2.

[2] Posamentier, Alfred. *The Pythagorean Theorem: The Story of Its Power and Beauty*, p. 23 (Prometheus Books 2010).

[3] George Johnston Allman (1889). *Greek Geometry from Thales to Euclid* (Reprinted by Kessinger Publishing LLC 2005 ed.). Hodges, Figgis, & Co. p. 26. ISBN 1-4326-0662-X. The discovery of the law of three squares, commonly called the "theorem of Pythagoras" is attributed to him by – amongst others – Vitruvius, Diogenes Laertius, Proclus, and Plutarch ...

[4] (Heath 1921, Vol I, p. 144)

[5] According to Heath 1921, Vol I, p. 147, Vitruvius says that Pythagoras first discovered the triangle (3,4,5); the fact that the latter is right-angled led to the theorem.

[6] Otto Neugebauer (1969). *The exact sciences in antiquity* (Republication of 1957 Brown University Press 2nd ed.). Courier Dover Publications. p. 36. ISBN 0-486-22332-9.. For a different view, see Dick Teresi (2003). *Lost Discoveries: The Ancient Roots of Modern Science*. Simon and Schuster. p. 52. ISBN 0-7432-4379-X., where the speculation is made that the first column of tablet 322 in the Plimpton collection supports a Babylonian knowledge of some elements of trigonometry. That notion is pretty much laid to rest, however, by Eleanor Robson (2002). "Words and Pictures: New Light on Plimpton 322". *The American Mathematical Monthly* (Mathematical Association of America) **109** (2): 105–120. doi:10.2307/2695324. JSTOR 2695324. (pdf file). The generally accepted view today is that the Babylonians had no awareness of trigonometric functions. See also Abdulrahman A. Abdulaziz (2010). "The Plimpton 322 Tablet and the Babylonian Method of Generating Pythagorean Triples". arXiv:1004.0025 [math.HO]. §2, page 7.

[7] Mario Livio (2003). *The golden ratio: the story of phi, the world's most astonishing number*. Random House, Inc. p. 25. ISBN 0-7679-0816-3.

[8] Benson, Donald. *The Moment of Proof : Mathematical Epiphanies*, pp. 172–173 (Oxford University Press, 1999).

[9] Maor, Eli. *The Pythagorean Theorem: A 4,000-year History*, p. 61 (Princeton University Press, 2007).

[10] (Loomis 1968)

[11] (Maor 2007, p. 39) page 39

[12] Stephen W. Hawking (2005). *God created the integers: the mathematical breakthroughs that changed history*. Philadelphia: Running Press Book Publishers. p. 12. ISBN 0-7624-1922-9. This proof first appeared after a computer program was set to check Euclidean proofs.

[13] See for example Pythagorean theorem by shear mapping, Saint Louis University website Java applet

[14] Jan Gullberg (1997). *Mathematics: from the birth of numbers*. W. W. Norton & Company. p. 435. ISBN 0-393-04002-X.

[15] Elements 1.47 by Euclid. Retrieved 19 December 2006.

[16] Euclid's Elements, Book I, Proposition 47: web page version using Java applets from Euclid's Elements by Prof. David E. Joyce, Clark University

[17] The proof by Pythagoras probably was not a general one, as the theory of proportions was developed only two centuries after Pythagoras; see (Maor 2007, p. 25) page 25

[18] Alexander Bogomolny. "Pythagorean Theorem, proof number 10". *Cut the Knot*. Retrieved 27 February 2010.

[19] (Loomis 1968, Geometric proof 22 and Figure 123, page= 113)

[20] Alexander Bogomolny. "Cut-the-knot.org: Pythagorean theorem and its many proofs, Proof #3". *Cut the Knot*. Retrieved 4 November 2010.

[21] Alexander Bogomolny. "Cut-the-knot.org: Pythagorean theorem and its many proofs, Proof #4". *Cut the Knot*. Retrieved 4 November 2010.

[22] Published in a weekly mathematics column: James A Garfield (1876). *The New England Journal of Education* **3**: 161. Missing or empty |title= (help) as noted in William Dunham (1997). *The mathematical universe: An alphabetical journey through the great proofs, problems, and personalities*. Wiley. p. 96. ISBN 0-471-17661-3. and in A calendar of mathematical dates: April 1, 1876 by V. Frederick Rickey

[23] Prof. David Lantz' animation from his web site of animated proofs

[24] Mike Staring (1996). "The Pythagorean proposition: A proof by means of calculus". *Mathematics Magazine* (Mathematical Association of America) **69** (1): 45–46. doi:10.2307/2691395. JSTOR 2691395.

[25] Bogomolny, Alexander. "Pythagorean Theorem". *Interactive Mathematics Miscellany and Puzzles*. Alexander Bogomolny. Retrieved 2010-05-09. External link in |work= (help)

[26] Bruce C. Berndt (1988). "Ramanujan—100 years old (fashioned) or 100 years new (fangled)?". *The Mathematical Intelligencer* **10** (3): 24. doi:10.1007/BF03026638.

[27] Judith D. Sally, Paul J. Sally Jr. (2007-12-21). "Theorem 2.4 (Converse of the Pythagorean Theorem).". *Roots to Research*. American Mathematical Society. pp. 54–55. ISBN 0-8218-4403-2.

[28] Euclid's Elements, Book I, Proposition 48 From D.E. Joyce's web page at Clark University

[29] Casey, Stephen, "The converse of the theorem of Pythagoras", *Mathematical Gazette* 92, July 2008, 309–313.

[30] Mitchell, Douglas W., "Feedback on 92.47", *Mathematical Gazette* 93, March 2009, 156.

[31] Ernest Julius Wilczynski, Herbert Ellsworth Slaught (1914). "Theorem 1 and Theorem 2". *Plane trigonometry and applications*. Allyn and Bacon. p. 85.

[32] "Dijkstra's generalization" (PDF).

[33] Law, Henry (1853). "Corollary 5 of Proposition XLVII (*Pythagoras's Theorem*)". *The Elements of Euclid: with many additional propositions, and explanatory notes, to which is prefixed an introductory essay on logic*. John Weale. p. 49.

[34] Shaughan Lavine (1994). *Understanding the infinite*. Harvard University Press. p. 13. ISBN 0-674-92096-1.

[35] (Heath 1921, Vol I, pp. 65); Hippasus was on a voyage at the time, and his fellows cast him overboard. See James R. Choike (1980). "The pentagram and the discovery of an irrational number". *The College Mathematics Journal* **11**: 312–316.

[36] A careful discussion of Hippasus's contributions is found in Kurt Von Fritz (Apr 1945). "The Discovery of Incommensurability by Hippasus of Metapontum". *Annals of Mathematics*. Second Series (Annals of Mathematics) **46** (2): 242–264. doi:10.2307/1969021. JSTOR 1969021.

[37] Jon Orwant, Jarkko Hietaniemi, John Macdonald (1999). "Euclidean distance". *Mastering algorithms with Perl*. O'Reilly Media, Inc. p. 426. ISBN 1-56592-398-7.

[38] Wentworth, George (2009). *Plane Trigonometry and Tables*. BiblioBazaar, LLC. p. 116. ISBN 1-103-07998-0., Exercises, page 116

[39] Lawrence S. Leff (2005). *PreCalculus the Easy Way* (7th ed.). Barron's Educational Series. p. 296. ISBN 0-7641-2892-2.

[40] WS Massey (Dec 1983). "Cross products of vectors in higher-dimensional Euclidean spaces". *The American Mathematical Monthly* (Mathematical Association of America) **90** (10): 697–701. doi:10.2307/2323537. JSTOR 2323537.

[41] Pertti Lounesto (2001). "§7.4 Cross product of two vectors". *Clifford algebras and spinors* (2nd ed.). Cambridge University Press. p. 96. ISBN 0-521-00551-5.

[42] Francis Begnaud Hildebrand (1992). *Methods of applied mathematics* (Reprint of Prentice-Hall 1965 2nd ed.). Courier Dover Publications. p. 24. ISBN 0-486-67002-3.

[43] Heath, T. L., *A History of Greek Mathematics*, Oxford University Press, 1921; reprinted by Dover, 1981.

[44] Euclid's *Elements*: Book VI, Proposition VI 31: "In right-angled triangles the figure on the side subtending the right angle is equal to the similar and similarly described figures on the sides containing the right angle."

[45] Putz, John F. and Sipka, Timothy A. "On generalizing the Pythagorean theorem", *The College Mathematics Journal* 34 (4), September 2003, pp. 291–295.

[46] Lawrence S. Leff (2005-05-01). cited work. Barron's Educational Series. p. 326. ISBN 0-7641-2892-2.

[47] Howard Whitley Eves (1983). "§4.8:...generalization of Pythagorean theorem". *Great moments in mathematics (before 1650)*. Mathematical Association of America. p. 41. ISBN 0-88385-310-8.

[48] Tâbit ibn Qorra (full name Thābit ibn Qurra ibn Marwan Al-Ṣābiʾ al-Ḥarrānī) (826–901 AD) was a physician living in Baghdad who wrote extensively on Euclid's Elements and other mathematical subjects.

[49] Aydin Sayili (Mar 1960). "Thâbit ibn Qurra's Generalization of the Pythagorean Theorem". *Isis* **51** (1): 35–37. doi:10.1086/348837. JSTOR 227603.

[50] Judith D. Sally, Paul Sally (2007-12-21). "Exercise 2.10 (ii)". *Roots to Research: A Vertical Development of Mathematical Problems*. p. 62. ISBN 0-8218-4403-2.

[51] For the details of such a construction, see George Jennings (1997). "Figure 1.32: The generalized Pythagorean theorem". *Modern geometry with applications: with 150 figures* (3rd ed.). Springer. p. 23. ISBN 0-387-94222-X.

[52] Claudi Alsina, Roger B. Nelsen: *Charming Proofs: A Journey Into Elegant Mathematics*. MAA, 2010, ISBN 9780883853481, pp. 77–78 (*excerpt*, p. 77, at Google Books)

[53] Rajendra Bhatia (1997). *Matrix analysis*. Springer. p. 21. ISBN 0-387-94846-5.

[54] For an extended discussion of this generalization, see, for example, Willie W. Wong 2002, *A generalized n-dimensional Pythagorean theorem*.

[55] Ferdinand van der Heijden, Dick de Ridder (2004). *Classification, parameter estimation, and state estimation*. Wiley. p. 357. ISBN 0-470-09013-8.

[56] Qun Lin, Jiafu Lin (2006). *Finite element methods: accuracy and improvement*. Elsevier. p. 23. ISBN 7-03-016656-6.

[57] Howard Anton, Chris Rorres (2010). *Elementary Linear Algebra: Applications Version* (10th ed.). Wiley. p. 336. ISBN 0-470-43205-5.

[58] Karen Saxe (2002). "Theorem 1.2". *Beginning functional analysis*. Springer. p. 7. ISBN 0-387-95224-1.

[59] Douglas, Ronald G. (1998). *Banach Algebra Techniques in Operator Theory, 2nd edition*. New York, New York: Springer-Verlag New York, Inc. pp. 60–1. ISBN 978-0-387-98377-6.

[60] Stephen W. Hawking (2005). cited work. p. 4. ISBN 0-7624-1922-9.

[61] Eric W. Weisstein (2003). *CRC concise encyclopedia of mathematics* (2nd ed.). p. 2147. ISBN 1-58488-347-2. The parallel postulate is equivalent to the *Equidistance postulate*, *Playfair axiom*, *Proclus axiom*, the *Triangle postulate* and the *Pythagorean theorem*.

[62] Alexander R. Pruss (2006). *The principle of sufficient reason: a reassessment*. Cambridge University Press. p. 11. ISBN 0-521-85959-X. We could include...the parallel postulate and derive the Pythagorean theorem. Or we could instead make the Pythagorean theorem among the other axioms and derive the parallel postulate.

[63] Victor Pambuccian (December 2010). "Maria Teresa Calapso's Hyperbolic Pythagorean Theorem". *The Mathematical Intelligencer* **32** (4): 2. doi:10.1007/s00283-010-9169-0.

[64] Barrett O'Neill (2006). "Exercise 4". *Elementary differential geometry* (2nd ed.). Academic Press. p. 441. ISBN 0-12-088735-5.

[65] Saul Stahl (1993). "Theorem 8.3". *The Poincaré half-plane: a gateway to modern geometry*. Jones & Bartlett Learning. p. 122. ISBN 0-86720-298-X.

[66] Jane Gilman (1995). "Hyperbolic triangles". *Two-generator discrete subgroups of PSL(2,R)*. American Mathematical Society Bookstore. ISBN 0-8218-0361-1.

[67] Tai L. Chow (2000). *Mathematical methods for physicists: a concise introduction*. Cambridge University Press. p. 52. ISBN 0-521-65544-7.

[68] Friberg, Joran. "Methods and Traditions of Babylonian Mathematics", Historia Mathematica, Vol. 8, pp. 277, 306 (1991).

[69] Lederman, Leon and Teresi, Dick. The God Particle: If the Universe Is the Answer, What Is the Question?, p. 80 (Houghton Mifflin Harcourt 2006).

[70] Teresi, Dick. Lost Discoveries: The Ancient Roots of Modern Science—from the Babylonians to the Maya, p. 8 (Simon and Schuster 2010).

[71] (van_der_Waerden 1983, p. 5) See also Frank Swetz, T. I. Kao (1977). *Was Pythagoras Chinese?: An examination of right triangle theory in ancient China*. Penn State Press. p. 12. ISBN 0-271-01238-2.

[72] Carl Benjamin Boyer (1968). "China and India". *A history of mathematics*. Wiley. p. 229. we find rules for the construction of right angles by means of triples of cords the lengths of which form Pythagorean triages, such as 3, 4, and 5, or 5, 12, and 13, or 8, 15, and 17, or 12, 35, and 37. However all of these triads are easily derived from the old Babylonian rule; hence, Mesopotamian influence in the *Sulvasūtras* is not unlikely. Aspastamba knew that the square on the diagonal of a rectangle is equal to the sum of the squares on the two adjacent sides, but this form of the Pythagorean theorem also may have been derived from Mesopotamia. [...] So conjectural are the origin and period of the *Sulvasūtras* that we cannot tell whether or not the rules are related to early Egyptian surveying or to the later Greek problem of altar doubling. They are variously dated within an interval of almost a thousand years stretching from the eighth century B.C. to the second century of our era.; See also Carl B. Boyer , Uta C. Merzbach (2010). *A History of Mathematics* (3rd ed.). Wiley. ISBN 0-470-52548-7.

[73] Robert P. Crease (2008). *The great equations: breakthroughs in science from Pythagoras to Heisenberg*. W W Norton & Co. p. 25. ISBN 0-393-06204-X.

[74] A rather extensive discussion of the origins of the various texts in the Zhou Bi is provided by Christopher Cullen (2007). *Astronomy and Mathematics in Ancient China: The 'Zhou Bi Suan Jing'*. Cambridge University Press. pp. 139 *f.* ISBN 0-521-03537-6.

[75] This work is a compilation of 246 problems, some of which survived the book burning of 213 BC, and was put in final form before 100 AD. It was extensively commented upon by Liu Hui in 263 AD. Philip D Straffin, Jr. (2004). "Liu Hui and the first golden age of Chinese mathematics". In Marlow Anderson, Victor J. Katz, Robin J. Wilson. *Sherlock Holmes in Babylon: and other tales of mathematical history*. Mathematical Association of America. pp. 69 *ff.* ISBN 0-88385-546-1. See particularly §3: *Nine chapters on the mathematical art*, pps. 71 *ff.*

[76] Kangshen Shen, John N. Crossley, Anthony Wah-Cheung Lun (1999). *The nine chapters on the mathematical art: companion and commentary*. Oxford University Press. p. 488. ISBN 0-19-853936-3.

[77] In particular, Li Jimin; see *Centaurus, Volume 39*. Copenhagen: Munksgaard. 1997. pp. 193 & 205.

[78] Chen, Cheng-Yih (1996). "§3.3.4 Chén Zǐ's formula and the Chóng-Chã method; Figure 40". *Early Chinese work in natural science: a re-examination of the physics of motion, acoustics, astronomy and scientific thoughts*. Hong Kong University Press. p. 142. ISBN 962-209-385-X.

[79] Wen-tsün Wu (2008). "The Gougu theorem". *Selected works of Wen-tsün Wu*. World Scientific. p. 158. ISBN 981-279-107-8.

[80] (Euclid 1956, p. 351) page 351

[81] An extensive discussion of the historical evidence is provided in (Euclid 1956, p. 351) page=351

[82] Asger Aaboe (1997). *Episodes from the early history of mathematics*. Mathematical Association of America. p. 51. ISBN 0-88385-613-1. ...it is not until Euclid that we find a logical sequence of general theorems with proper proofs.

[83] H.C. Andersen: *Formens evige Magie* (1831), visithcandersen.dk

[84] Maor (2007), p. 47.

[85] "The Scarecrow's Formula". *Internet Movie Data Base*. Retrieved 2010-05-12.

[86] Singh, Simon (2013). *The Simpsons and Their Mathematical Secrets*. New York: Bloomsbury. ISBN 9781620402771.

[87] "Uganda 2000 Shillings". Numismatic Guaranty Corporation. Retrieved 2014-06-13.

[88] "Le Saviez-vous?".

[89] Miller, Jeff (2007-08-03). "Images of Mathematicians on Postage Stamps". Retrieved 2010-07-18.

9.11   References

- Bell, John L. (1999). *The Art of the Intelligible: An Elementary Survey of Mathematics in its Conceptual Development*. Kluwer. ISBN 0-7923-5972-0.

- Euclid (1956). Translated by Johan Ludvig Heiberg with an introduction and commentary by Sir Thomas L. Heath, ed. *The Elements (3 vols.)*. Vol. 1 (Books I and II) (Reprint of 1908 ed.). Dover. ISBN 0-486-60088-2. On-line text at Euclid

- Heath, Sir Thomas (1921). "The 'Theorem of Pythagoras'". *A History of Greek Mathematics (2 Vols.)* (Dover Publications, Inc. (1981) ed.). Clarendon Press, Oxford. p. 144 *ff*. ISBN 0-486-24073-8.

- Libeskind, Shlomo (2008). *Euclidean and transformational geometry: a deductive inquiry*. Jones & Bartlett Learning. ISBN 0-7637-4366-6. This high-school geometry text covers many of the topics in this WP article.

- Loomis, Elisha Scott (1968). *The Pythagorean proposition* (2nd ed.). The National Council of Teachers of Mathematics. ISBN 978-0-87353-036-1. For full text of 2nd edition of 1940, see Elisha Scott Loomis. "The Pythagorean proposition: its demonstrations analyzed and classified, and bibliography of sources for data of the four kinds of proofs" (PDF). *Education Resources Information Center*. Institute of Education Sciences (IES) of the U.S. Department of Education. Retrieved 2010-05-04. Originally published in 1940 and reprinted in 1968 by National Council of Teachers of Mathematics, isbn=0-87353-036-5.

- Maor, Eli (2007). *The Pythagorean Theorem: A 4,000-Year History*. Princeton, New Jersey: Princeton University Press. ISBN 978-0-691-12526-8.

- Stillwell, John (1989). *Mathematics and Its History*. Springer-Verlag. ISBN 0-387-96981-0. Also ISBN 3-540-96981-0.

- Swetz, Frank; Kao, T. I. (1977). *Was Pythagoras Chinese?: An Examination of Right Triangle Theory in Ancient China*. Pennsylvania State University Press. ISBN 0-271-01238-2.

- van der Waerden, Bartel Leendert (1983). *Geometry and Algebra in Ancient Civilizations*. Springer. ISBN 3-540-12159-5.

9.12   External links

- Euclid, David E. Joyce, ed. (1997) [c. 300 BC]. *Elements*. Retrieved 2006-08-30. In HTML with Java-based interactive figures.

- Hazewinkel, Michiel, ed. (2001), "Pythagorean theorem", *Encyclopedia of Mathematics*, Springer, ISBN 978-1-55608-010-4

- History topic: Pythagoras's theorem in Babylonian mathematics

- Interactive links:
  - Interactive proof in Java of The Pythagorean Theorem
  - Another interactive proof in Java of The Pythagorean Theorem
  - Pythagorean theorem with interactive animation
  - Animated, Non-Algebraic, and User-Paced Pythagorean Theorem

- Pythagorean theorem water demo @YouTube

- Pythagorean Theorem (more than 70 proofs from cut-the-knot)

- Weisstein, Eric W., "Pythagorean theorem", *MathWorld*.

Chapter 10

Parallelogram law

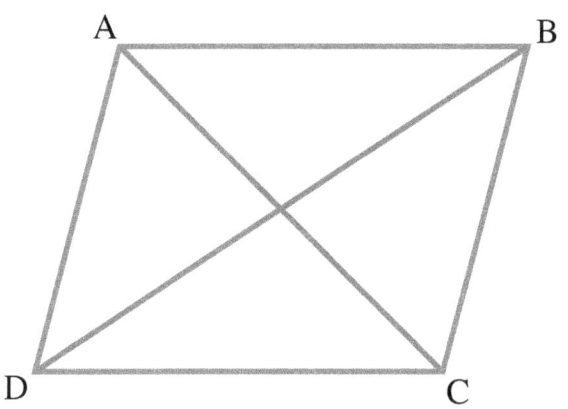

A parallelogram. The sides are shown in blue and the diagonals in red.

In mathematics, the simplest form of the **parallelogram law** (also called the **parallelogram identity**) belongs to elementary geometry. It states that the sum of the squares of the lengths of the four sides of a parallelogram equals the sum of the squares of the lengths of the two diagonals. Using the notation in the diagram on the right, the sides are (AB), (BC), (CD), (DA). But since in Euclidean geometry a parallelogram necessarily has opposite sides equal, or $(AB) = (CD)$ and $(BC) = (DA)$, the law can be stated as,

$$2(AB)^2 + 2(BC)^2 = (AC)^2 + (BD)^2$$

In case the parallelogram is a rectangle, the two diagonals are of equal lengths $(AC) = (BD)$ so,

$$2(AB)^2 + 2(BC)^2 = 2(AC)^2$$

and the statement reduces to the Pythagorean theorem. For the general quadrilateral with four sides not necessarily equal,

$$(AB)^2 + (BC)^2 + (CD)^2 + (DA)^2 = (AC)^2 + (BD)^2 + 4x^2.$$

where x is the length of the line joining the midpoints of the diagonals. It can be seen from the diagram that, for a parallelogram, $x = 0$, and the general formula is equivalent to the parallelogram law.

10.1 The parallelogram law in inner product spaces

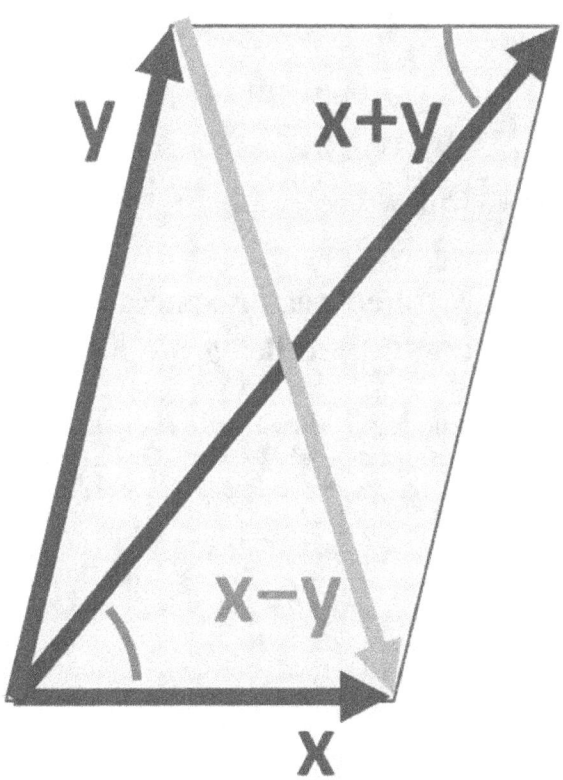

Vectors involved in the parallelogram law.

In a normed space, the statement of the parallelogram law is an equation relating norms:

83

$$2\|x\|^2 + 2\|y\|^2 = \|x+y\|^2 + \|x-y\|^2.$$

In an inner product space, the norm is determined using the inner product:

$$\|x\|^2 = \langle x, x \rangle.$$

As a consequence of this definition, in an inner product space the parallelogram law is an algebraic identity, readily established using the properties of the inner product:

$$\|x+y\|^2 = \langle x+y, x+y \rangle = \langle x,x \rangle + \langle x,y \rangle + \langle y,x \rangle + \langle y,y \rangle,$$

$$\|x-y\|^2 = \langle x-y, x-y \rangle = \langle x,x \rangle - \langle x,y \rangle - \langle y,x \rangle + \langle y,y \rangle.$$

Adding these two expressions:

$$\|x+y\|^2 + \|x-y\|^2 = 2\langle x,x \rangle + 2\langle y,y \rangle = 2\|x\|^2 + 2\|y\|^2,$$

as required.

If x is orthogonal to y, then $\langle x, y \rangle = 0$ and the above equation for the norm of a sum becomes:

$$\|x+y\|^2 = \langle x,x \rangle + \langle x,y \rangle + \langle y,x \rangle + \langle y,y \rangle = \|x\|^2 + \|y\|^2,$$

which is Pythagoras' theorem.

10.2 Normed vector spaces satisfying the parallelogram law

Most real and complex normed vector spaces do not have inner products, but all normed vector spaces have norms (by definition). For example, a commonly used norm is the p-norm:

$$\|x\|_p = \left(\sum_{i=1}^{n} |x_i|^p \right)^{1/p},$$

where the x_i are the components of vector x .

Given a norm, one can evaluate both sides of the parallelogram law above. A remarkable fact is that if the parallelogram law holds, then the norm must arise in the usual way from some inner product. In particular, it holds for the p-norm if and only if $p = 2$, the so-called *Euclidean* norm or *standard* norm.[1][2]

For any norm satisfying the parallelogram law (which necessarily is an inner product norm), the inner product generating the norm is unique as a consequence of the polarization identity. In the real case, the polarization identity is given by:

$$\langle x, y \rangle = \frac{\|x+y\|^2 - \|x-y\|^2}{4},$$

or, equivalently, by:

$$\frac{\|x+y\|^2 - \|x\|^2 - \|y\|^2}{2} \text{ or } \frac{\|x\|^2 + \|y\|^2 - \|x-y\|^2}{2}.$$

In the complex case it is given by:

$$\langle x, y \rangle = \frac{\|x+y\|^2 - \|x-y\|^2}{4} + i\frac{\|ix-y\|^2 - \|ix+y\|^2}{4}.$$

For example, using the p-norm with $p = 2$ and real vectors x, y , the evaluation of the inner product proceeds as follows:

$$\begin{aligned}
\langle x, y \rangle &= \frac{\|x+y\|^2 - \|x-y\|^2}{4} \\
&= \frac{1}{4}\left[\sum |x_i + y_i|^2 - \sum |x_i - y_i|^2 \right] \\
&= \frac{1}{4}\left[4\sum x_i y_i \right] \\
&= (x \cdot y),
\end{aligned}$$

which is the standard dot product of two vectors.

10.3 Notes and in-line references

[1] Cyrus D. Cantrell (2000). *Modern mathematical methods for physicists and engineers*. Cambridge University Press. p. 535. ISBN 0-521-59827-3. if $p \neq 2$, there is no inner product such that $\sqrt{\langle x, x \rangle} = \|x\|_p$ because the p-norm violates the parallelogram law.

[2] Karen Saxe (2002). *Beginning functional analysis*. Springer. p. 10. ISBN 0-387-95224-1.

10.4 See also

- Commutative property

- Inner product space

- Normed vector space

- Polarization identity

10.5 External links

- Weisstein, Eric W., "Parallelogram Law", *MathWorld*.

- The Parallelogram Law Proven Simply at Dreamshire blog

- The Parallelogram Law: A Proof Without Words at cut-the-knot

- A generalization of the "Parallelogram Law/Identity" to a Parallelo-hexagon and to 2n-gons in General - Relations between the sides and diagonals of 2n-gons (Douglas' Theorem) at Dynamic Geometry Sketches, an interactive dynamic geometry sketch.

Chapter 11

Altitude (triangle)

"Orthocenter" and "Orthocentre" redirect here. For the orthocentric system, see Orthocentric system.

In geometry, an **altitude** of a triangle is a line segment

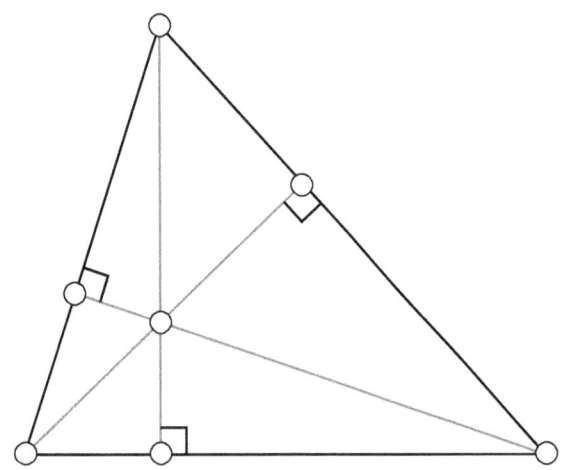

Three altitudes intersecting at the orthocenter

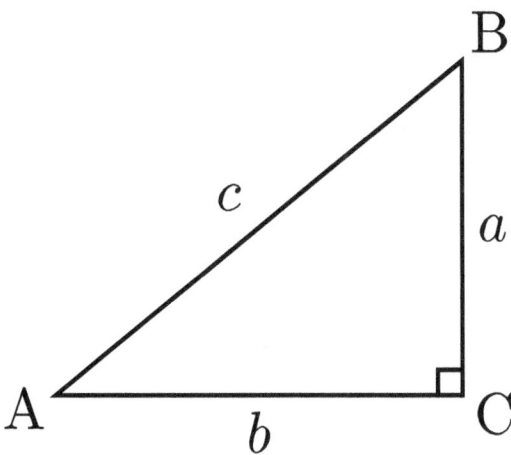

A right triangle, in which the altitude from each acute angle coincides with a leg and intersects the opposite side at (has its foot at) the right-angled vertex.

through a vertex and perpendicular to (i.e. forming a right angle with) a line containing the base (the opposite side of the triangle). This line containing the opposite side is called the *extended base* of the altitude. The intersection between the extended base and the altitude is called the *foot* of the altitude. The length of the altitude, often simply called the altitude, is the distance between the extended base and the vertex. The process of drawing the altitude from the vertex to the foot is known as *dropping the altitude* of that vertex. It is a special case of orthogonal projection.

Altitudes can be used to compute the area of a triangle: one half of the product of an altitude's length and its base's length equals the triangle's area. Thus the longest altitude is perpendicular to the shortest side of the triangle. The altitudes are also related to the sides of the triangle through the trigonometric functions.

In an isosceles triangle (a triangle with two congruent sides), the altitude having the incongruent side as its base will have the midpoint of that side as its foot. Also the altitude having

the incongruent side as its base will form the angle bisector of the vertex.

It is common to mark the altitude with the letter h (as in *height*), often subscripted with the name of the side the altitude comes from.

In a right triangle, the altitude with the hypotenuse c as base divides the hypotenuse into two lengths p and q. If we denote the length of the altitude by hc, we then have the relation

$$h_c = \sqrt{pq} \text{ (Geometric mean theorem)}$$

For acute and right triangles the feet of the altitudes all fall on the triangle's interior or edge. In an obtuse triangle (one with an obtuse angle), the foot of the altitude to the obtuse-angled vertex falls on the opposite side, but the feet of the altitudes to the acute-angled vertices fall on the opposite extended side, exterior to the triangle. This is illustrated in the diagram to the right: in this obtuse triangle, an altitude

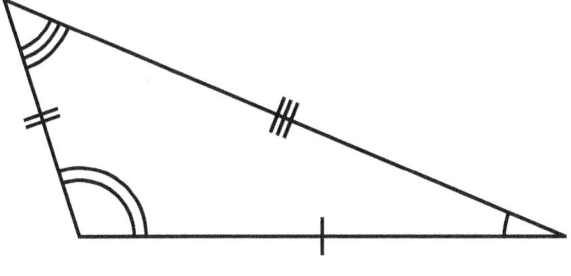

dropped perpendicularly from the top vertex, which has an acute angle, intersects the extended horizontal side outside the triangle.

11.1   Orthocenter

See also: Orthocentric system

The three altitudes intersect in a single point, called the

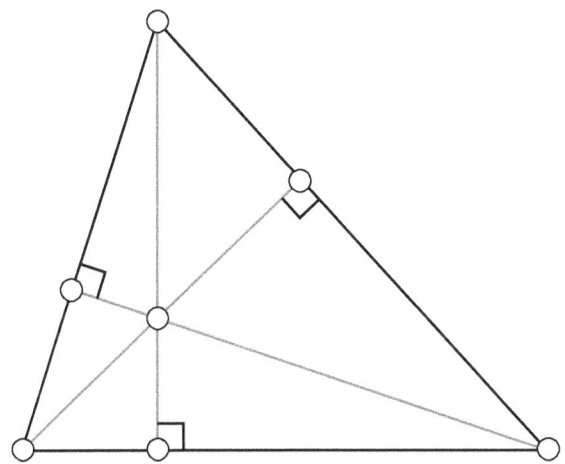

Three altitudes intersecting at the orthocenter

orthocenter of the triangle. The orthocenter lies inside the triangle if and only if the triangle is acute (i.e. does not have an angle greater than or equal to a right angle). If one angle is a right angle, the orthocenter coincides with the vertex of the right angle.

The product of the distances from the orthocenter to a vertex and to the foot of the corresponding altitude is the same for all three altitudes.[1]:p. 176 This product is the squared radius of the triangle's polar circle.

The orthocenter H, the centroid G, the circumcenter O, and the center N of the nine-point circle all lie on a single line, known as the Euler line. The center of the nine-point circle lies at the midpoint between the orthocenter and the circumcenter, and the distance between the centroid and the

circumcenter is half that between the centroid and the orthocenter:

$$OH = 2NH,$$

$$2OG = GH.$$

The orthocenter is closer to the incenter I than it is to the centroid, and the orthocenter is farther than the incenter is from the centroid:

$$HI < HG,$$

$$HG > IG.$$

In terms of the sides a, b, c, inradius r and circumradius R,[2]

$$OH^2 = R^2 - 8R^2 \cos A \cos B \cos C = 9R^2 - (a^2 + b^2 + c^2),\text{ [3]:p. 449}$$

$$HI^2 = 2r^2 - 4R^2 \cos A \cos B \cos C.$$

The isogonal conjugate and also the complement of the orthocenter is the circumcenter.

Four points in the plane such that one of them is the orthocenter of the triangle formed by the other three are called an orthocentric system or orthocentric quadrangle.

Let A, B, C denote the angles of the reference triangle, and let $a = |BC|$, $b = |CA|$, $c = |AB|$ be the sidelengths. The orthocenter has trilinear coordinates[4]

$$\sec A : \sec B : \sec C = \cos A - \sin B \sin C : \cos B - \sin C \sin A : \cos C - \sin A \sin B,$$

and barycentric coordinates

$$(a^2+b^2-c^2)(a^2-b^2+c^2) : (a^2+b^2-c^2)(-a^2+b^2+c^2) : (a^2-b^2+c^2)(-a^2+b^2+c^2)$$

$$= \tan A : \tan B : \tan C.$$

Since barycentric coordinates are all positive for a point in a triangle's interior but at least one is negative for a point in the exterior, and two of the barycentric coordinates are zero for a vertex point, the barycentric coordinates given for the orthocenter show that the orthocenter is in an acute triangle's interior, on the right-angled vertex of a right triangle, and exterior to an obtuse triangle.

Denote the vertices of a triangle as A, B, and C and the orthocenter as H, and let D, E, and F denote the feet of the altitudes from A, B, and C respectively. Then:

- The sum of the ratios on the three altitudes of the distance of the orthocenter from the base to the length of the altitude is 1:[5] (This property and the next one are applications of a more general property of any interior point and the three cevians through it.)

$$\frac{HD}{AD} + \frac{HE}{BE} + \frac{HF}{CF} = 1.$$

- The sum of the ratios on the three altitudes of the distance of the orthocenter from the vertex to the length of the altitude is 2:[5]

$$\frac{AH}{AD} + \frac{BH}{BE} + \frac{CH}{CF} = 2.$$

- The product of the lengths of the segments that the orthocenter divides an altitude into is the same for all three altitudes:[6]

$$AH \cdot HD = BH \cdot HE = CH \cdot HF.$$

- If any altitude, say AD, is extended to intersect the circumcircle at P, so that AP is a chord of the circumcircle, then the foot D bisects segment HP:[6]

$$HD = DP.$$

Denote the orthocenter of triangle ABC as H, denote the sidelengths as a, b, and c, and denote the circumradius of the triangle as R. Then[7][8]

$$a^2 + b^2 + c^2 + AH^2 + BH^2 + CH^2 = 12R^2.$$

In addition, denoting r as the radius of the triangle's incircle, ra, rb, and rc as the radii if its excircles, and R again as the radius of its circumcircle, the following relations hold regarding the distances of the orthocenter from the vertices:[9]

$$r_a + r_b + r_c + r = AH + BH + CH + 2R,$$

$$r_a^2 + r_b^2 + r_c^2 + r^2 = AH^2 + BH^2 + CH^2 + (2R)^2.$$

The directrices of all parabolas that are externally tangent to one side of a triangle and tangent to the extensions of the other sides pass through the orthocenter.[10]

A circumconic passing through the orthocenter of a triangle is a rectangular hyperbola.[11]

In any acute triangle, the inscribed triangle with the smallest perimeter is the pedal triangle of the orthocenter (the triangle whose vertices are the feet of the perpendiculars from the orthocenter to the sides).[1]:p.168 The sides of the pedal triangle of the orthocenter are parallel to the tangents to the circumcircle at the original triangle's vertices.[1]:p.172

11.2 Orthic triangle

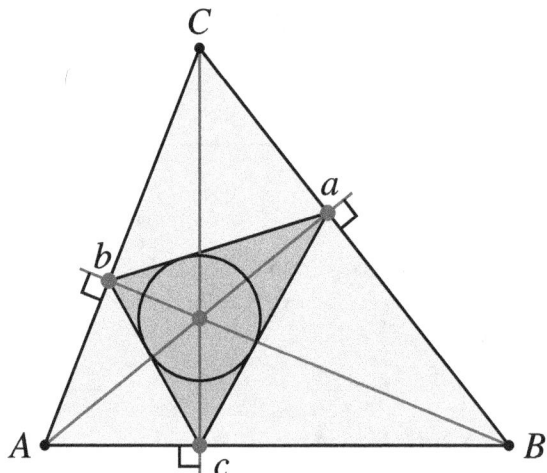

Triangle abc *is the orthic triangle of triangle* ABC

If the triangle ABC is oblique (not right-angled), the points of intersection of the altitudes with the sides of the triangle form another triangle, A'B'C', called the **orthic triangle** or **altitude triangle**. It is the pedal triangle of the orthocenter of the original triangle. Also, the incenter (that is, the center for the inscribed circle) of the orthic triangle is the orthocenter of the original triangle.[12]

The sides of the orthic triangle meet the sides of its reference triangle at three collinear points.[13]:p. 165[14]

The orthic triangle is closely related to the tangential triangle, constructed as follows: let LA be the line tangent to the circumcircle of triangle ABC at vertex A, and define LB and LC analogously. Let $A'' = LB \cap LC$, $B'' = LC \cap LA$, $C'' = LC \cap LA$. The tangential triangle is $A''B''C''$, whose sides are the tangents to the reference triangle's circumcircle at its vertices; it is homothetic to the orthic triangle. The circumcenter of the tangential triangle, and the center of similitude of the orthic and tangential triangles, are on the Euler line.[3]:p. 447

The orthic triangle provides the solution to Fagnano's problem, posed in 1775, of finding for the minimum perimeter triangle inscribed in a given acute-angle triangle.

The orthic triangle of an acute triangle gives a triangular light route.[15]

Trilinear coordinates for the vertices of the orthic triangle are given by

- A' = 0 : sec B : sec C
- B' = sec A : 0 : sec C
- C' = sec A : sec B : 0

Trilinear coordinates for the vertices of the tangential triangle are given by

- $A" = -a : b : c$

- $B" = a : -b : c$

- $C" = a : b : -c$

For more information on the orthic triangle, see here.

11.3 Some additional altitude theorems

11.3.1 Altitude in terms of the sides

For any triangle with sides a, b, c and semiperimeter $s = (a+b+c) / 2$, the altitude from side a is given by

$$h_a = \frac{2\sqrt{s(s-a)(s-b)(s-c)}}{a}.$$

This follows from combining Heron's formula for the area of a triangle in terms of the sides with the area formula (1/2)×base×height, where the base is taken as side a and the height is the altitude from a.

11.3.2 Inradius theorems

Consider an arbitrary triangle with sides a, b, c and with corresponding altitudes ha, hb, and hc. The altitudes and the incircle radius r are related by

$$\frac{1}{r} = \frac{1}{h_a} + \frac{1}{h_b} + \frac{1}{h_c}.$$

11.3.3 Circumradius theorem

Denoting the altitude from one side of a triangle as ha, the other two sides as b and c, and the triangle's circumradius (radius of the triangle's circumscribed circle) as R, the altitude is given by[1]:p. 71

$$h_a = \frac{bc}{2R}.$$

11.3.4 Interior point

If p_1, p_2, and p_3 are the perpendicular distances from any point P to the sides, and h_1, h_2, and h_3 are the altitudes to the respective sides, then[1]:p. 74

$$\frac{p_1}{h_1} + \frac{p_2}{h_2} + \frac{p_3}{h_3} = 1.$$

11.3.5 Area theorem

Denoting the altitudes of any triangle from sides a, b, and c respectively as h_a , h_b , and h_c ,and denoting the semi-sum of the reciprocals of the altitudes as $H = (h_a^{-1} + h_b^{-1} + h_c^{-1})/2$ we have[16]

$$\text{Area}^{-1} = 4\sqrt{H(H - h_a^{-1})(H - h_b^{-1})(H - h_c^{-1})}.$$

11.3.6 General point on an altitude

If E is any point on an altitude AD of any triangle ABC, then[17]:77–78

$$AC^2 + EB^2 = AB^2 + CE^2.$$

11.3.7 Feet of the altitudes

The lines connecting the feet of the altitudes intersect the opposite sides at collinear points.[1]:p.199

11.3.8 Special case triangles

Equilateral triangle

For any point P within an equilateral triangle, the sum of the perpendiculars to the three sides is equal to the altitude of the triangle. This is Viviani's theorem.

Right triangle

In a right triangle the three altitudes ha, hb, and hc (the first two of which equal the leg lengths b and a respectively) are related according to[18][19]

$$\frac{1}{h_a^2} + \frac{1}{h_b^2} = \frac{1}{h_c^2}.$$

11.4   See also

- Triangle center

- Median (geometry)

11.5   Notes

[1] Johnson, Roger A. *Advanced Euclidean Geometry*, Dover Publications, 2007.

[2] Marie-Nicole Gras, "Distances between the circumcenter of the extouch triangle and the classical centers", *Forum Geometricorum* 14 (2014), 51-61. http://forumgeom.fau.edu/FG2014volume14/FG201405index.html

[3] Smith, Geoff, and Leversha, Gerry, "Euler and triangle geometry", *Mathematical Gazette* 91, November 2007, 436–452.

[4] Clark Kimberling's Encyclopedia of Triangle Centers http://faculty.evansville.edu/ck6/encyclopedia/ETC.html

[5] Panapoi,Ronnachai, "Some properties of the orthocenter of a triangle", University of Georgia.

[6] "Orthocenter of a triangle"

[7] Weisstein, Eric W. "Orthocenter." From MathWorld--A Wolfram Web Resource.

[8] Altshiller-Court, Nathan, *College Geometry*, Dover Publications, 2007 (orig. Barnes & Noble 1952), p. 102.

[9] Bell, Amy, "Hansen's right triangle theorem, its converse and a generalization", *Forum Geometricorum* 6, 2006, 335–342.

[10] Weisstein, Eric W. "Kiepert Parabola." From MathWorld--A Wolfram Web Resource. http://mathworld.wolfram.com/KiepertParabola.html

[11] Weisstein, Eric W. "Jerabek Hyperbola." From MathWorld--A Wolfram Web Resource. http://mathworld.wolfram.com/JerabekHyperbola.htm

[12] William H. Barker, Roger Howe (2007). "§ VI.2: The classical coincidences". *Continuous symmetry: from Euclid to Klein*. American Mathematical Society Bookstore. p. 292. ISBN 0-8218-3900-4. See also: Corollary 5.5, p. 318.

[13] Altshiller-Court, Nathan, *College Geometry*, Dover Publications, 2007 (orig. 1952)

[14] William H. Barker, Roger Howe (2007). "§ VI.2: The classical coincidences". *Continuous symmetry: from Euclid to Klein*. American Mathematical Society Bookstore. p. 292. ISBN 0-8218-3900-4. See also: Corollary 5.5, p. 318.

[15] Bryant, V., and Bradley, H., "Triangular Light Routes," *Mathematical Gazette* 82, July 1998, 298-299.

[16] Mitchell, Douglas W., "A Heron-type formula for the reciprocal area of a triangle," *Mathematical Gazette* 89, November 2005, 494.

[17] Alfred S. Posamentier and Charles T. Salkind, *Challenging Problems in Geometry*, Dover Publishing Co., second revised edition, 1996.

[18] Voles, Roger, "Integer solutions of $a^{-2} + b^{-2} = d^{-2}$," *Mathematical Gazette* 83, July 1999, 269–271.

[19] Richinick, Jennifer, "The upside-down Pythagorean Theorem," *Mathematical Gazette* 92, July 2008, 313–317.

11.6   References

- Weisstein, Eric W., "Altitude", *MathWorld*.

11.7   External links

- Orthocenter of a triangle With interactive animation

- Animated demonstration of orthocenter construction Compass and straightedge.

- An interactive Java applet for the orthocenter

- Fagnano's Problem by Jay Warendorff, Wolfram Demonstrations Project.

Chapter 12

Vector space

This article is about linear (vector) spaces. For the structure in incidence geometry, see Linear space (geometry).

A **vector space** (also called a **linear space**) is a collection

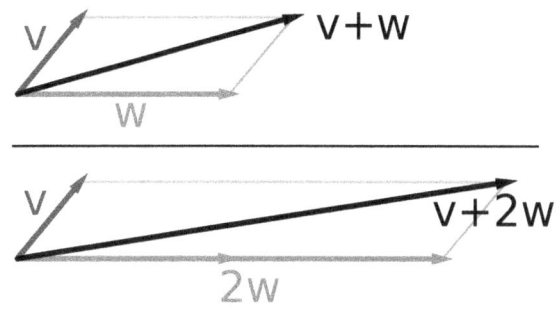

*Vector addition and scalar multiplication: a vector **v** (blue) is added to another vector **w** (red, upper illustration). Below, **w** is stretched by a factor of 2, yielding the sum **v** + 2**w**.*

of objects called **vectors**, which may be added together and multiplied ("scaled") by numbers, called *scalars* in this context. Scalars are often taken to be real numbers, but there are also vector spaces with scalar multiplication by complex numbers, rational numbers, or generally any field. The operations of vector addition and scalar multiplication must satisfy certain requirements, called *axioms*, listed below.

Euclidean vectors are an example of a vector space. They represent physical quantities such as forces: any two forces (of the same type) can be added to yield a third, and the multiplication of a force vector by a real multiplier is another force vector. In the same vein, but in a more geometric sense, vectors representing displacements in the plane or in three-dimensional space also form vector spaces. Vectors in vector spaces do not necessarily have to be arrow-like objects as they appear in the mentioned examples: vectors are regarded as abstract mathematical objects with particular properties, which in some cases can be visualized as arrows.

Vector spaces are the subject of linear algebra and are well understood from this point of view since vector spaces are characterized by their dimension, which, roughly speaking, specifies the number of independent directions in the space. A vector space may be endowed with additional structure, such as a norm or inner product. Such spaces arise naturally in mathematical analysis, mainly in the guise of infinite-dimensional function spaces whose vectors are functions. Analytical problems call for the ability to decide whether a sequence of vectors converges to a given vector. This is accomplished by considering vector spaces with additional structure, mostly spaces endowed with a suitable topology, thus allowing the consideration of proximity and continuity issues. These topological vector spaces, in particular Banach spaces and Hilbert spaces, have a richer theory.

Historically, the first ideas leading to vector spaces can be traced back as far as the 17th century's analytic geometry, matrices, systems of linear equations, and Euclidean vectors. The modern, more abstract treatment, first formulated by Giuseppe Peano in 1888, encompasses more general objects than Euclidean space, but much of the theory can be seen as an extension of classical geometric ideas like lines, planes and their higher-dimensional analogs.

Today, vector spaces are applied throughout mathematics, science and engineering. They are the appropriate linear-algebraic notion to deal with systems of linear equations; offer a framework for Fourier expansion, which is employed in image compression routines; or provide an environment that can be used for solution techniques for partial differential equations. Furthermore, vector spaces furnish an abstract, coordinate-free way of dealing with geometrical and physical objects such as tensors. This in turn allows the examination of local properties of manifolds by linearization techniques. Vector spaces may be generalized in several ways, leading to more advanced notions in geometry and abstract algebra.

12.1 Introduction and definition

The concept of vector space will first be explained by describing two particular examples:

12.1.1 First example: arrows in the plane

The first example of a vector space consists of arrows in a fixed plane, starting at one fixed point. This is used in physics to describe forces or velocities. Given any two such arrows, **v** and **w**, the parallelogram spanned by these two arrows contains one diagonal arrow that starts at the origin, too. This new arrow is called the *sum* of the two arrows and is denoted **v** + **w**. In the special case of two arrows on the same line, their sum is the arrow on this line whose length is the sum or the difference of the lengths, depending on whether the arrows have the same direction. Another operation that can be done with arrows is scaling: given any positive real number a, the arrow that has the same direction as **v**, but is dilated or shrunk by multiplying its length by a, is called *multiplication* of **v** by a. It is denoted a**v**. When a is negative, a**v** is defined as the arrow pointing in the opposite direction, instead.

The following shows a few examples: if $a = 2$, the resulting vector a**w** has the same direction as **w**, but is stretched to the double length of **w** (right image below). Equivalently 2**w** is the sum **w** + **w**. Moreover, (-1)**v** = $-$**v** has the opposite direction and the same length as **v** (blue vector pointing down in the right image).

12.1.2 Second example: ordered pairs of numbers

A second key example of a vector space is provided by pairs of real numbers x and y. (The order of the components x and y is significant, so such a pair is also called an ordered pair.) Such a pair is written as (x, y). The sum of two such pairs and multiplication of a pair with a number is defined as follows:

$$(x_1, y_1) + (x_2, y_2) = (x_1 + x_2, y_1 + y_2)$$

and

$$a\,(x, y) = (ax, ay).$$

The first example above reduces to this one if the arrows are represented by the pair of Cartesian coordinates of their end points.

12.1.3 Definition

A vector space over a field F is a set V together with two operations that satisfy the eight axioms listed below. Elements of V are commonly called *vectors*. Elements of F are commonly called *scalars*. The first operation, called *vector addition* or simply *addition*, takes any two vectors **v** and **w** and assigns to them a third vector which is commonly written as **v** + **w**, and called the sum of these two vectors. The second operation, called *scalar multiplication* takes any scalar a and any vector **v** and gives another vector a**v**.

In this article, vectors are distinguished from scalars by boldface.[nb 1] In the two examples above, the field is the field of the real numbers and the set of the vectors consists of the planar arrows with fixed starting point and of pairs of real numbers, respectively.

To qualify as a vector space, the set V and the operations of addition and multiplication must adhere to a number of requirements called axioms.[1] In the list below, let **u**, **v** and **w** be arbitrary vectors in V, and a and b scalars in F.

These axioms generalize properties of the vectors introduced in the above examples. Indeed, the result of addition of two ordered pairs (as in the second example above) does not depend on the order of the summands:

$$(x_v, y_v) + (x_w, y_w) = (x_w, y_w) + (x_v, y_v).$$

Likewise, in the geometric example of vectors as arrows, **v** + **w** = **w** + **v** since the parallelogram defining the sum of the vectors is independent of the order of the vectors. All other axioms can be checked in a similar manner in both examples. Thus, by disregarding the concrete nature of the particular type of vectors, the definition incorporates these two and many more examples in one notion of vector space.

Subtraction of two vectors and division by a (non-zero) scalar can be defined as

$$\mathbf{v} - \mathbf{w} = \mathbf{v} + (-\mathbf{w}),$$
$$\mathbf{v}/a = (1/a)\mathbf{v}.$$

When the scalar field F is the real numbers **R**, the vector space is called a *real vector space*. When the scalar field is the complex numbers, it is called a *complex vector space*. These two cases are the ones used most often in engineering. The general definition of a vector space allows scalars to be elements of any fixed field F. The notion is then known as an *F-vector spaces* or a *vector space over F*. A field is, essentially, a set of numbers possessing addition, subtraction, multiplication and division operations.[nb 3] For example, rational numbers also form a field.

In contrast to the intuition stemming from vectors in the plane and higher-dimensional cases, there is, in general vector spaces, no notion of nearness, angles or distances. To deal with such matters, particular types of vector spaces are introduced; see below.

12.1.4 Alternative formulations and elementary consequences

The requirement that vector addition and scalar multiplication be binary operations includes (by definition of binary operations) a property called closure: that $\mathbf{u} + \mathbf{v}$ and $a\mathbf{v}$ are in V for all a in F, and $\mathbf{u}$, $\mathbf{v}$ in V. Some older sources mention these properties as separate axioms.[2]

In the parlance of abstract algebra, the first four axioms can be subsumed by requiring the set of vectors to be an abelian group under addition. The remaining axioms give this group an F-module structure. In other words, there is a ring homomorphism f from the field F into the endomorphism ring of the group of vectors. Then scalar multiplication $a\mathbf{v}$ is defined as $(f(a))(\mathbf{v})$.[3]

There are a number of direct consequences of the vector space axioms. Some of them derive from elementary group theory, applied to the additive group of vectors: for example the zero vector $\mathbf{0}$ of V and the additive inverse $-\mathbf{v}$ of any vector $\mathbf{v}$ are unique. Other properties follow from the distributive law, for example $a\mathbf{v}$ equals $\mathbf{0}$ if and only if a equals 0 or $\mathbf{v}$ equals $\mathbf{0}$.

12.2 History

Vector spaces stem from affine geometry via the introduction of coordinates in the plane or three-dimensional space. Around 1636, Descartes and Fermat founded analytic geometry by equating solutions to an equation of two variables with points on a plane curve.[4] In 1884, to achieve geometric solutions without using coordinates, Bolzano introduced certain operations on points, lines and planes, which are predecessors of vectors.[5] His work was then used in the conception of barycentric coordinates by Möbius in 1827.[6] The definition of vectors was founded on Bellavitis' notion of the bipoint, an oriented segment of which one end is the origin and the other a target, then further elaborated with the presentation of complex numbers by Argand and Hamilton and the introduction of quaternions and biquaternions by the latter.[7] They are elements in $\mathbf{R}^2$, $\mathbf{R}^4$, and $\mathbf{R}^8$; their treatment as linear combinations can be traced back to Laguerre in 1867, who also defined systems of linear equations.

In 1857, Cayley introduced matrix notation, which allows for a harmonization and simplification of linear maps. Around the same time, Grassmann studied the barycen-

tric calculus initiated by Möbius. He envisaged sets of abstract objects endowed with operations.[8] In his work, the concepts of linear independence and dimension, as well as scalar products, are present. In fact, Grassmann's 1844 work exceeds the framework of vector spaces, since his consideration of multiplication led him to what are today called algebras. Peano was the first to give the modern definition of vector spaces and linear maps in 1888.[9]

An important development of vector spaces is due to the construction of function spaces by Lebesgue. This was later formalized by Banach and Hilbert, around 1920.[10] At that time, algebra and the new field of functional analysis began to interact, notably with key concepts such as spaces of p-integrable functions and Hilbert spaces.[11] Vector spaces, including infinite-dimensional ones, then became a firmly established notion, and many mathematical branches started making use of this concept.

12.3 Examples

Main article: Examples of vector spaces

12.3.1 Coordinate spaces

Main article: Coordinate space

The most simple example of a vector space over a field F is the field itself, equipped with its standard addition and multiplication. More generally, a vector space can be composed of n-tuples (sequences of length n) of elements of F, such as

$$(a_1, a_2, ..., an), \text{ where each } ai \text{ is an element of } F.\text{[12]}$$

A vector space composed of all the n-tuples of a field F is known as a *coordinate space*, usually denoted F^n. The case $n = 1$ is the above-mentioned simplest example, in which the field F is also regarded as a vector space over itself. The case $F = \mathbf{R}$ and $n = 2$ was discussed in the introduction above.

12.3.2 Complex numbers and other field extensions

The set of complex numbers $\mathbf{C}$, i.e., numbers that can be written in the form $x + iy$ for real numbers x and y where i is the imaginary unit, form a vector space over the reals with the usual addition and multiplication: $(x + iy) + (a + ib) = (x + a) + i(y + b)$ and $c \cdot (x + iy) = (c \cdot x) + i(c \cdot y)$

for real numbers x, y, a, b and c. The various axioms of a vector space follow from the fact that the same rules hold for complex number arithmetic.

In fact, the example of complex numbers is essentially the same (i.e., it is *isomorphic*) to the vector space of ordered pairs of real numbers mentioned above: if we think of the complex number $x + i\,y$ as representing the ordered pair (x, y) in the complex plane then we see that the rules for sum and scalar product correspond exactly to those in the earlier example.

More generally, field extensions provide another class of examples of vector spaces, particularly in algebra and algebraic number theory: a field F containing a smaller field E is an E-vector space, by the given multiplication and addition operations of F.[13] For example, the complex numbers are a vector space over **R**, and the field extension $\mathbf{Q}(i\sqrt{5})$ is a vector space over **Q**.

12.3.3   Function spaces

Functions from any fixed set Ω to a field F also form vector spaces, by performing addition and scalar multiplication pointwise. That is, the sum of two functions f and g is the function $(f + g)$ given by

$$(f + g)(w) = f(w) + g(w),$$

and similarly for multiplication. Such function spaces occur in many geometric situations, when Ω is the real line or an interval, or other subsets of **R**. Many notions in topology and analysis, such as continuity, integrability or differentiability are well-behaved with respect to linearity: sums and scalar multiples of functions possessing such a property still have that property.[14] Therefore, the set of such functions are vector spaces. They are studied in greater detail using the methods of functional analysis, see below. Algebraic constraints also yield vector spaces: the vector space $F[\mathrm{x}]$ is given by polynomial functions:

$f(x) = r_0 + r_1 x + \ldots + r_{n-1}x^{n-1} + r_n x^n$, where the coefficients $r_0, \ldots, r_n$ are in F.[15]

12.3.4   Linear equations

Main articles: Linear equation, Linear differential equation and Systems of linear equations

Systems of homogeneous linear equations are closely tied to vector spaces.[16] For example, the solutions of

are given by triples with arbitrary a, $b = a/2$, and $c = -5a/2$. They form a vector space: sums and scalar multiples of such triples still satisfy the same ratios of the three variables; thus they are solutions, too. Matrices can be used to condense multiple linear equations as above into one vector equation, namely

$$A\mathbf{x} = \mathbf{0},$$

where $A = \begin{bmatrix} 1 & 3 & 1 \\ 4 & 2 & 2 \end{bmatrix}$ is the matrix containing the coefficients of the given equations, $\mathbf{x}$ is the vector (a, b, c), $A\mathbf{x}$ denotes the matrix product, and $\mathbf{0} = (0, 0)$ is the zero vector. In a similar vein, the solutions of homogeneous *linear differential equations* form vector spaces. For example,

$$f''(x) + 2f'(x) + f(x) = 0$$

yields $f(x) = a\,e^{-x} + bx\,e^{-x}$, where a and b are arbitrary constants, and e^x is the natural exponential function.

12.4   Basis and dimension

Main articles: Basis and Dimension
Bases allow to represent vectors by a sequence of scalars

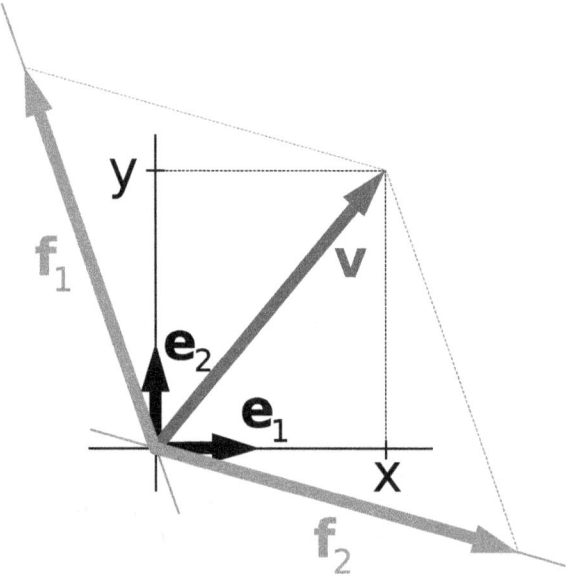

*A vector **v** in $\mathbf{R}^2$ (blue) expressed in terms of different bases: using the standard basis of $\mathbf{R}^2$ $\mathbf{v} = \mathrm{x}\mathbf{e}_1 + \mathrm{y}\mathbf{e}_2$ (black), and using a different, non-orthogonal basis: $\mathbf{v} = \mathbf{f}_1 + \mathbf{f}_2$ (red).*

called *coordinates* or *components*. A basis is a (finite or infinite) set $B = \{\mathbf{b}i\}i \in I$ of vectors $\mathbf{b}i$, for convenience often indexed by some index set I, that spans the whole space and is linearly independent. "Spanning the whole space" means that any vector **v** can be expressed as a finite sum (called a *linear combination*) of the basis elements:

where the a_k are scalars, called the coordinates (or the components) of the vector **v** with respect to the basis B, and $\mathbf{b}_{ik}$ ($k = 1, ..., n$) elements of B. Linear independence means that the coordinates a_k are uniquely determined for any vector in the vector space.

For example, the coordinate vectors $\mathbf{e}_1 = (1, 0, ..., 0)$, $\mathbf{e}_2 = (0, 1, 0, ..., 0)$, to $\mathbf{e}n = (0, 0, ..., 0, 1)$, form a basis of F^n, called the standard basis, since any vector $(x_1, x_2, ..., xn)$ can be uniquely expressed as a linear combination of these vectors:

$$(x_1, x_2, ..., xn) = x_1(1, 0, ..., 0) + x_2(0, 1, 0, ..., 0)$$
$$+ ... + xn(0, ..., 0, 1) = x_1\mathbf{e}_1 + x_2\mathbf{e}_2 + ... + xn\mathbf{e}n.$$

The corresponding coordinates x_1, x_2, ..., xn are just the Cartesian coordinates of the vector.

Every vector space has a basis. This follows from Zorn's lemma, an equivalent formulation of the Axiom of Choice.[17] Given the other axioms of Zermelo–Fraenkel set theory, the existence of bases is equivalent to the axiom of choice.[18] The ultrafilter lemma, which is weaker than the axiom of choice, implies that all bases of a given vector space have the same number of elements, or cardinality (cf. *Dimension theorem for vector spaces*).[19] It is called the *dimension* of the vector space, denoted dim V. If the space is spanned by finitely many vectors, the above statements can be proven without such fundamental input from set theory.[20]

The dimension of the coordinate space F^n is n, by the basis exhibited above. The dimension of the polynomial ring $F[x]$ introduced above is countably infinite, a basis is given by 1, x, x^2, ... A fortiori, the dimension of more general function spaces, such as the space of functions on some (bounded or unbounded) interval, is infinite.[nb 4] Under suitable regularity assumptions on the coefficients involved, the dimension of the solution space of a homogeneous ordinary differential equation equals the degree of the equation.[21] For example, the solution space for the above equation is generated by e^{-x} and xe^{-x}. These two functions are linearly independent over **R**, so the dimension of this space is two, as is the degree of the equation.

A field extension over the rationals **Q** can be thought of as a vector space over **Q** (by defining vector addition as field addition, defining scalar multiplication as field multiplication by elements of **Q**, and otherwise ignoring the field multiplication). The dimension (or degree) of the field extension

$\mathbf{Q}(\alpha)$ over **Q** depends on α. If α satisfies some polynomial equation

$$qn\alpha^n + qn_{-1}\alpha^{n-1} + ... + q_0 = 0, \text{ with rational coefficients } qn, ..., q_0.$$

("α is algebraic"), the dimension is finite. More precisely, it equals the degree of the minimal polynomial having α as a root.[22] For example, the complex numbers **C** are a two-dimensional real vector space, generated by 1 and the imaginary unit i. The latter satisfies $i^2 + 1 = 0$, an equation of degree two. Thus, **C** is a two-dimensional **R**-vector space (and, as any field, one-dimensional as a vector space over itself, **C**). If α is not algebraic, the dimension of $\mathbf{Q}(\alpha)$ over **Q** is infinite. For instance, for $\alpha = \pi$ there is no such equation, in other words π is transcendental.[23]

12.5 Linear maps and matrices

Main article: Linear map

The relation of two vector spaces can be expressed by *linear map* or *linear transformation*. They are functions that reflect the vector space structure—i.e., they preserve sums and scalar multiplication:

$$f(\mathbf{x} + \mathbf{y}) = f(\mathbf{x}) + f(\mathbf{y}) \text{ and } f(a \cdot \mathbf{x}) = a \cdot f(\mathbf{x}) \text{ for all } \mathbf{x} \text{ and } \mathbf{y} \text{ in } V, \text{ all } a \text{ in } F.[24]$$

An *isomorphism* is a linear map $f : V \rightarrow W$ such that there exists an inverse map $g : W \rightarrow V$, which is a map such that the two possible compositions $f \circ g : W \rightarrow W$ and $g \circ f : V \rightarrow V$ are identity maps. Equivalently, f is both one-to-one (injective) and onto (surjective).[25] If there exists an isomorphism between V and W, the two spaces are said to be *isomorphic*; they are then essentially identical as vector spaces, since all identities holding in V are, via f, transported to similar ones in W, and vice versa via g.

For example, the "arrows in the plane" and "ordered pairs of numbers" vector spaces in the introduction are isomorphic: a planar arrow **v** departing at the origin of some (fixed) coordinate system can be expressed as an ordered pair by considering the x- and y-component of the arrow, as shown in the image at the right. Conversely, given a pair (x, y), the arrow going by x to the right (or to the left, if x is negative), and y up (down, if y is negative) turns back the arrow **v**.

Linear maps $V \rightarrow W$ between two vector spaces form a vector space Hom$F(V, W)$, also denoted L(V, W).[26] The space of linear maps from V to F is called the *dual vector space*, denoted V^*.[27] Via the injective natural map $V \rightarrow V^{**}$, any vector space can be embedded into its *bidual*;

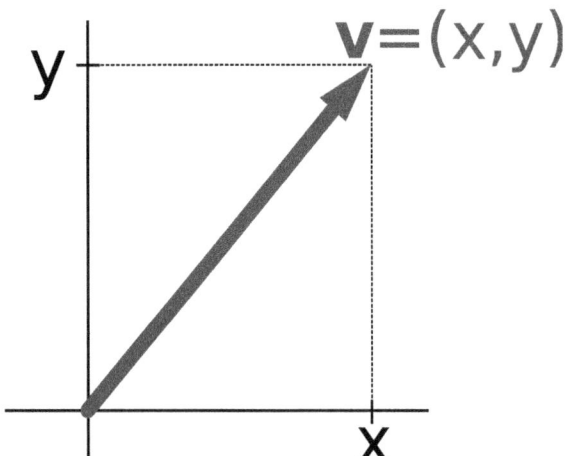

*Describing an arrow vector **v** by its coordinates x and y yields an isomorphism of vector spaces.*

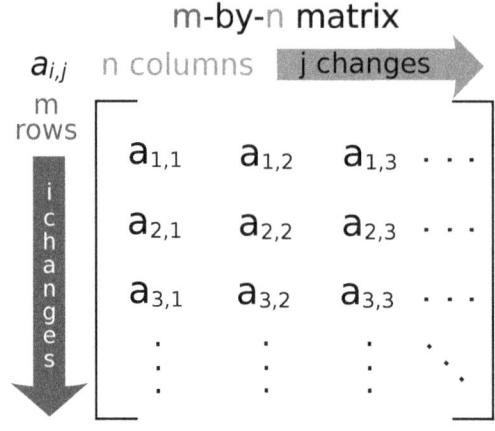

A typical matrix

the map is an isomorphism if and only if the space is finite-dimensional.[28]

Once a basis of V is chosen, linear maps $f : V \to W$ are completely determined by specifying the images of the basis vectors, because any element of V is expressed uniquely as a linear combination of them.[29] If dim $V =$ dim W, a 1-to-1 correspondence between fixed bases of V and W gives rise to a linear map that maps any basis element of V to the corresponding basis element of W. It is an isomorphism, by its very definition.[30] Therefore, two vector spaces are isomorphic if their dimensions agree and vice versa. Another way to express this is that any vector space is *completely classified* (up to isomorphism) by its dimension, a single number. In particular, any n-dimensional F-vector space V is isomorphic to F^n. There is, however, no "canonical" or preferred isomorphism; actually an isomorphism $\varphi : F^n \to V$ is equivalent to the choice of a basis of V, by mapping the standard basis of F^n to V, via φ. The freedom of choosing a convenient basis is particularly useful in the infinite-dimensional context, see below.

12.5.1 Matrices

Main articles: Matrix and Determinant
Matrices are a useful notion to encode linear maps.[31] They are written as a rectangular array of scalars as in the image at the right. Any m-by-n matrix A gives rise to a linear map from F^n to F^m, by the following

$$\mathbf{x} \quad = \quad (x_1, x_2, \cdots, x_n) \quad \mapsto$$
$$\left(\sum_{j=1}^{n} a_{1j} x_j, \sum_{j=1}^{n} a_{2j} x_j, \cdots, \sum_{j=1}^{n} a_{mj} x_j \right)$$
, where $\sum$ denotes summation,

or, using the matrix multiplication of the matrix A with the coordinate vector $\mathbf{x}$:

$$\mathbf{x} \mapsto A\mathbf{x}.$$

Moreover, after choosing bases of V and W, *any* linear map $f : V \to W$ is uniquely represented by a matrix via this assignment.[32]

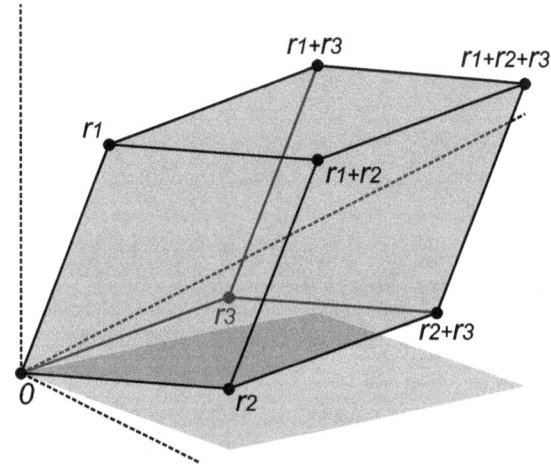

The volume of this parallelepiped is the absolute value of the determinant of the 3-by-3 matrix formed by the vectors r_1, r_2, and r_3.

The determinant det (A) of a square matrix A is a scalar that tells whether the associated map is an isomorphism or not: to be so it is sufficient and necessary that the determinant is nonzero.[33] The linear transformation of $\mathbf{R}^n$ corresponding to a real n-by-n matrix is orientation preserving if and only if its determinant is positive.

12.5.2 Eigenvalues and eigenvectors

Main article: Eigenvalues and eigenvectors

Endomorphisms, linear maps $f : V \to V$, are particularly important since in this case vectors $\mathbf{v}$ can be compared with their image under f, $f(\mathbf{v})$. Any nonzero vector $\mathbf{v}$ satisfying $\lambda\mathbf{v} = f(\mathbf{v})$, where λ is a scalar, is called an *eigenvector* of f with *eigenvalue* λ.[nb 5][34] Equivalently, $\mathbf{v}$ is an element of the kernel of the difference $f - \lambda \cdot \mathrm{Id}$ (where Id is the identity map $V \to V$). If V is finite-dimensional, this can be rephrased using determinants: f having eigenvalue λ is equivalent to

$$\det(f - \lambda \cdot \mathrm{Id}) = 0.$$

By spelling out the definition of the determinant, the expression on the left hand side can be seen to be a polynomial function in λ, called the characteristic polynomial of f.[35] If the field F is large enough to contain a zero of this polynomial (which automatically happens for F algebraically closed, such as $F = \mathbf{C}$) any linear map has at least one eigenvector. The vector space V may or may not possess an eigenbasis, a basis consisting of eigenvectors. This phenomenon is governed by the Jordan canonical form of the map.[nb 6] The set of all eigenvectors corresponding to a particular eigenvalue of f forms a vector space known as the *eigenspace* corresponding to the eigenvalue (and f) in question. To achieve the spectral theorem, the corresponding statement in the infinite-dimensional case, the machinery of functional analysis is needed, see below.

12.6 Basic constructions

In addition to the above concrete examples, there are a number of standard linear algebraic constructions that yield vector spaces related to given ones. In addition to the definitions given below, they are also characterized by universal properties, which determine an object X by specifying the linear maps from X to any other vector space.

12.6.1 Subspaces and quotient spaces

Main articles: Linear subspace and Quotient vector space
A nonempty subset W of a vector space V that is closed under addition and scalar multiplication (and therefore contains the $\mathbf{0}$-vector of V) is called a *subspace* of V.[36] Subspaces of V are vector spaces (over the same field) in their own right. The intersection of all subspaces containing a given set S of vectors is called its span, and it is the smallest subspace of V containing the set S. Expressed in terms

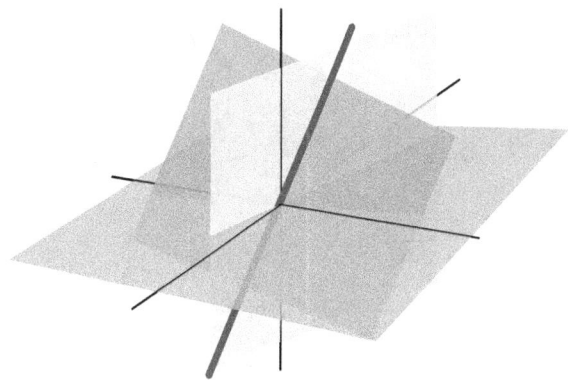

A line passing through the origin (blue, thick) in $\mathbf{R}^3$ is a linear subspace. It is the intersection of two planes (green and yellow).

of elements, the span is the subspace consisting of all the linear combinations of elements of S.[37]

The counterpart to subspaces are *quotient vector spaces*.[38] Given any subspace $W \subset V$, the quotient space V/W ("V modulo W") is defined as follows: as a set, it consists of $\mathbf{v} + W = \{\mathbf{v} + \mathbf{w} : \mathbf{w} \in W\}$, where $\mathbf{v}$ is an arbitrary vector in V. The sum of two such elements $\mathbf{v}_1 + W$ and $\mathbf{v}_2 + W$ is $(\mathbf{v}_1 + \mathbf{v}_2) + W$, and scalar multiplication is given by $a \cdot (\mathbf{v} + W) = (a \cdot \mathbf{v}) + W$. The key point in this definition is that $\mathbf{v}_1 + W = \mathbf{v}_2 + W$ if and only if the difference of $\mathbf{v}_1$ and $\mathbf{v}_2$ lies in W.[nb 7] This way, the quotient space "forgets" information that is contained in the subspace W.

The kernel $\ker(f)$ of a linear map $f : V \to W$ consists of vectors $\mathbf{v}$ that are mapped to $\mathbf{0}$ in W.[39] Both kernel and image $\mathrm{im}(f) = \{f(\mathbf{v}) : \mathbf{v} \in V\}$ are subspaces of V and W, respectively.[40] The existence of kernels and images is part of the statement that the category of vector spaces (over a fixed field F) is an abelian category, i.e. a corpus of mathematical objects and structure-preserving maps between them (a category) that behaves much like the category of abelian groups.[41] Because of this, many statements such as the first isomorphism theorem (also called rank–nullity theorem in matrix-related terms)

$$V / \ker(f) \equiv \mathrm{im}(f).$$

and the second and third isomorphism theorem can be formulated and proven in a way very similar to the corresponding statements for groups.

An important example is the kernel of a linear map $\mathbf{x} \mapsto A\mathbf{x}$ for some fixed matrix A, as above. The kernel of this map is the subspace of vectors $\mathbf{x}$ such that $A\mathbf{x} = 0$, which is precisely the set of solutions to the system of homogeneous linear equations belonging to A. This concept also extends to linear differential equations

$a_0 f + a_1 \frac{df}{dx} + a_2 \frac{d^2 f}{dx^2} + \cdots + a_n \frac{d^n f}{dx^n} = 0$, where the coefficients a_i are functions in x, too.

In the corresponding map

$$f \mapsto D(f) = \sum_{i=0}^{n} a_i \frac{d^i f}{dx^i}$$

the derivatives of the function f appear linearly (as opposed to $f''(x)^2$, for example). Since differentiation is a linear procedure (i.e., $(f + g)' = f' + g'$ and $(c \cdot f)' = c \cdot f'$ for a constant c) this assignment is linear, called a linear differential operator. In particular, the solutions to the differential equation $D(f) = 0$ form a vector space (over **R** or **C**).

12.6.2 Direct product and direct sum

Main articles: Direct product and Direct sum of modules

The *direct product* of vector spaces and the *direct sum* of vector spaces are two ways of combining an indexed family of vector spaces into a new vector space.

The *direct product* $\prod_{i \in I} V_i$ of a family of vector spaces V_i consists of the set of all tuples $(v_i)_{i} \in I$, which specify for each index i in some index set I an element v_i of V_i.[42] Addition and scalar multiplication is performed componentwise. A variant of this construction is the *direct sum* $\oplus_{i \in I} V_i$ (also called coproduct and denoted $\coprod_{i \in I} V_i$), where only tuples with finitely many nonzero vectors are allowed. If the index set I is finite, the two constructions agree, but in general they are different.

12.6.3 Tensor product

Main article: Tensor product of vector spaces

The *tensor product* $V \otimes_F W$, or simply $V \otimes W$, of two vector spaces V and W is one of the central notions of multilinear algebra which deals with extending notions such as linear maps to several variables. A map $g : V \times W \to X$ is called bilinear if g is linear in both variables **v** and **w**. That is to say, for fixed **w** the map $\mathbf{v} \mapsto g(\mathbf{v}, \mathbf{w})$ is linear in the sense above and likewise for fixed **v**.

The tensor product is a particular vector space that is a *universal* recipient of bilinear maps g, as follows. It is defined as the vector space consisting of finite (formal) sums of symbols called tensors

$$\mathbf{v}_1 \otimes \mathbf{w}_1 + \mathbf{v}_2 \otimes \mathbf{w}_2 + \ldots + \mathbf{v}n \otimes \mathbf{w}n,$$

subject to the rules

$a \cdot (\mathbf{v} \otimes \mathbf{w}) = (a \cdot \mathbf{v}) \otimes \mathbf{w} = \mathbf{v} \otimes (a \cdot \mathbf{w})$, where a is a scalar,

$(\mathbf{v}_1 + \mathbf{v}_2) \otimes \mathbf{w} = \mathbf{v}_1 \otimes \mathbf{w} + \mathbf{v}_2 \otimes \mathbf{w}$, and

$\mathbf{v} \otimes (\mathbf{w}_1 + \mathbf{w}_2) = \mathbf{v} \otimes \mathbf{w}_1 + \mathbf{v} \otimes \mathbf{w}_2$.[43]

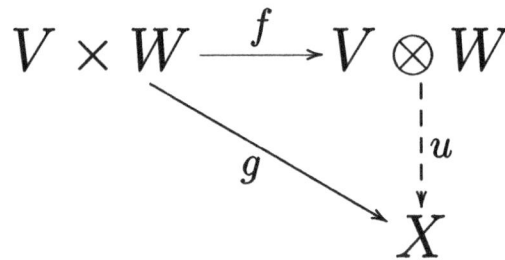

Commutative diagram depicting the universal property of the tensor product.

These rules ensure that the map f from the $V \times W$ to $V \otimes W$ that maps a tuple $(\mathbf{v}, \mathbf{w})$ to $\mathbf{v} \otimes \mathbf{w}$ is bilinear. The universality states that given *any* vector space X and *any* bilinear map $g : V \times W \to X$, there exists a unique map u, shown in the diagram with a dotted arrow, whose composition with f equals g: $u(\mathbf{v} \otimes \mathbf{w}) = g(\mathbf{v}, \mathbf{w})$.[44] This is called the universal property of the tensor product, an instance of the method—much used in advanced abstract algebra—to indirectly define objects by specifying maps from or to this object.

12.7 Vector spaces with additional structure

From the point of view of linear algebra, vector spaces are completely understood insofar as any vector space is characterized, up to isomorphism, by its dimension. However, vector spaces *per se* do not offer a framework to deal with the question—crucial to analysis—whether a sequence of functions converges to another function. Likewise, linear algebra is not adapted to deal with infinite series, since the addition operation allows only finitely many terms to be added. Therefore, the needs of functional analysis require considering additional structures.

A vector space may be given a partial order $\leq$, under which some vectors can be compared.[45] For example, n-dimensional real space **R**^{n} can be ordered by comparing its vectors componentwise. Ordered vector spaces, for example Riesz spaces, are fundamental to Lebesgue integration, which relies on the ability to express a function as a difference of two positive functions

$$f = f^+ - f^-,$$

where f^+ denotes the positive part of f and f^- the negative part.[46]

12.7.1 Normed vector spaces and inner product spaces

Main articles: Normed vector space and Inner product space

"Measuring" vectors is done by specifying a norm, a datum which measures lengths of vectors, or by an inner product, which measures angles between vectors. Norms and inner products are denoted $|\mathbf{v}|$ and $\langle \mathbf{v}, \mathbf{w} \rangle$, respectively. The datum of an inner product entails that lengths of vectors can be defined too, by defining the associated norm $|\mathbf{v}| := \sqrt{\langle \mathbf{v}, \mathbf{v} \rangle}$. Vector spaces endowed with such data are known as *normed vector spaces* and *inner product spaces*, respectively.[47]

Coordinate space F^n can be equipped with the standard dot product:

$$\langle \mathbf{x}, \mathbf{y} \rangle = \mathbf{x} \cdot \mathbf{y} = x_1 y_1 + \cdots + x_n y_n.$$

In $\mathbf{R}^2$, this reflects the common notion of the angle between two vectors $\mathbf{x}$ and $\mathbf{y}$, by the law of cosines:

$$\mathbf{x} \cdot \mathbf{y} = \cos\left(\angle(\mathbf{x}, \mathbf{y})\right) \cdot |\mathbf{x}| \cdot |\mathbf{y}|.$$

Because of this, two vectors satisfying $\langle \mathbf{x}, \mathbf{y} \rangle = 0$ are called orthogonal. An important variant of the standard dot product is used in Minkowski space: $\mathbf{R}^4$ endowed with the Lorentz product

$$\langle \mathbf{x} | \mathbf{y} \rangle = x_1 y_1 + x_2 y_2 + x_3 y_3 - x_4 y_4. \quad [48]$$

In contrast to the standard dot product, it is not positive definite: $\langle \mathbf{x} | \mathbf{x} \rangle$ also takes negative values, for example for $\mathbf{x} = (0, 0, 0, 1)$. Singling out the fourth coordinate—corresponding to time, as opposed to three space-dimensions—makes it useful for the mathematical treatment of special relativity.

12.7.2 Topological vector spaces

Main article: Topological vector space

Convergence questions are treated by considering vector spaces V carrying a compatible topology, a structure that allows one to talk about elements being close to each other.[49][50] Compatible here means that addition and scalar multiplication have to be continuous maps. Roughly, if $\mathbf{x}$ and $\mathbf{y}$ in V, and a in F vary by a bounded amount, then so do $\mathbf{x} + \mathbf{y}$ and $a\mathbf{x}$.[nb 8] To make sense of specifying the amount a scalar changes, the field F also has to carry a topology in this context; a common choice are the reals or the complex numbers.

In such *topological vector spaces* one can consider series of vectors. The infinite sum

$$\sum_{i=0}^{\infty} f_i$$

denotes the limit of the corresponding finite partial sums of the sequence $(f_i)_{i \in \mathbf{N}}$ of elements of V. For example, the f_i could be (real or complex) functions belonging to some function space V, in which case the series is a function series. The mode of convergence of the series depends on the topology imposed on the function space. In such cases, pointwise convergence and uniform convergence are two prominent examples.

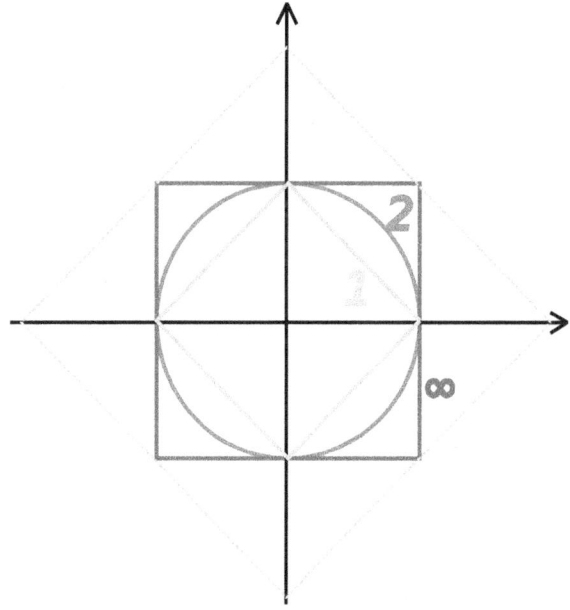

Unit "spheres" in $\mathbf{R}^2$ consist of plane vectors of norm 1. Depicted are the unit spheres in different p-norms, for p = 1, 2, and ∞. The bigger diamond depicts points of 1-norm equal to $\sqrt{2}$.

A way to ensure the existence of limits of certain infinite series is to restrict attention to spaces where any Cauchy sequence has a limit; such a vector space is called complete. Roughly, a vector space is complete provided that it contains all necessary limits. For example, the vector space of polynomials on the unit interval [0,1], equipped with the topology of uniform convergence is not complete because any continuous function on [0,1] can be uniformly approximated by a sequence of polynomials, by the Weierstrass approximation theorem.[51] In contrast, the space of *all* continuous functions on [0,1] with the same topology is complete.[52] A norm gives rise to a topology by defining that a sequence of vectors **v**n converges to **v** if and only if

$$\lim_{n \to \infty} |\mathbf{v}_n - \mathbf{v}| = 0.$$

Banach and Hilbert spaces are complete topological vector spaces whose topologies are given, respectively, by a norm and an inner product. Their study—a key piece of functional analysis—focusses on infinite-dimensional vector spaces, since all norms on finite-dimensional topological vector spaces give rise to the same notion of convergence.[53] The image at the right shows the equivalence of the 1-norm and ∞-norm on **R**^{2}: as the unit "balls" enclose each other, a sequence converges to zero in one norm if and only if it so does in the other norm. In the infinite-dimensional case, however, there will generally be inequivalent topologies, which makes the study of topological vector spaces richer than that of vector spaces without additional data.

From a conceptual point of view, all notions related to topological vector spaces should match the topology. For example, instead of considering all linear maps (also called functionals) $V \to W$, maps between topological vector spaces are required to be continuous.[54] In particular, the (topological) dual space V^* consists of continuous functionals $V \to \mathbf{R}$ (or to **C**). The fundamental Hahn–Banach theorem is concerned with separating subspaces of appropriate topological vector spaces by continuous functionals.[55]

Banach spaces

Main article: Banach space

Banach spaces, introduced by Stefan Banach, are complete normed vector spaces.[56] A first example is the vector space ℓ^p consisting of infinite vectors with real entries **x** = $(x_1, x_2, \ldots)$ whose p-norm $(1 \le p \le \infty)$ given by

$$|\mathbf{x}|_p := \left(\sum_i |x_i|^p \right)^{1/p} \text{ for } p < \infty \text{ and } |\mathbf{x}|_\infty := \sup_i |x_i|$$

is finite. The topologies on the infinite-dimensional space ℓ^p are inequivalent for different p. E.g. the sequence of vectors **x**n = $(2^{-n}, 2^{-n}, \ldots, 2^{-n}, 0, 0, \ldots)$, i.e. the first 2^n components are 2^{-n}, the following ones are 0, converges to the zero vector for $p = \infty$, but does not for $p = 1$:

$|x_n|_\infty = \sup(2^{-n}, 0) = 2^{-n} \to 0$, but $|x_n|_1 = \sum_{i=1}^{2^n} 2^{-n} = 2^n \cdot 2^{-n} = 1.$

More generally than sequences of real numbers, functions $f \colon \Omega \to \mathbf{R}$ are endowed with a norm that replaces the above sum by the Lebesgue integral

$$|f|_p := \left(\int_\Omega |f(x)|^p \, dx \right)^{1/p}.$$

The space of integrable functions on a given domain Ω (for example an interval) satisfying $|f|p < \infty$, and equipped with this norm are called Lebesgue spaces, denoted $L^p(\Omega)$.[nb 9] These spaces are complete.[57] (If one uses the Riemann integral instead, the space is *not* complete, which may be seen as a justification for Lebesgue's integration theory.[nb 10]) Concretely this means that for any sequence of Lebesgue-integrable functions $f_1, f_2, \ldots$ with $|f_n|p < \infty$, satisfying the condition

$$\lim_{k, n \to \infty} \int_\Omega |f_k(x) - f_n(x)|^p \, dx = 0$$

there exists a function $f(x)$ belonging to the vector space $L^p(\Omega)$ such that

$$\lim_{k \to \infty} \int_\Omega |f(x) - f_k(x)|^p \, dx = 0.$$

Imposing boundedness conditions not only on the function, but also on its derivatives leads to Sobolev spaces.[58]

Hilbert spaces

Main article: Hilbert space
Complete inner product spaces are known as *Hilbert spaces*,

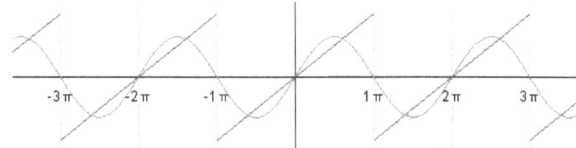

The succeeding snapshots show summation of 1 to 5 terms in approximating a periodic function (blue) by finite sum of sine functions (red).

in honor of David Hilbert.[59] The Hilbert space $L^2(\Omega)$, with inner product given by

$$\langle f, \, g \rangle = \int_\Omega f(x)\overline{g(x)} \, dx,$$

where $\overline{g(x)}$ denotes the complex conjugate of $g(x)$,[60][nb 11] is a key case.

By definition, in a Hilbert space any Cauchy sequence converges to a limit. Conversely, finding a sequence of functions f_n with desirable properties that approximates a given limit function, is equally crucial. Early analysis, in the guise of the Taylor approximation, established an approximation of differentiable functions f by polynomials.[61] By the Stone–Weierstrass theorem, every continuous function on $[a, b]$ can be approximated as closely as desired by a polynomial.[62] A similar approximation technique by trigonometric functions is commonly called Fourier expansion, and is much applied in engineering, see below. More generally, and more conceptually, the theorem yields a simple description of what "basic functions", or, in abstract Hilbert spaces, what basic vectors suffice to generate a Hilbert space H, in the sense that the *closure* of their span (i.e., finite linear combinations and limits of those) is the whole space. Such a set of functions is called a *basis* of H, its cardinality is known as the Hilbert space dimension.[nb 12] Not only does the theorem exhibit suitable basis functions as sufficient for approximation purposes, but together with the Gram-Schmidt process, it enables one to construct a basis of orthogonal vectors.[63] Such orthogonal bases are the Hilbert space generalization of the coordinate axes in finite-dimensional Euclidean space.

The solutions to various differential equations can be interpreted in terms of Hilbert spaces. For example, a great many fields in physics and engineering lead to such equations and frequently solutions with particular physical properties are used as basis functions, often orthogonal.[64] As an example from physics, the time-dependent Schrödinger equation in quantum mechanics describes the change of physical properties in time by means of a partial differential equation, whose solutions are called wavefunctions.[65] Definite values for physical properties such as energy, or momentum, correspond to eigenvalues of a certain (linear) differential operator and the associated wavefunctions are called eigenstates. The spectral theorem decomposes a linear compact operator acting on functions in terms of these eigenfunctions and their eigenvalues.[66]

12.7.3 Algebras over fields

Main articles: Algebra over a field and Lie algebra
General vector spaces do not possess a multiplication between vectors. A vector space equipped with an additional bilinear operator defining the multiplication of two vectors is an *algebra over a field*.[67] Many algebras stem from functions on some geometrical object: since functions with values in a given field can be multiplied pointwise, these entities form algebras. The Stone–Weierstrass theorem mentioned above, for example, relies on Banach algebras which are both Banach spaces and algebras.

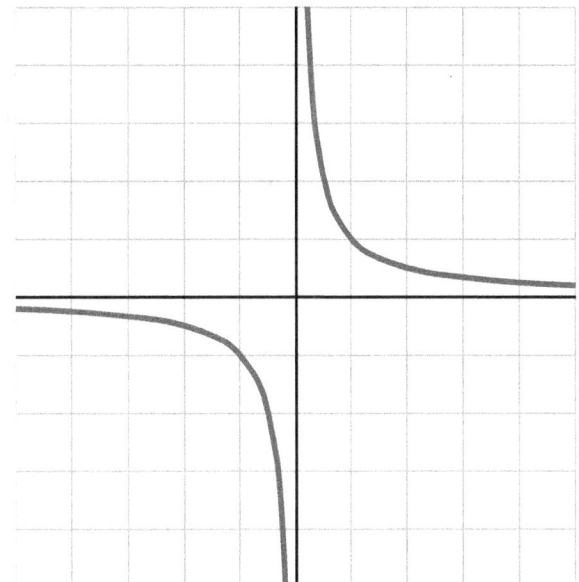

A hyperbola, given by the equation x ⋅ y = 1. The coordinate ring of functions on this hyperbola is given by $\mathbf{R}[x, y] / (x \cdot y - 1)$, an infinite-dimensional vector space over $\mathbf{R}$.

tioned above, for example, relies on Banach algebras which are both Banach spaces and algebras.

Commutative algebra makes great use of rings of polynomials in one or several variables, introduced above. Their multiplication is both commutative and associative. These rings and their quotients form the basis of algebraic geometry, because they are rings of functions of algebraic geometric objects.[68]

Another crucial example are *Lie algebras*, which are neither commutative nor associative, but the failure to be so is limited by the constraints ($[x, y]$ denotes the product of x and y):

- $[x, y] = -[y, x]$ (anticommutativity), and

- $[x, [y, z]] + [y, [z, x]] + [z, [x, y]] = 0$ (Jacobi identity).[69]

Examples include the vector space of n-by-n matrices, with $[x, y] = xy - yx$, the commutator of two matrices, and $\mathbf{R}^3$, endowed with the cross product.

The tensor algebra $T(V)$ is a formal way of adding products to any vector space V to obtain an algebra.[70] As a vector space, it is spanned by symbols, called simple tensors

$$\mathbf{v}_1 \otimes \mathbf{v}_2 \otimes \ldots \otimes \mathbf{v}_n,$$ where the degree n varies.

The multiplication is given by concatenating such symbols, imposing the distributive law under addition, and requiring that scalar multiplication commute with the tensor product $\otimes$, much the same way as with the tensor product of two vector spaces introduced above. In general, there are no relations between $\mathbf{v}_1 \otimes \mathbf{v}_2$ and $\mathbf{v}_2 \otimes \mathbf{v}_1$. Forcing two such elements to be equal leads to the symmetric algebra, whereas forcing $\mathbf{v}_1 \otimes \mathbf{v}_2 = -\mathbf{v}_2 \otimes \mathbf{v}_1$ yields the exterior algebra.[71]

When a field, F is explicitly stated, a common term used is F-algebra.

12.8 Applications

Vector spaces have manifold applications as they occur in many circumstances, namely wherever functions with values in some field are involved. They provide a framework to deal with analytical and geometrical problems, or are used in the Fourier transform. This list is not exhaustive: many more applications exist, for example in optimization. The minimax theorem of game theory stating the existence of a unique payoff when all players play optimally can be formulated and proven using vector spaces methods.[72] Representation theory fruitfully transfers the good understanding of linear algebra and vector spaces to other mathematical domains such as group theory.[73]

12.8.1 Distributions

Main article: Distribution

A *distribution* (or *generalized function*) is a linear map assigning a number to each "test" function, typically a smooth function with compact support, in a continuous way: in the above terminology the space of distributions is the (continuous) dual of the test function space.[74] The latter space is endowed with a topology that takes into account not only f itself, but also all its higher derivatives. A standard example is the result of integrating a test function f over some domain Ω:

$$I(f) = \int_\Omega f(x)\,dx.$$

When $\Omega = \{p\}$, the set consisting of a single point, this reduces to the Dirac distribution, denoted by δ, which associates to a test function f its value at the p: $\delta(f) = f(p)$. Distributions are a powerful instrument to solve differential equations. Since all standard analytic notions such as derivatives are linear, they extend naturally to the space of distributions. Therefore, the equation in question can be transferred to a distribution space, which is bigger than the underlying function space, so that more flexible methods are

available for solving the equation. For example, Green's functions and fundamental solutions are usually distributions rather than proper functions, and can then be used to find solutions of the equation with prescribed boundary conditions. The found solution can then in some cases be proven to be actually a true function, and a solution to the original equation (e.g., using the Lax–Milgram theorem, a consequence of the Riesz representation theorem).[75]

12.8.2 Fourier analysis

Main article: Fourier analysis
Resolving a periodic function into a sum of trigonometric

The heat equation describes the dissipation of physical properties over time, such as the decline of the temperature of a hot body placed in a colder environment (yellow depicts colder regions than red).

functions forms a *Fourier series*, a technique much used in physics and engineering.[nb 13][76] The underlying vector space is usually the Hilbert space $L^2(0, 2\pi)$, for which the functions $\sin mx$ and $\cos mx$ (m an integer) form an orthogonal basis.[77] The Fourier expansion of an L^2 function f is

$$\frac{a_0}{2} + \sum_{m=1}^{\infty} \left[a_m \cos(mx) + b_m \sin(mx) \right].$$

The coefficients a_m and b_m are called Fourier coefficients of f, and are calculated by the formulas[78]

$$a_m = \tfrac{1}{\pi} \int_0^{2\pi} f(t) \cos(mt)\,dt \ , \ b_m = \tfrac{1}{\pi} \int_0^{2\pi} f(t) \sin(mt)\,dt.$$

In physical terms the function is represented as a superposition of sine waves and the coefficients give information about the function's frequency spectrum.[79] A complex-number form of Fourier series is also commonly used.[78] The concrete formulae above are consequences of a more general mathematical duality called Pontryagin duality.[80] Applied to the group **R**, it yields the classical Fourier transform; an application in physics are reciprocal lattices, where the underlying group is a finite-dimensional real vector space endowed with the additional datum of a lattice encoding positions of atoms in crystals.[81]

Fourier series are used to solve boundary value problems in partial differential equations.[82] In 1822, Fourier first used this technique to solve the heat equation.[83] A discrete version of the Fourier series can be used in sampling applications where the function value is known only at a finite number of equally spaced points. In this case the Fourier series is finite and its value is equal to the sampled values at all points.[84] The set of coefficients is known as the discrete Fourier transform (DFT) of the given sample sequence. The DFT is one of the key tools of digital signal processing, a field whose applications include radar, speech encoding, image compression.[85] The JPEG image format is an application of the closely related discrete cosine transform.[86]

The fast Fourier transform is an algorithm for rapidly computing the discrete Fourier transform.[87] It is used not only for calculating the Fourier coefficients but, using the convolution theorem, also for computing the convolution of two finite sequences.[88] They in turn are applied in digital filters[89] and as a rapid multiplication algorithm for polynomials and large integers (Schönhage-Strassen algorithm).[90][91]

12.8.3 Differential geometry

Main article: Tangent space
 The tangent plane to a surface at a point is naturally a vec-

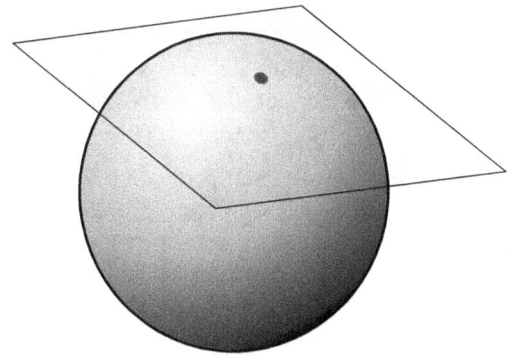

The tangent space to the 2-sphere at some point is the infinite plane touching the sphere in this point.

tor space whose origin is identified with the point of contact. The tangent plane is the best linear approximation, or linearization, of a surface at a point.[nb 14] Even in a three-dimensional Euclidean space, there is typically no natural way to prescribe a basis of the tangent plane, and so it is conceived of as an abstract vector space rather than a real coordinate space. The *tangent space* is the generalization to higher-dimensional differentiable manifolds.[92]

Riemannian manifolds are manifolds whose tangent spaces are endowed with a suitable inner product.[93] Derived therefrom, the Riemann curvature tensor encodes all curvatures of a manifold in one object, which finds applications in general relativity, for example, where the Einstein curvature tensor describes the matter and energy content of space-time.[94][95] The tangent space of a Lie group can be given naturally the structure of a Lie algebra and can be used to classify compact Lie groups.[96]

12.9 Generalizations

12.9.1 Vector bundles

Main articles: Vector bundle and Tangent bundle
 A *vector bundle* is a family of vector spaces parametrized

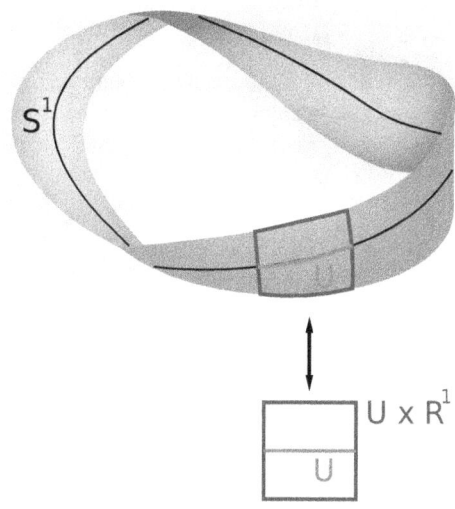

A Möbius strip. Locally, it looks like U × **R**.

continuously by a topological space X.[92] More precisely, a vector bundle over X is a topological space E equipped with a continuous map

$$\pi : E \to X$$

such that for every x in X, the fiber $\pi^{-1}(x)$ is a vector space. The case dim $V = 1$ is called a line bundle. For any vector space V, the projection $X \times V \to X$ makes the product $X \times$

V into a "trivial" vector bundle. Vector bundles over X are required to be locally a product of X and some (fixed) vector space V: for every x in X, there is a neighborhood U of x such that the restriction of π to $\pi^{-1}(U)$ is isomorphic[nb 15] to the trivial bundle $U \times V \to U$. Despite their locally trivial character, vector bundles may (depending on the shape of the underlying space X) be "twisted" in the large (i.e., the bundle need not be (globally isomorphic to) the trivial bundle $X \times V$). For example, the Möbius strip can be seen as a line bundle over the circle S^1 (by identifying open intervals with the real line). It is, however, different from the cylinder $S^1 \times \mathbf{R}$, because the latter is orientable whereas the former is not.[97]

Properties of certain vector bundles provide information about the underlying topological space. For example, the tangent bundle consists of the collection of tangent spaces parametrized by the points of a differentiable manifold. The tangent bundle of the circle S^1 is globally isomorphic to $S^1 \times \mathbf{R}$, since there is a global nonzero vector field on S^1.[nb 16] In contrast, by the hairy ball theorem, there is no (tangent) vector field on the 2-sphere S^2 which is everywhere nonzero.[98] K-theory studies the isomorphism classes of all vector bundles over some topological space.[99] In addition to deepening topological and geometrical insight, it has purely algebraic consequences, such as the classification of finite-dimensional real division algebras: $\mathbf{R}$, $\mathbf{C}$, the quaternions $\mathbf{H}$ and the octonions.

The cotangent bundle of a differentiable manifold consists, at every point of the manifold, of the dual of the tangent space, the cotangent space. Sections of that bundle are known as differential one-forms.

12.9.2 Modules

Main article: Module

Modules are to rings what vector spaces are to fields. The very same axioms, applied to a ring R instead of a field F yield modules.[100] The theory of modules, compared to that of vector spaces, is complicated by the presence of ring elements that do not have multiplicative inverses. For example, modules need not have bases, as the $\mathbf{Z}$-module (i.e., abelian group) $\mathbf{Z}/2\mathbf{Z}$ shows; those modules that do (including all vector spaces) are known as free modules. Nevertheless, a vector space can be compactly defined as a module over a ring which is a field with the elements being called vectors. Some authors use the term *vector space* to mean modules over a division ring.[101] The algebro-geometric interpretation of commutative rings via their spectrum allows the development of concepts such as locally free modules, the algebraic counterpart to vector bundles.

12.9.3 Affine and projective spaces

Main articles: Affine space and Projective space
Roughly, *affine spaces* are vector spaces whose origins are

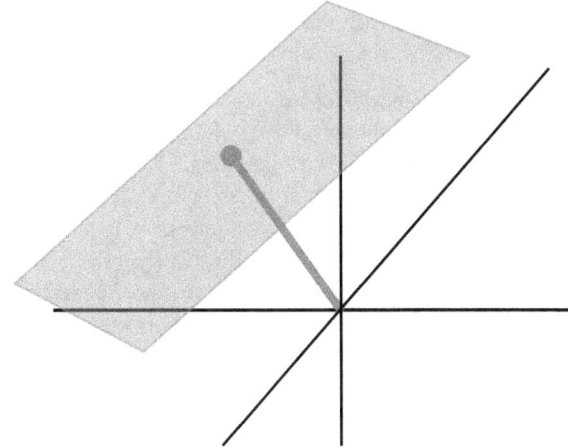

An affine plane (light blue) in $\mathbf{R}^3$. It is a two-dimensional subspace shifted by a vector $\mathbf{x}$ (red).

not specified.[102] More precisely, an affine space is a set with a free transitive vector space action. In particular, a vector space is an affine space over itself, by the map

$$V \times V \to V, (\mathbf{v}, \mathbf{a}) \mapsto \mathbf{a} + \mathbf{v}.$$

If W is a vector space, then an affine subspace is a subset of W obtained by translating a linear subspace V by a fixed vector $\mathbf{x} \in W$; this space is denoted by $\mathbf{x} + V$ (it is a coset of V in W) and consists of all vectors of the form $\mathbf{x} + \mathbf{v}$ for $\mathbf{v} \in V$. An important example is the space of solutions of a system of inhomogeneous linear equations

$$A\mathbf{x} = \mathbf{b}$$

generalizing the homogeneous case $\mathbf{b} = 0$ above.[103] The space of solutions is the affine subspace $\mathbf{x} + V$ where $\mathbf{x}$ is a particular solution of the equation, and V is the space of solutions of the homogeneous equation (the nullspace of A).

The set of one-dimensional subspaces of a fixed finite-dimensional vector space V is known as *projective space*; it may be used to formalize the idea of parallel lines intersecting at infinity.[104] Grassmannians and flag manifolds generalize this by parametrizing linear subspaces of fixed dimension k and flags of subspaces, respectively.

12.10   See also

- Vector (mathematics and physics), for a list of various kinds of vectors

12.11 Notes

[1] It is also common, especially in physics, to denote vectors with an arrow on top: $\vec{v}$.

[2] This axiom refers to two different operations: scalar multiplication: $b\mathbf{v}$; and field multiplication: ab. It does not assert the associativity of either operation. More formally, scalar multiplication is the *semigroup action* of the scalars on the vector space. Combined with the axiom of the identity element of scalar multiplication, it is a *monoid action*.

[3] Some authors (such as Brown 1991) restrict attention to the fields **R** or **C**, but most of the theory is unchanged for an arbitrary field.

[4] The indicator functions of intervals (of which there are infinitely many) are linearly independent, for example.

[5] The nomenclature derives from German "eigen", which means own or proper.

[6] Roman 2005, ch. 8, p. 140. See also Jordan–Chevalley decomposition.

[7] Some authors (such as Roman 2005) choose to start with this equivalence relation and derive the concrete shape of V/W from this.

[8] This requirement implies that the topology gives rise to a uniform structure, Bourbaki 1989, ch. II

[9] The triangle inequality for $|-|p$ is provided by the Minkowski inequality. For technical reasons, in the context of functions one has to identify functions that agree almost everywhere to get a norm, and not only a seminorm.

[10] "Many functions in L^2 of Lebesgue measure, being unbounded, cannot be integrated with the classical Riemann integral. So spaces of Riemann integrable functions would not be complete in the L^2 norm, and the orthogonal decomposition would not apply to them. This shows one of the advantages of Lebesgue integration.", Dudley 1989, §5.3, p. 125

[11] For $p \neq 2$, $L^p(\Omega)$ is not a Hilbert space.

[12] A basis of a Hilbert space is not the same thing as a basis in the sense of linear algebra above. For distinction, the latter is then called a Hamel basis.

[13] Although the Fourier series is periodic, the technique can be applied to any L^2 function on an interval by considering the function to be continued periodically outside the interval. See Kreyszig 1988, p. 601

[14] That is to say (BSE-3 2001), the plane passing through the point of contact P such that the distance from a point P_1 on the surface to the plane is infinitesimally small compared to the distance from P_1 to P in the limit as P_1 approaches P along the surface.

[15] That is, there is a homeomorphism from $\pi^{-1}(U)$ to $V \times U$ which restricts to linear isomorphisms between fibers.

[16] A line bundle, such as the tangent bundle of S^1 is trivial if and only if there is a section that vanishes nowhere, see Husemoller 1994, Corollary 8.3. The sections of the tangent bundle are just vector fields.

12.12 Footnotes

[1] Roman 2005, ch. 1, p. 27

[2] van der Waerden 1993, Ch. 19

[3] Bourbaki 1998, §II.1.1. Bourbaki calls the group homomorphisms $f(a)$ *homotheties*.

[4] Bourbaki 1969, ch. "Algèbre linéaire et algèbre multilinéaire", pp. 78–91

[5] Bolzano 1804

[6] Möbius 1827

[7] Hamilton 1853

[8] Grassmann 2000

[9] Peano 1888, ch. IX

[10] Banach 1922

[11] Dorier 1995, Moore 1995

[12] Lang 1987, ch. I.1

[13] Lang 2002, ch. V.1

[14] e.g. Lang 1993, ch. XII.3., p. 335

[15] Lang 1987, ch. IX.1

[16] Lang 1987, ch. VI.3.

[17] Roman 2005, Theorem 1.9, p. 43

[18] Blass 1984

[19] Halpern 1966, pp. 670–673

[20] Artin 1991, Theorem 3.3.13

[21] Braun 1993, Th. 3.4.5, p. 291

[22] Stewart 1975, Proposition 4.3, p. 52

[23] Stewart 1975, Theorem 6.5, p. 74

[24] Roman 2005, ch. 2, p. 45

[25] Lang 1987, ch. IV.4, Corollary, p. 106

[26] Lang 1987, Example IV.2.6

[27] Lang 1987, ch. VI.6

[28] Halmos 1974, p. 28, Ex. 9

[29] Lang 1987, Theorem IV.2.1, p. 95

[30] Roman 2005, Th. 2.5 and 2.6, p. 49

[31] Lang 1987, ch. V.1

[32] Lang 1987, ch. V.3., Corollary, p. 106

[33] Lang 1987, Theorem VII.9.8, p. 198

[34] Roman 2005, ch. 8, p. 135–156

[35] Lang 1987, ch. IX.4

[36] Roman 2005, ch. 1, p. 29

[37] Roman 2005, ch. 1, p. 35

[38] Roman 2005, ch. 3, p. 64

[39] Lang 1987, ch. IV.3.

[40] Roman 2005, ch. 2, p. 48

[41] Mac Lane 1998

[42] Roman 2005, ch. 1, pp. 31–32

[43] Lang 2002, ch. XVI.1

[44] Roman 2005, Th. 14.3. See also Yoneda lemma.

[45] Schaefer & Wolff 1999, pp. 204–205

[46] Bourbaki 2004, ch. 2, p. 48

[47] Roman 2005, ch. 9

[48] Naber 2003, ch. 1.2

[49] Treves 1967

[50] Bourbaki 1987

[51] Kreyszig 1989, §4.11-5

[52] Kreyszig 1989, §1.5-5

[53] Choquet 1966, Proposition III.7.2

[54] Treves 1967, p. 34–36

[55] Lang 1983, Cor. 4.1.2, p. 69

[56] Treves 1967, ch. 11

[57] Treves 1967, Theorem 11.2, p. 102

[58] Evans 1998, ch. 5

[59] Treves 1967, ch. 12

[60] Dennery 1996, p.190

[61] Lang 1993, Th. XIII.6, p. 349

[62] Lang 1993, Th. III.1.1

[63] Choquet 1966, Lemma III.16.11

[64] Kreyszig 1999, Chapter 11

[65] Griffiths 1995, Chapter 1

[66] Lang 1993, ch. XVII.3

[67] Lang 2002, ch. III.1, p. 121

[68] Eisenbud 1995, ch. 1.6

[69] Varadarajan 1974

[70] Lang 2002, ch. XVI.7

[71] Lang 2002, ch. XVI.8

[72] Luenberger 1997, §7.13

[73] See representation theory and group representation.

[74] Lang 1993, Ch. XI.1

[75] Evans 1998, Th. 6.2.1

[76] Folland 1992, p. 349 ff

[77] Gasquet & Witomski 1999, p. 150

[78] Gasquet & Witomski 1999, §4.5

[79] Gasquet & Witomski 1999, p. 57

[80] Loomis 1953, Ch. VII

[81] Ashcroft & Mermin 1976, Ch. 5

[82] Kreyszig 1988, p. 667

[83] Fourier 1822

[84] Gasquet & Witomski 1999, p. 67

[85] Ifeachor & Jervis 2002, pp. 3–4, 11

[86] Wallace Feb 1992

[87] Ifeachor & Jervis 2002, p. 132

[88] Gasquet & Witomski 1999, §10.2

[89] Ifeachor & Jervis 2002, pp. 307–310

[90] Gasquet & Witomski 1999, §10.3

[91] Schönhage & Strassen 1971

[92] Spivak 1999, ch. 3

[93] Jost 2005. See also Lorentzian manifold.

[94] Misner, Thorne & Wheeler 1973, ch. 1.8.7, p. 222 and ch. 2.13.5, p. 325

[95] Jost 2005, ch. 3.1

[96] Varadarajan 1974, ch. 4.3, Theorem 4.3.27

[97] Kreyszig 1991, §34, p. 108

[98] Eisenberg & Guy 1979

[99] Atiyah 1989

[100] Artin 1991, ch. 12

[101] Grillet, Pierre Antoine. Abstract algebra. Vol. 242. Springer Science & Business Media, 2007.

[102] Meyer 2000, Example 5.13.5, p. 436

[103] Meyer 2000, Exercise 5.13.15–17, p. 442

[104] Coxeter 1987

12.13 References

12.13.1 Algebra

- Artin, Michael (1991), *Algebra*, Prentice Hall, ISBN 978-0-89871-510-1

- Blass, Andreas (1984), "Existence of bases implies the axiom of choice", *Axiomatic set theory (Boulder, Colorado, 1983)*, Contemporary Mathematics **31**, Providence, R.I.: American Mathematical Society, pp. 31–33, MR 763890

- Brown, William A. (1991), *Matrices and vector spaces*, New York: M. Dekker, ISBN 978-0-8247-8419-5

- Lang, Serge (1987), *Linear algebra*, Berlin, New York: Springer-Verlag, ISBN 978-0-387-96412-6

- Lang, Serge (2002), *Algebra*, Graduate Texts in Mathematics **211** (Revised third ed.), New York: Springer-Verlag, ISBN 978-0-387-95385-4, MR 1878556

- Mac Lane, Saunders (1999), *Algebra* (3rd ed.), pp. 193–222, ISBN 0-8218-1646-2

- Meyer, Carl D. (2000), *Matrix Analysis and Applied Linear Algebra*, SIAM, ISBN 978-0-89871-454-8

- Roman, Steven (2005), *Advanced Linear Algebra*, Graduate Texts in Mathematics **135** (2nd ed.), Berlin, New York: Springer-Verlag, ISBN 978-0-387-24766-3

- Spindler, Karlheinz (1993), *Abstract Algebra with Applications: Volume 1: Vector spaces and groups*, CRC, ISBN 978-0-8247-9144-5

- van der Waerden, Bartel Leendert (1993), *Algebra* (in German) (9th ed.), Berlin, New York: Springer-Verlag, ISBN 978-3-540-56799-8

12.13.2 Analysis

- Bourbaki, Nicolas (1987), *Topological vector spaces*, Elements of mathematics, Berlin, New York: Springer-Verlag, ISBN 978-3-540-13627-9

- Bourbaki, Nicolas (2004), *Integration I*, Berlin, New York: Springer-Verlag, ISBN 978-3-540-41129-1

- Braun, Martin (1993), *Differential equations and their applications: an introduction to applied mathematics*, Berlin, New York: Springer-Verlag, ISBN 978-0-387-97894-9

- BSE-3 (2001), "Tangent plane", in Hazewinkel, Michiel, *Encyclopedia of Mathematics*, Springer, ISBN 978-1-55608-010-4

- Choquet, Gustave (1966), *Topology*, Boston, MA: Academic Press

- Dennery, Philippe; Krzywicki, Andre (1996), *Mathematics for Physicists*, Courier Dover Publications, ISBN 978-0-486-69193-0

- Dudley, Richard M. (1989), *Real analysis and probability*, The Wadsworth & Brooks/Cole Mathematics Series, Pacific Grove, CA: Wadsworth & Brooks/Cole Advanced Books & Software, ISBN 978-0-534-10050-6

- Dunham, William (2005), *The Calculus Gallery*, Princeton University Press, ISBN 978-0-691-09565-3

- Evans, Lawrence C. (1998), *Partial differential equations*, Providence, R.I.: American Mathematical Society, ISBN 978-0-8218-0772-9

- Folland, Gerald B. (1992), *Fourier Analysis and Its Applications*, Brooks-Cole, ISBN 978-0-534-17094-3

- Gasquet, Claude; Witomski, Patrick (1999), *Fourier Analysis and Applications: Filtering, Numerical Computation, Wavelets*, Texts in Applied Mathematics, New York: Springer-Verlag, ISBN 0-387-98485-2

- Ifeachor, Emmanuel C.; Jervis, Barrie W. (2001), *Digital Signal Processing: A Practical Approach* (2nd ed.), Harlow, Essex, England: Prentice-Hall (published 2002), ISBN 0-201-59619-9

- Krantz, Steven G. (1999), *A Panorama of Harmonic Analysis*, Carus Mathematical Monographs, Washington, DC: Mathematical Association of America, ISBN 0-88385-031-1

- Kreyszig, Erwin (1988), *Advanced Engineering Mathematics* (6th ed.), New York: John Wiley & Sons, ISBN 0-471-85824-2

- Kreyszig, Erwin (1989), *Introductory functional analysis with applications*, Wiley Classics Library, New York: John Wiley & Sons, ISBN 978-0-471-50459-7, MR 992618

- Lang, Serge (1983), *Real analysis*, Addison-Wesley, ISBN 978-0-201-14179-5

- Lang, Serge (1993), *Real and functional analysis*, Berlin, New York: Springer-Verlag, ISBN 978-0-387-94001-4

- Loomis, Lynn H. (1953), *An introduction to abstract harmonic analysis*, Toronto-New York–London: D. Van Nostrand Company, Inc., pp. x+190

- Schaefer, Helmut H.; Wolff, M.P. (1999), *Topological vector spaces* (2nd ed.), Berlin, New York: Springer-Verlag, ISBN 978-0-387-98726-2

- Treves, François (1967), *Topological vector spaces, distributions and kernels*, Boston, MA: Academic Press

12.13.3   Historical references

- Banach, Stefan (1922), "Sur les opérations dans les ensembles abstraits et leur application aux équations intégrales (On operations in abstract sets and their application to integral equations)" (PDF), *Fundamenta Mathematicae* (in French) **3**, ISSN 0016-2736

- Bolzano, Bernard (1804), *Betrachtungen über einige Gegenstände der Elementargeometrie (Considerations of some aspects of elementary geometry)* (in German)

- Bourbaki, Nicolas (1969), *Éléments d'histoire des mathématiques (Elements of history of mathematics)* (in French), Paris: Hermann

- Dorier, Jean-Luc (1995), "A general outline of the genesis of vector space theory", *Historia Mathematica* **22** (3): 227–261, doi:10.1006/hmat.1995.1024, MR 1347828

- Fourier, Jean Baptiste Joseph (1822), *Théorie analytique de la chaleur* (in French), Chez Firmin Didot, père et fils

- Grassmann, Hermann (1844), *Die Lineale Ausdehnungslehre - Ein neuer Zweig der Mathematik* (in German), O. Wigand, reprint: Hermann Grassmann. Translated by Lloyd C. Kannenberg. (2000), Kannenberg, L.C., ed., *Extension Theory*, Providence, R.I.: American Mathematical Society, ISBN 978-0-8218-2031-5

- Hamilton, William Rowan (1853), *Lectures on Quaternions*, Royal Irish Academy

- Möbius, August Ferdinand (1827), *Der Barycentrische Calcul : ein neues Hülfsmittel zur analytischen Behandlung der Geometrie (Barycentric calculus: a new utility for an analytic treatment of geometry)* (in German)

- Moore, Gregory H. (1995), "The axiomatization of linear algebra: 1875–1940", *Historia Mathematica* **22** (3): 262–303, doi:10.1006/hmat.1995.1025

- Peano, Giuseppe (1888), *Calcolo Geometrico secondo l'Ausdehnungslehre di H. Grassmann preceduto dalle Operazioni della Logica Deduttiva* (in Italian), Turin

12.13.4   Further references

- Ashcroft, Neil; Mermin, N. David (1976), *Solid State Physics*, Toronto: Thomson Learning, ISBN 978-0-03-083993-1

- Atiyah, Michael Francis (1989), *K-theory*, Advanced Book Classics (2nd ed.), Addison-Wesley, ISBN 978-0-201-09394-0, MR 1043170

- Bourbaki, Nicolas (1998), *Elements of Mathematics : Algebra I Chapters 1-3*, Berlin, New York: Springer-Verlag, ISBN 978-3-540-64243-5

- Bourbaki, Nicolas (1989), *General Topology. Chapters 1-4*, Berlin, New York: Springer-Verlag, ISBN 978-3-540-64241-1

- Coxeter, Harold Scott MacDonald (1987), *Projective Geometry* (2nd ed.), Berlin, New York: Springer-Verlag, ISBN 978-0-387-96532-1

- Eisenberg, Murray; Guy, Robert (1979), "A proof of the hairy ball theorem", *The American Mathematical Monthly* (Mathematical Association of America) **86** (7): 572–574, doi:10.2307/2320587, JSTOR 2320587

- Eisenbud, David (1995), *Commutative algebra*, Graduate Texts in Mathematics **150**, Berlin, New York: Springer-Verlag, ISBN 978-0-387-94269-8, MR 1322960

- Goldrei, Derek (1996), *Classic Set Theory: A guided independent study* (1st ed.), London: Chapman and Hall, ISBN 0-412-60610-0

- Griffiths, David J. (1995), *Introduction to Quantum Mechanics*, Upper Saddle River, NJ: Prentice Hall, ISBN 0-13-124405-1

- Halmos, Paul R. (1974), *Finite-dimensional vector spaces*, Berlin, New York: Springer-Verlag, ISBN 978-0-387-90093-3

- Halpern, James D. (Jun 1966), "Bases in Vector Spaces and the Axiom of Choice", *Proceedings of the American Mathematical Society* (American Mathematical Society) **17** (3): 670–673, doi:10.2307/2035388, JSTOR 2035388

- Husemoller, Dale (1994), *Fibre Bundles* (3rd ed.), Berlin, New York: Springer-Verlag, ISBN 978-0-387-94087-8

- Jost, Jürgen (2005), *Riemannian Geometry and Geometric Analysis* (4th ed.), Berlin, New York: Springer-Verlag, ISBN 978-3-540-25907-7

- Kreyszig, Erwin (1991), *Differential geometry*, New York: Dover Publications, pp. xiv+352, ISBN 978-0-486-66721-8

- Kreyszig, Erwin (1999), *Advanced Engineering Mathematics* (8th ed.), New York: John Wiley & Sons, ISBN 0-471-15496-2

- Luenberger, David (1997), *Optimization by vector space methods*, New York: John Wiley & Sons, ISBN 978-0-471-18117-0

- Mac Lane, Saunders (1998), *Categories for the Working Mathematician* (2nd ed.), Berlin, New York: Springer-Verlag, ISBN 978-0-387-98403-2

- Misner, Charles W.; Thorne, Kip; Wheeler, John Archibald (1973), *Gravitation*, W. H. Freeman, ISBN 978-0-7167-0344-0

- Naber, Gregory L. (2003), *The geometry of Minkowski spacetime*, New York: Dover Publications, ISBN 978-0-486-43235-9, MR 2044239

- Schönhage, A.; Strassen, Volker (1971), "Schnelle Multiplikation großer Zahlen (Fast multiplication of big numbers)" (PDF), *Computing* (in German) **7**: 281–292, doi:10.1007/bf02242355, ISSN 0010-485X

- Spivak, Michael (1999), *A Comprehensive Introduction to Differential Geometry (Volume Two)*, Houston, TX: Publish or Perish

- Stewart, Ian (1975), *Galois Theory*, Chapman and Hall Mathematics Series, London: Chapman and Hall, ISBN 0-412-10800-3

- Varadarajan, V. S. (1974), *Lie groups, Lie algebras, and their representations*, Prentice Hall, ISBN 978-0-13-535732-3

- Wallace, G.K. (Feb 1992), "The JPEG still picture compression standard", *IEEE Transactions on Consumer Electronics* **38** (1): xviii–xxxiv, doi:10.1109/30.125072, ISSN 0098-3063

- Weibel, Charles A. (1994), *An introduction to homological algebra*, Cambridge Studies in Advanced Mathematics **38**, Cambridge University Press, ISBN 978-0-521-55987-4, OCLC 36131259, MR 1269324

12.14 External links

- Hazewinkel, Michiel, ed. (2001), "Vector space", *Encyclopedia of Mathematics*, Springer, ISBN 978-1-55608-010-4

- A lecture about fundamental concepts related to vector spaces (given at MIT)

- A graphical simulator for the concepts of span, linear dependency, base and dimension

Chapter 13

Lebesgue integration

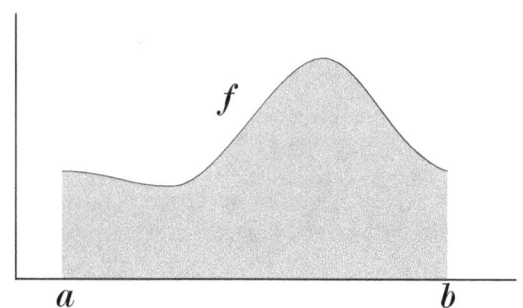

The integral of a positive function can be interpreted as the area under a curve.

In mathematics, the integral of a non-negative function can be regarded, in the simplest case, as the area between the graph of that function and the x-axis. The **Lebesgue integral** extends the integral to a larger class of functions. It also extends the domains on which these functions can be defined.

Mathematicians had long understood that for non-negative functions with a smooth enough graph—such as continuous functions on closed bounded intervals—the *area under the curve* could be defined as the integral, and computed using approximation techniques on the region by polygons. However, as the need to consider more irregular functions arose—e.g., as a result of the limiting processes of mathematical analysis and the mathematical theory of probability—it became clear that more careful approximation techniques were needed to define a suitable integral. Also, we might wish to integrate on spaces more general than the real line. The Lebesgue integral provides the right abstractions needed to do this important job.

The Lebesgue integral plays an important role in the branch of mathematics called real analysis, and in many other mathematical sciences fields. It is named after Henri Lebesgue (1875–1941), who introduced the integral (Lebesgue 1904). It is also a pivotal part of the axiomatic theory of probability.

The term *Lebesgue integration* can mean either the general theory of integration of a function with respect to a general measure, as introduced by Lebesgue—or the specific case of integration of a function defined on a sub-domain of the real line with respect to Lebesgue measure.

13.1 Introduction

The integral of a function f between limits a and b can be interpreted as the area under the graph of f. This is easy to understand for familiar functions such as polynomials, but what does it mean for more exotic functions? In general, for which class of functions does "area under the curve" make sense? The answer to this question has great theoretical and practical importance.

As part of a general movement toward rigour in mathematics in the nineteenth century, mathematicians attempted to put integral calculus on a firm foundation. The Riemann integral—proposed by Bernhard Riemann (1826–1866)—is a broadly successful attempt to provide such a foundation. Riemann's definition starts with the construction of a sequence of easily calculated areas that converge to the integral of a given function. This definition is successful in the sense that it gives the expected answer for many already-solved problems, and gives useful results for many other problems.

However, Riemann integration does not interact well with taking limits of sequences of functions, making such limiting processes difficult to analyze. This is important, for instance, in the study of Fourier series, Fourier transforms and other topics. The Lebesgue integral is better able to describe how and when it is possible to take limits under the integral sign (via the powerful monotone convergence theorem and dominated convergence theorem).

The Lebesgue definition considers a different class of easily calculated areas than the Riemann definition—which is the main reason the Lebesgue integral behaves better. The Lebesgue definition also makes it possible to calculate integrals for a broader class of functions. For example, the Dirichlet function, which is 0 where its argument is irrational and 1 otherwise, has a Lebesgue integral, but does not have a Riemann integral.

Lebesgue summarized his approach to integration in a letter to Paul Montel:

> I have to pay a certain sum, which I have collected in my pocket. I take the bills and coins out of my pocket and give them to the creditor in the order I find them until I have reached the total sum. This is the Riemann integral. But I can proceed differently. After I have taken all the money out of my pocket I order the bills and coins according to identical values and then I pay the several heaps one after the other to the creditor. This is my integral.
>
> — *Source*: (Siegmund-Schultze 2008)

The insight is that one should be able to rearrange the values of a function freely, while preserving the value of the integral. This process of rearrangement can convert a very pathological function into one that is "nice" from the point of view of integration—and thus let such pathological functions be integrated.

13.1.1 Intuitive interpretation

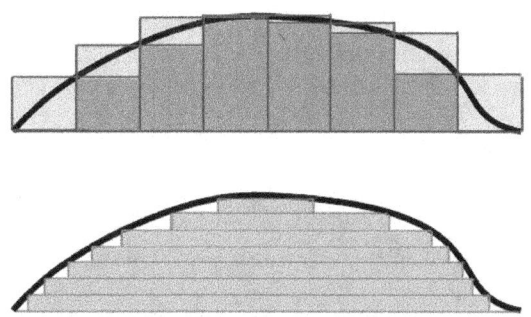

Riemann-Darboux's integration (in blue) and Lebesgue integration (in red).

To get some intuition about the different approaches to integration, let's imagine we want to find a mountain's volume (above sea level).

The Riemann-Darboux approach Divide the base of the mountain into a grid of 1 meter squares. Measure the altitude of the mountain at the center of each square. The volume on a single grid square is approximately 1

m² × (that square's altitude), so the total volume is 1 m² times the sum of the altitudes.

The Lebesgue approach Draw a contour map of the mountain, where adjacent contours are 1 meter of altitude apart. The volume of earth a single contour contains is approximately 1 m × (that contour's area), so the total volume is the sum of these areas times 1 m.

Folland summarizes the difference between the Riemann and Lebesgue approaches thus: "to compute the Riemann integral of f, one partitions the domain $[a, b]$ into subintervals", while in the Lebesgue integral, "one is in effect partitioning the range of f ."[1]

13.1.2 Towards a formal definition

To define the Lebesgue integral requires the formal notion of a measure that, roughly, associates to each set A of real numbers a nonnegative number $\mu(A)$ representing the "size" of A. This notion of "size" should agree with the usual length of an interval or disjoint union of intervals. Suppose that $f : \mathbb{R} \to \mathbb{R}^+$ is a non-negative real-valued function. Using the "partitioning the range of f " philosophy, the integral of f should be the sum over t of the elementary area contained in the thin horizontal strip between $y = t$ and $y = t + dt$. This elementary area is just

$$\mu\left(\{x \mid f(x) > t\}\right) dt.$$

Let

$$f^*(t) = \mu\left(\{x \mid f(x) > t\}\right).$$

The Lebesgue integral of f is then defined by[2]

$$\int f \, d\mu = \int_0^\infty f^*(t) \, dt$$

where the integral on the right is an ordinary improper Riemann integral (note that f^* is a non-negative decreasing function, and therefore has a well-defined improper Riemann integral). For a suitable class of functions (the measurable functions), this defines the Lebesgue integral.

A general (not necessarily positive) function f is Lebesgue integrable if the area between the graph of f and the x-axis is finite:

$$\int |f| \, d\mu < +\infty.$$

In that case, the integral is, as in the Riemannian case, the difference between the area above the x-axis and the area below the x-axis:

$$\int f\, d\mu = \int f^+\, d\mu - \int f^-\, d\mu$$

where $f = f^+ - f^-$ is the unique decomposition of f into the difference of two non-negative functions, given explicitly by

$$f^+(x) = \max\{f(x), 0\} \quad = \begin{cases} f(x), & \text{if } f(x) > 0, \\ 0, & \text{otherwise} \end{cases}$$

$$f^-(x) = \max\{-f(x), 0\} \quad = \begin{cases} -f(x), & \text{if } f(x) < 0, \\ 0, & \text{otherwise.} \end{cases}$$

13.2   Construction

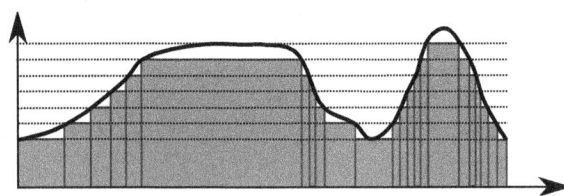

Approximating a function by simple functions.

The discussion that follows parallels the most common expository approach to the Lebesgue integral. In this approach, the theory of integration has two distinct parts:

1. A theory of measurable sets and measures on these sets

2. A theory of measurable functions and integrals on these functions

The function whose integral is to be found is then approximated by certain so-called simple functions, whose integrals can be written in terms of the measure. The integral of the original function is then the limit of the integral of the simple functions.

13.2.1   Measure theory

Further information: Measure (mathematics)

Measure theory was initially created to provide a useful abstraction of the notion of length of subsets of the real line—and, more generally, area and volume of subsets of Euclidean spaces. In particular, it provided a systematic answer to the question of which subsets of $\mathbb{R}$ have a length. As later set theory developments showed (see non-measurable set), it is actually impossible to assign a length to all subsets of $\mathbb{R}$ in a way that preserves some natural additivity and translation invariance properties. This suggests that picking out a suitable class of *measurable* subsets is an essential prerequisite.

The Riemann integral uses the notion of length explicitly. Indeed, the element of calculation for the Riemann integral is the rectangle $[a, b] \times [c, d]$, whose area is calculated to be $(b - a)(d - c)$. The quantity $b - a$ is the length of the base of the rectangle and $d - c$ is the height of the rectangle. Riemann could only use planar rectangles to approximate the area under the curve, because there was no adequate theory for measuring more general sets.

In the development of the theory in most modern textbooks (after 1950), the approach to measure and integration is *axiomatic*. This means that a measure is any function μ defined on a certain class X of subsets of a set E, which satisfies a certain list of properties. These properties can be shown to hold in many different cases.

13.2.2   Integration

We start with a measure space (E, X, μ) where E is a set, X is a σ-algebra of subsets of E, and μ is a (non-negative) measure on E defined on the sets of X.

For example, E can be Euclidean n-space $\mathbb{R}^n$ or some Lebesgue measurable subset of it, X is the σ-algebra of all Lebesgue measurable subsets of E, and μ is the Lebesgue measure. In the mathematical theory of probability, we confine our study to a probability measure μ, which satisfies $\mu(E) = 1$.

Lebesgue's theory defines integrals for a class of functions called measurable functions. A real-valued function f on E is measurable if the pre-image of every interval of the form (t, ∞) is in X:

$$\{x \mid f(x) > t\} \in X \quad \text{all for } t \in \mathbb{R}.$$

We can show that this is equivalent to requiring that the pre-image of any Borel subset of $\mathbb{R}$ be in X. The set of measurable functions is closed under algebraic operations, but more importantly it is closed under various kinds of pointwise sequential limits:

$$\sup_{k \in \mathbb{N}} f_k, \quad \liminf_{k \in \mathbb{N}} f_k, \quad \limsup_{k \in \mathbb{N}} f_k$$

are measurable if the original sequence $(fk)k$, where $k \in \mathbb{N}$, consists of measurable functions.

We build up an integral

$$\int_E f \, d\mu = \int_E f(x) \, \mu(dx)$$

for measurable real-valued functions f defined on E in stages:

Indicator functions: To assign a value to the integral of the indicator function $1S$ of a measurable set S consistent with the given measure μ, the only reasonable choice is to set:

$$\int 1_S \, d\mu = \mu(S).$$

Notice that the result may be equal to $+\infty$, unless μ is a *finite* measure.

Simple functions: A finite linear combination of indicator functions

$$\sum_k a_k 1_{S_k}$$

where the coefficients ak are real numbers and the sets Sk are measurable, is called a measurable simple function. We extend the integral by linearity to *non-negative* measurable simple functions. When the coefficients ak are non-negative, we set

$$\int \left(\sum_k a_k 1_{S_k} \right) d\mu = \sum_k a_k \int 1_{S_k} \, d\mu = \sum_k a_k \, \mu(S_k).$$

The convention $0 \times \infty = 0$ must be used, and the result may be infinite. Even if a simple function can be written in many ways as a linear combination of indicator functions, the integral is always the same. This can be shown using the additivity property of measures.

Some care is needed when defining the integral of a *real-valued* simple function, to avoid the undefined expression $\infty - \infty$: one assumes that the representation

$$f = \sum_k a_k 1_{S_k}$$

is such that $\mu(Sk) < \infty$ whenever $ak \neq 0$. Then the above formula for the integral of f makes sense, and the result does not depend upon the particular representation of f satisfying the assumptions.

If B is a measurable subset of E and s is a measurable simple function one defines

$$\int_B s \, d\mu = \int 1_B \, s \, d\mu = \sum_k a_k \, \mu(S_k \cap B).$$

Non-negative functions: Let f be a non-negative measurable function on E, which we allow to attain the value $+\infty$, in other words, f takes non-negative values in the extended real number line. We define

$$\int_E f \, d\mu = \sup \left\{ \int_E s \, d\mu : 0 \le s \le f, \, s \text{ simple} \right\}.$$

We need to show this integral coincides with the preceding one, defined on the set of simple functions, when E is a segment $[a, b]$. There is also the question of whether this corresponds in any way to a Riemann notion of integration. It is possible to prove that the answer to both questions is yes.

We have defined the integral of f for any non-negative extended real-valued measurable function on E. For some functions, this integral $\int E f \, d\mu$ is infinite.

Signed functions: To handle signed functions, we need a few more definitions. If f is a measurable function of the set E to the reals (including $\pm\infty$), then we can write

$$f = f^+ - f^-,$$

where

$$f^+(x) = \begin{cases} f(x) & \text{if } f(x) > 0 \\ 0 & \text{otherwise} \end{cases}$$

$$f^-(x) = \begin{cases} -f(x) & \text{if } f(x) < 0 \\ 0 & \text{otherwise} \end{cases}$$

Note that both f^+ and f^- are non-negative measurable functions. Also note that

$$|f| = f^+ + f^-.$$

We say that the Lebesgue integral of the measurable function f *exists*, or *is defined* if at least one of $\int f^+ \, d\mu$ and $\int f^- \, d\mu$ is finite:

$$\min \left(\int f^+ \, d\mu, \int f^- \, d\mu \right) < \infty.$$

In this case we *define*

$$\int f \, d\mu = \int f^+ \, d\mu - \int f^- \, d\mu.$$

If

$$\int |f| \, d\mu < \infty,$$

we say that f is *Lebesgue integrable*.

It turns out that this definition gives the desirable properties of the integral.

Complex valued functions can be similarly integrated, by considering the real part and the imaginary part separately.

If $h = f + ig$ for real-valued integrable functions f, g, then the integral of h is defined by

$$\int h \, d\mu = \int f \, d\mu + i \int g \, d\mu.$$

13.2.3   Example

Consider the indicator function of the rational numbers, $1_{\mathbf{Q}}$. This function is nowhere continuous.

- $1_{\mathbf{Q}}$ **is not Riemann-integrable on** $[0, 1]$: No matter how the set $[0, 1]$ is partitioned into subintervals, each partition contains at least one rational and at least one irrational number, because rationals and irrationals are both dense in the reals. Thus the upper Darboux sums are all one, and the lower Darboux sums are all zero.

- $1_{\mathbf{Q}}$ **is Lebesgue-integrable on** $[0, 1]$ using the Lebesgue measure: Indeed, it is the indicator function of the rationals so by definition

$$\int_{[0,1]} 1_{\mathbf{Q}} \, d\mu = \mu(\mathbf{Q} \cap [0, 1]) = 0,$$

because $\mathbf{Q}$ is countable.

13.2.4   Domain of integration

A technical issue in Lebesgue integration is that the domain of integration is defined as a *set* (a subset of a measure space), with no notion of orientation. In elementary calculus, one defines integration with respect to an orientation:

$$\int_b^a f := - \int_a^b f.$$

Generalizing this to higher dimensions yields integration of differential forms. By contrast, Lebesgue integration provides an alternative generalization, integrating over subsets with respect to a measure; this can be notated as

$$\int_A f \, d\mu = \int_{[a,b]} f \, d\mu$$

to indicate integration over a subset A. For details on the relation between these generalizations, see Differential form: Relation with measures.

13.3   Limitations of the Riemann integral

Here we discuss the limitations of the Riemann integral and the greater scope offered by the Lebesgue integral. This discussion presumes a working understanding of the Riemann integral.

With the advent of Fourier series, many analytical problems involving integrals came up whose satisfactory solution required interchanging limit processes and integral signs. However, the conditions under which the integrals

$$\sum_k \int f_k(x)dx, \quad \int \left[\sum_k f_k(x) \right] dx$$

are equal proved quite elusive in the Riemann framework. There are some other technical difficulties with the Riemann integral. These are linked with the limit-taking difficulty discussed above.

Failure of monotone convergence. As shown above, the indicator function $1_{\mathbf{Q}}$ on the rationals is not Riemann integrable. In particular, the Monotone convergence theorem fails. To see why, let $\{ak\}$ be an enumeration of all the rational numbers in $[0, 1]$ (they are countable so this can be done.) Then let

$$g_k(x) = \begin{cases} 1 & \text{if } x = a_j, j \le k \\ 0 & \text{otherwise} \end{cases}$$

The function gk is zero everywhere, except on a finite set of points. Hence its Riemann integral is zero. Each gk is non-negative, and this sequence of functions is monotonically increasing, but its limit as $k \to \infty$ is $1_{\mathbf{Q}}$, which is not Riemann integrable.

Unsuitability for unbounded intervals. The Riemann integral can only integrate functions on a bounded interval. It can however be extended to unbounded intervals by taking limits, so long as this doesn't yield an answer such as $\infty - \infty$.

Integrating on structures other than Euclidean space. The Riemann integral is inextricably linked to the order structure of the real line.

13.4 Basic theorems of the Lebesgue integral

The Lebesgue integral does not distinguish between functions that differ only on a set of μ-measure zero. To make this precise, functions f and g are said to be equal almost everywhere (a.e.) if

$$\mu(\{x \in E : f(x) \neq g(x)\}) = 0.$$

- If f, g are non-negative measurable functions (possibly assuming the value $+\infty$) such that $f = g$ almost everywhere, then

$$\int f \, d\mu = \int g \, d\mu.$$

To wit, the integral respects the equivalence relation of almost-everywhere equality.

- If f, g are functions such that $f = g$ almost everywhere, then f is Lebesgue integrable if and only if g is Lebesgue integrable, and the integrals of f and g are the same if they exist.

The Lebesgue integral has the following properties:

Linearity: If f and g are Lebesgue integrable functions and a and b are real numbers, then $af + bg$ is Lebesgue integrable and

$$\int (af + bg) \, d\mu = a \int f \, d\mu + b \int g \, d\mu.$$

Monotonicity: If $f \leq g$, then

$$\int f \, d\mu \leq \int g \, d\mu.$$

Monotone convergence theorem: Suppose $\{fk\}k \in \mathbb{N}$ is a sequence of non-negative measurable functions such that

$$f_k(x) \leq f_{k+1}(x) \quad \forall k \in \mathbb{N}, \forall x \in E.$$

Then, the pointwise limit f of fk is Lebesgue integrable and

$$\lim_k \int f_k \, d\mu = \int f \, d\mu.$$

The value of any of the integrals is allowed to be infinite.

Fatou's lemma: If $\{fk\}k \in \mathbf{N}$ is a sequence of non-negative measurable functions, then

$$\int \liminf_k f_k \, d\mu \leq \liminf_k \int f_k \, d\mu.$$

Again, the value of any of the integrals may be infinite.

Dominated convergence theorem: Suppose $\{fk\}k \in \mathbf{N}$ is a sequence of complex measurable functions with pointwise limit f, and there is a Lebesgue integrable function g (i.e., g belongs to the space L^1) such that $|fk| \leq g$ for all k.

Then, f is Lebesgue integrable and

$$\lim_k \int f_k \, d\mu = \int f \, d\mu.$$

13.5 Proof techniques

To illustrate some of the proof techniques used in Lebesgue integration theory, we sketch a proof of the above-mentioned Lebesgue monotone convergence theorem. Let $\{fk\}k \in \mathbf{N}$ be a non-decreasing sequence of non-negative measurable functions and put

$$f = \sup_{k \in \mathbb{N}} f_k = \lim_{k \in \mathbb{N}} f_k.$$

By the monotonicity property of the integral, it is immediate that:

$$\int f \, d\mu \geq \lim_k \int f_k \, d\mu$$

and the limit on the right exists, because the sequence is monotonic. We now prove the inequality in the other direction. It follows from the definition of integral that there is a non-decreasing sequence (gn) of non-negative simple functions such that $gn \leq f$ and

$$\lim_n \int g_n \, d\mu = \int f \, d\mu.$$

Therefore, it suffices to prove that for each $n \in \mathbb{N}$,

$$\int g_n \, d\mu \leq \lim_k \int f_k \, d\mu.$$

We will show that if g is a simple function and

$$\lim_k f_k(x) \geq g(x)$$

almost everywhere, then

$$\lim_k \int f_k \, d\mu \geq \int g \, d\mu.$$

By breaking up the function g into its constant value parts, this reduces to the case in which g is the indicator function of a set. The result we have to prove is then

> Suppose A is a measurable set and $\{ fk \}k \in \mathbb{N}$ is a nondecreasing sequence of non-negative measurable functions on E such that
>
> $$\lim_k f_k(x) \geq 1$$
>
> for almost all $x \in A$. Then
>
> $$\lim_k \int f_k \, d\mu \geq \mu(A).$$

To prove this result, fix $\varepsilon > 0$ and define the sequence of measurable sets

$$B_k = \{x \in A : f_k(x) \geq 1 - \varepsilon\}.$$

By monotonicity of the integral, it follows that for any $k \in \mathbb{N}$,

$$(1 - \varepsilon)\mu(B_k) = \int (1 - \varepsilon) 1_{B_k} \, d\mu \leq \int f_k \, d\mu$$

Because almost every x is in Bk for large enough k, we have

$$\bigcup_k B_k = A,$$

up to a set of measure 0. Thus by countable additivity of μ, and because Bk increases with k,

$$\mu(A) = \lim_k \mu(B_k) \leq \lim_k (1 - \varepsilon)^{-1} \int f_k \, d\mu.$$

As this is true for any positive ε the result follows.

For another Proof of the Monotone Convergence Theorem, we follow:[1]

Let (X, M, μ) be a measure space.

$\{ f_n \}$ is an increasing sequence of numbers, therefore its limit exists, even if it's equal to ∞. We know that

$$\int f_n \leq \int f$$

for all n, so that

$$\lim_{n \to \infty} \int f_n \leq \int f$$

Now we need to establish the reverse inequality. Fix $\alpha \in (0, 1)$, let φ be a simple function with $0 \leq \varphi \leq f$ and let

$$E_n = \{x : f_n(x) \geq \alpha\phi(x)\}$$

Then $\{E_n\}$ is an increasing sequence of measurable sets with $\bigcup\limits^{\infty} E_n = X$. We know that

$$\int f_n \geq \int_{E_n} f_n \geq \alpha \int_{E_n} \phi$$

This is true for all n, including the limit:

$$\lim \int_{E_n} \phi = \int \phi$$

Hence,

$$\lim \int f_n \geq \alpha \int \phi$$

This was true for all $\alpha \in (0, 1)$, so it remains true for $\alpha = 1$, and taking the supremum over simple $\varphi \leq f$ by the definition of integration in L^+,

$$\lim \int f_n \geq \int f$$

Now we have both inequalities, so we've shown the Monotone Convergence theorem:

$$\lim \int f_n = \int f$$

for $f_{\{n+1\}} \geq f_n$, and $f_n \to f$ pointwise, $\{f_n\} \in L^+$, the set of positive measurable functions from $X \to [0, \infty]$.

13.6 Alternative formulations

It is possible to develop the integral with respect to the Lebesgue measure without relying on the full machinery of measure theory. One such approach is provided by the Daniell integral.

There is also an alternative approach to developing the theory of integration via methods of functional analysis. The Riemann integral exists for any continuous function f of compact support defined on $\mathbb{R}^n$ (or a fixed open subset). Integrals of more general functions can be built starting from these integrals.

Let Cc be the space of all real-valued compactly supported continuous functions of $\mathbb{R}$. Define a norm on Cc by

$$\|f\| = \int |f(x)| \, dx.$$

Then Cc is a normed vector space (and in particular, it is a metric space.) All metric spaces have Hausdorff completions, so let L^1 be its completion. This space is isomorphic to the space of Lebesgue integrable functions modulo the subspace of functions with integral zero. Furthermore, the Riemann integral $\int$ is a uniformly continuous functional with respect to the norm on Cc, which is dense in L^1. Hence $\int$ has a unique extension to all of L^1. This integral is precisely the Lebesgue integral.

More generally, when the measure space on which the functions are defined is also a locally compact topological space (as is the case with the real numbers $\mathbb{R}$), measures compatible with the topology in a suitable sense (Radon measures, of which the Lebesgue measure is an example) an integral with respect to them can be defined in the same manner, starting from the integrals of continuous functions with compact support. More precisely, the compactly supported functions form a vector space that carries a natural topology, and a (Radon) measure is defined as a continuous linear functional on this space. The value of a measure at a compactly supported function is then also by definition the integral of the function. One then proceeds to expand the measure (the integral) to more general functions by continuity, and defines the measure of a set as the integral of its indicator function. This is the approach taken by Bourbaki (2004) and a certain number of other authors. For details see Radon measures.

13.7 Limitations of Lebesgue integral

The main purpose of Lebesgue integral is to provide an integral notion where limits of integrals hold under mild assumptions. There is no guarantee that every function is Lebesgue integrable. But it may happen that improper integrals exist for functions that are not Lebesgue integrable. One example would be

$$\frac{\sin(x)}{x}$$

over the entire real line. This function is not Lebesgue integrable, as

$$\int_{-\infty}^{\infty} \left| \frac{\sin(x)}{x} \right| \, d\mu = \infty.$$

On the other hand, $\int_{-\infty}^{\infty} \frac{\sin(x)}{x} \, d\mu$ exists as an improper integral and can be computed to be finite; it is twice the Dirichlet integral.

13.8 See also

- Henri Lebesgue, for a non-technical description of Lebesgue integration
- Null set
- Integration
- Measure
- Sigma-algebra
- Lebesgue space
- Lebesgue–Stieltjes integration
- Henstock–Kurzweil integral

13.9 Notes

[1] Folland, Gerald B. (1984). *Real Analysis: Modern Techniques and Their Applications*. Wiley. p. 56.

[2] Lieb & Loss 2001

13.10 References

- Bartle, Robert G. (1995). *The elements of integration and Lebesgue measure*. Wiley Classics Library. New York: John Wiley & Sons Inc. xii+179. ISBN 0-471-04222-6. MR 1312157.

- Bauer, Heinz (2001). *Measure and Integration Theory*. De Gruyter Studies in Mathematics 26. Berlin: De Gruyter. 236. ISBN 978-3-11-016719-1.

- Bourbaki, Nicolas (2004). *Integration. I. Chapters 1–6. Translated from the 1959, 1965 and 1967 French originals by Sterling K. Berberian*. Elements of Mathematics (Berlin). Berlin: Springer-Verlag. xvi+472. ISBN 3-540-41129-1. MR 2018901.

- Dudley, Richard M. (1989). *Real analysis and probability*. The Wadsworth & Brooks/Cole Mathematics Series. Pacific Grove, CA: Wadsworth & Brooks/Cole Advanced Books & Software. xii+436. ISBN 0-534-10050-3. MR 982264. Very thorough treatment, particularly for probabilists with good notes and historical references.

- Folland, Gerald B. (1999). *Real analysis: Modern techniques and their applications*. Pure and Applied Mathematics (New York) (Second ed.). New York: John Wiley & Sons Inc. xvi+386. ISBN 0-471-31716-0. MR 1681462.

- Halmos, Paul R. (1950). *Measure Theory*. New York, N. Y.: D. Van Nostrand Company, Inc. pp. xi+304. MR 0033869. A classic, though somewhat dated presentation.

- Hazewinkel, Michiel, ed. (2001), "Lebesgue integral", *Encyclopedia of Mathematics*, Springer, ISBN 978-1-55608-010-4

- Lebesgue, Henri (1904). "Leçons sur l'intégration et la recherche des fonctions primitives". Paris: Gauthier-Villars.

- Lebesgue, Henri (1972). *Oeuvres scientifiques (en cinq volumes)* (in French). Geneva: Institut de Mathématiques de l'Université de Genève. p. 405. MR 0389523.

- Lieb, Elliott; Loss, Michael (2001). *Analysis*. Graduate Studies in Mathematics 14 (2nd ed.). American Mathematical Society. ISBN 978-0821827833.

- Loomis, Lynn H. (1953). *An introduction to abstract harmonic analysis*. Toronto-New York-London: D. Van Nostrand Company, Inc. pp. x+190. MR 0054173. Includes a presentation of the Daniell integral.

- Munroe, M. E. (1953). *Introduction to measure and integration*. Cambridge, Mass.: Addison-Wesley Publishing Company Inc. pp. x+310. MR 0053186. Good treatment of the theory of outer measures.

- Royden, H. L. (1988). *Real analysis* (Third ed.). New York: Macmillan Publishing Company. pp. xx+444. ISBN 0-02-404151-3. MR 1013117.

- Rudin, Walter (1976). *Principles of mathematical analysis*. International Series in Pure and Applied Mathematics (Third ed.). New York: McGraw-Hill Book Co. pp. x+342. MR 0385023. Known as *Little Rudin*, contains the basics of the Lebesgue theory, but does not treat material such as Fubini's theorem.

- Rudin, Walter (1966). *Real and complex analysis*. New York: McGraw-Hill Book Co. pp. xi+412. MR 0210528. Known as *Big Rudin*. A complete and careful presentation of the theory. Good presentation of the Riesz extension theorems. However, there is a minor flaw (in the first edition) in the proof of one of the extension theorems, the discovery of which constitutes exercise 21 of Chapter 2.

- Saks, Stanisław (1937). "Theory of the Integral". Monografie Matematyczne 7 (2nd ed.). Warszawa-Lwów: G.E. Stechert & Co. pp. VI+347. JFM 63.0183.05. Zbl 0017.30004.. English translation by Laurence Chisholm Young, with two additional notes by Stefan Banach.

- Shilov, G. E.; Gurevich, B. L. (1977). *Integral, measure and derivative: a unified approach. Translated from the Russian and edited by Richard A. Silverman*. Dover Books on Advanced Mathematics. New York: Dover Publications Inc. xiv+233. ISBN 0-486-63519-8. MR 0466463. Emphasizes the Daniell integral.

- Siegmund-Schultze, Reinhard (2008), "Henri Lebesgue", in Timothy Gowers, June Barrow-Green, Imre Leader, *Princeton Companion to Mathematics*, Princeton University Press.

- Teschl, Gerald. *Topics in Real and Functional Analysis*. (lecture notes).

- Yeh, James (2006). *Real Analysis: Theory of Measure and Integral 2nd. Edition Paperback*. Singapore: World Scientific Publishing Company Pte. Ltd. p. 760. ISBN 978-981-256-6.

Chapter 14

Ergodic theory

Ergodic theory (Ancient Greek: *ergon* work, *hodos* way) is a branch of mathematics that studies dynamical systems with an invariant measure and related problems. Its initial development was motivated by problems of statistical physics.

A central concern of ergodic theory is the behavior of a dynamical system when it is allowed to run for a long time. The first result in this direction is the Poincaré recurrence theorem, which claims that almost all points in any subset of the phase space eventually revisit the set. More precise information is provided by various **ergodic theorems** which assert that, under certain conditions, the time average of a function along the trajectories exists almost everywhere and is related to the space average. Two of the most important theorems are those of Birkhoff (1931) and von Neumann which assert the existence of a time average along each trajectory. For the special class of **ergodic systems**, this time average is the same for almost all initial points: statistically speaking, the system that evolves for a long time "forgets" its initial state. Stronger properties, such as mixing and equidistribution, have also been extensively studied.

The problem of metric classification of systems is another important part of the abstract ergodic theory. An outstanding role in ergodic theory and its applications to stochastic processes is played by the various notions of entropy for dynamical systems.

The concepts of ergodicity and the ergodic hypothesis are central to applications of ergodic theory. The underlying idea is that for certain systems the time average of their properties is equal to the average over the entire space. Applications of ergodic theory to other parts of mathematics usually involve establishing ergodicity properties for systems of special kind. In geometry, methods of ergodic theory have been used to study the geodesic flow on Riemannian manifolds, starting with the results of Eberhard Hopf for Riemann surfaces of negative curvature. Markov chains form a common context for applications in probability theory. Ergodic theory has fruitful connections with harmonic analysis, Lie theory (representation theory,

lattices in algebraic groups), and number theory (the theory of diophantine approximations, L-functions).

14.1 Ergodic transformations

Main article: Ergodicity

Ergodic theory is often concerned with **ergodic transformations**. The intuition behind such transformations, which act on a given set, is that they do a thorough job "stirring" the elements of that set. (E.g., if the set is a quantity of hot oatmeal in a bowl, and if a spoon of syrup is dropped into the bowl, then iterations of the inverse of an ergodic transformation of the oatmeal will not allow the syrup to remain in a local subregion of the oatmeal, but will distribute the syrup evenly throughout. At the same time, these iterations will not compress or dilate any portion of the oatmeal: they preserve the measure that is density.) Here is the formal definition.

Let $T : X \to X$ be a measure-preserving transformation on a measure space (X, Σ, μ), with $\mu(X) = 1$. A measure-preserving transformation T as above is **ergodic** if for every E in Σ with $T^{-1}(E) = E$ either $\mu(E) = 0$ or $\mu(E) = 1$.

14.2 Examples

- An irrational rotation of the circle **R/Z**, $T: x \to x + \theta$, where θ is irrational, is ergodic. This transformation has even stronger properties of unique ergodicity, minimality, and equidistribution. By contrast, if $\theta = p/q$ is rational (in lowest terms) then T is periodic, with period q, and thus cannot be ergodic: for any interval I of length a, $0 < a < 1/q$, its orbit under T (that is, the union of I, $T(I)$, ..., $T^{q-1}(I)$, which contains the image of I under any number of applications of T) is a T-invariant mod 0 set that is a union of q intervals of

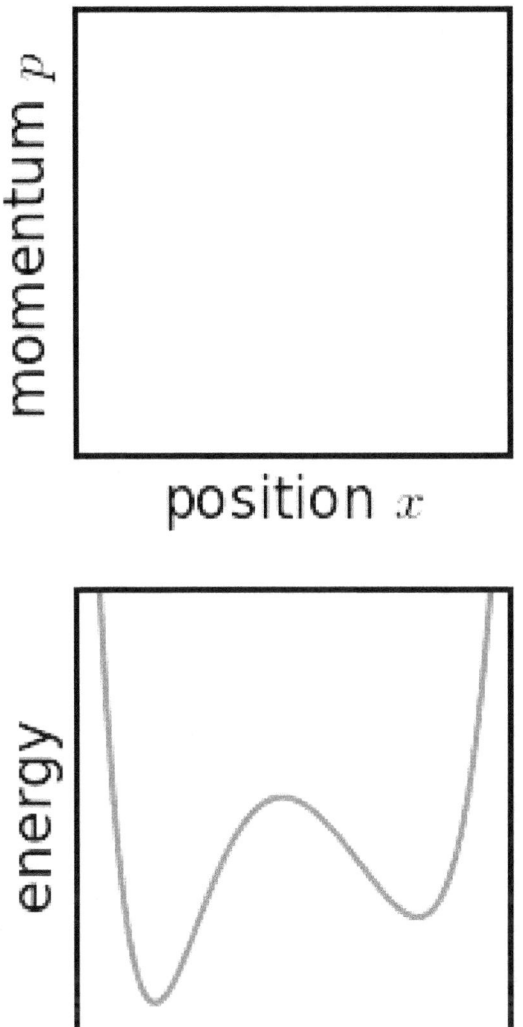

Evolution of an ensemble of classical systems in phase space (top). The systems are massive particles in a one-dimensional potential well (red curve, lower figure). The initially compact ensemble becomes swirled up over time and "spread around" phase space. This is however not *ergodic behaviour since the systems do not visit the left-hand potential well.*

length a, hence it has measure qa strictly between 0 and 1.

- Let G be a compact abelian group, μ the normalized Haar measure, and T a group automorphism of G. Let G^* be the Pontryagin dual group, consisting of the continuous characters of G, and T^* be the corresponding adjoint automorphism of G^*. The automorphism T is ergodic if and only if the equality $(T^*)^n(\chi)=\chi$ is possible only when $n = 0$ or χ is the trivial character of G. In particular, if G is the n-dimensional torus and the automorphism T is represented by a unimodular matrix A then T is ergodic if and only if no eigenvalue of A is a root of unity.

- A Bernoulli shift is ergodic. More generally, ergodicity of the shift transformation associated with a sequence of i.i.d. random variables and some more general stationary processes follows from Kolmogorov's zero-one law.

- Ergodicity of a continuous dynamical system means that its trajectories "spread around" the phase space. A system with a compact phase space which has a nonconstant first integral cannot be ergodic. This applies, in particular, to Hamiltonian systems with a first integral I functionally independent from the Hamilton function H and a compact level set $X = \{(p,q): H(p,q) = E\}$ of constant energy. Liouville's theorem implies the existence of a finite invariant measure on X, but the dynamics of the system is constrained to the level sets of I on X, hence the system possesses invariant sets of positive but less than full measure. A property of continuous dynamical systems that is the opposite of ergodicity is complete integrability.

14.3 Ergodic theorems

Let $T\colon X \to X$ be a measure-preserving transformation on a measure space (X, Σ, μ) and suppose f is a μ-integrable function, i.e. $f \in L^1(\mu)$. Then we define the following *averages*:

Time average: This is defined as the average (if it exists) over iterations of T starting from some initial point x:

$$\hat{f}(x) = \lim_{n\to\infty} \frac{1}{n} \sum_{k=0}^{n-1} f\left(T^k x\right).$$

Space average: If $\mu(X)$ is finite and nonzero, we can consider the *space* or *phase* average of f:

$$\bar{f} = \frac{1}{\mu(X)} \int f \, d\mu. \quad \text{space, probability a (For } \mu(X) = 1.)$$

In general the time average and space average may be different. But if the transformation is ergodic, and the measure is invariant, then the time average is equal to the space average almost everywhere. This is the celebrated ergodic theorem, in an abstract form due to George David Birkhoff.

(Actually, Birkhoff's paper considers not the abstract general case but only the case of dynamical systems arising from differential equations on a smooth manifold.) The equidistribution theorem is a special case of the ergodic theorem, dealing specifically with the distribution of probabilities on the unit interval.

More precisely, the **pointwise** or **strong ergodic theorem** states that the limit in the definition of the time average of f exists for almost every x and that the (almost everywhere defined) limit function $\hat{f}$ is integrable:

$$\hat{f} \in L^1(\mu).$$

Furthermore, $\hat{f}$ is T-invariant, that is to say

$$\hat{f} \circ T = \hat{f}$$

holds almost everywhere, and if $\mu(X)$ is finite, then the normalization is the same:

$$\int \hat{f} \, d\mu = \int f \, d\mu.$$

In particular, if T is ergodic, then $\hat{f}$ must be a constant (almost everywhere), and so one has that

$$\bar{f} = \hat{f}$$

almost everywhere. Joining the first to the last claim and assuming that $\mu(X)$ is finite and nonzero, one has that

$$\lim_{n \to \infty} \frac{1}{n} \sum_{k=0}^{n-1} f\left(T^k x\right) = \frac{1}{\mu(X)} \int f \, d\mu$$

for almost all x, i.e., for all x except for a set of measure zero.

For an ergodic transformation, the time average equals the space average almost surely.

As an example, assume that the measure space (X, Σ, μ) models the particles of a gas as above, and let $f(x)$ denote the velocity of the particle at position x. Then the pointwise ergodic theorems says that the average velocity of all particles at some given time is equal to the average velocity of one particle over time.

A generalization of Birkhoff's theorem is Kingman's subadditive ergodic theorem.

14.4 Probabilistic formulation: Birkhoff–Khinchin theorem

Birkhoff–Khinchin theorem. Let f be measurable, $E(|f|) < \infty$, and T be a measure-preserving map. Then with probability 1:

$$\lim_{n \to \infty} \frac{1}{n} \sum_{k=0}^{n-1} f\left(T^k x\right) = E(f|\mathcal{C}),$$

where $E(f|\mathcal{C})$ is the conditional expectation given the σ-algebra $\mathcal{C}$ of invariant sets of T.

Corollary (Pointwise Ergodic Theorem): In particular, if T is also ergodic, then $\mathcal{C}$ is the trivial σ-algebra, and thus with probability 1:

$$\lim_{n \to \infty} \frac{1}{n} \sum_{k=0}^{n-1} f\left(T^k x\right) = E(f).$$

14.5 Mean ergodic theorem

Von Neumann's mean ergodic theorem, holds in Hilbert spaces.[1]

Let U be a unitary operator on a Hilbert space H; more generally, an isometric linear operator (that is, a not necessarily surjective linear operator satisfying $\|Ux\| = \|x\|$ for all x in H, or equivalently, satisfying $U*U = I$, but not necessarily $UU* = I$). Let P be the orthogonal projection onto $\{\psi \in H| \, U\psi = \psi\} = \text{Ker}(I - U)$.

Then, for any x in H, we have:

$$\lim_{N \to \infty} \frac{1}{N} \sum_{n=0}^{N-1} U^n x = Px,$$

where the limit is with respect to the norm on H. In other words, the sequence of averages

$$\frac{1}{N} \sum_{n=0}^{N-1} U^n$$

converges to P in the strong operator topology.

Indeed, it is not difficult to see that in this case any $x \in H$ admits an orthogonal decomposition into parts from $\ker(I - U)$ and $\overline{\text{Ran}(I - U)}$ respectively. The former part is invariant in all the partial sums as N grows, while for the latter part, from the telescoping series one would have:

$$\lim_{N\to\infty} \frac{1}{N} \sum_{n=0}^{N-1} U^n(I-U) = \lim_{N\to\infty} \frac{1}{N}(I-U^N) = 0$$

This theorem specializes to the case in which the Hilbert space H consists of L^2 functions on a measure space and U is an operator of the form

$$Uf(x) = f(Tx)$$

where T is a measure-preserving endomorphism of X, thought of in applications as representing a time-step of a discrete dynamical system.[2] The ergodic theorem then asserts that the average behavior of a function f over sufficiently large time-scales is approximated by the orthogonal component of f which is time-invariant.

In another form of the mean ergodic theorem, let Ut be a strongly continuous one-parameter group of unitary operators on H. Then the operator

$$\frac{1}{T} \int_0^T U_t\, dt$$

converges in the strong operator topology as $T \to \infty$. In fact, this result also extends to the case of strongly continuous one-parameter semigroup of contractive operators on a reflexive space.

Remark: Some intuition for the mean ergodic theorem can be developed by considering the case where complex numbers of unit length are regarded as unitary transformations on the complex plane (by left multiplication). If we pick a single complex number of unit length (which we think of as U), it is intuitive that its powers will fill up the circle. Since the circle is symmetric around 0, it makes sense that the averages of the powers of U will converge to 0. Also, 0 is the only fixed point of U, and so the projection onto the space of fixed points must be the zero operator (which agrees with the limit just described).

14.6 Convergence of the ergodic means in the L^p norms

Let (X, Σ, μ) be as above a probability space with a measure preserving transformation T, and let $1 \le p \le \infty$. The conditional expectation with respect to the sub-σ-algebra ΣT of the T-invariant sets is a linear projector ET of norm 1 of the Banach space $L^p(X, \Sigma, \mu)$ onto its closed subspace $L^p(X, \Sigma T, \mu)$ The latter may also be characterized as the space of

all T-invariant L^p-functions on X. The ergodic means, as linear operators on $L^p(X, \Sigma, \mu)$ also have unit operator norm; and, as a simple consequence of the Birkhoff–Khinchin theorem, converge to the projector ET in the strong operator topology of L^p if $1 \le p \le \infty$, and in the weak operator topology if $p = \infty$. More is true if $1 < p \le \infty$ then the Wiener–Yoshida–Kakutani ergodic dominated convergence theorem states that the ergodic means of $f \in L^p$ are dominated in L^p; however, if $f \in L^1$, the ergodic means may fail to be equidominated in L^p. Finally, if f is assumed to be in the Zygmund class, that is $|f| \log^+(|f|)$ is integrable, then the ergodic means are even dominated in L^1.

14.7 Sojourn time

Let (X, Σ, μ) be a measure space such that $\mu(X)$ is finite and nonzero. The time spent in a measurable set A is called the **sojourn time**. An immediate consequence of the ergodic theorem is that, in an ergodic system, the relative measure of A is equal to the mean sojourn time:

$$\frac{\mu(A)}{\mu(X)} = \frac{1}{\mu(X)} \int \chi_A\, d\mu = \lim_{n\to\infty} \frac{1}{n} \sum_{k=0}^{n-1} \chi_A\left(T^k x\right)$$

for all x except for a set of measure zero, where χA is the indicator function of A.

The **occurrence times** of a measurable set A is defined as the set $k_1, k_2, k_3, ...,$ of times k such that $T^k(x)$ is in A, sorted in increasing order. The differences between consecutive occurrence times $Ri = ki - ki_{-1}$ are called the **recurrence times** of A. Another consequence of the ergodic theorem is that the average recurrence time of A is inversely proportional to the measure of A, assuming that the initial point x is in A, so that $k_0 = 0$.

$$\frac{R_1 + \cdots + R_n}{n} \to \frac{\mu(X)}{\mu(A)} \quad \text{(almost surely)}$$

(See almost surely.) That is, the smaller A is, the longer it takes to return to it.

14.8 Ergodic flows on manifolds

The ergodicity of the geodesic flow on compact Riemann surfaces of variable negative curvature and on compact manifolds of constant negative curvature of any dimension was proved by Eberhard Hopf in 1939, although special cases had been studied earlier: see for example, Hadamard's billiards (1898) and Artin billiard (1924). The relation between geodesic flows on Riemann surfaces and one-parameter subgroups on SL(2, **R**) was described in 1952 by S. V. Fomin and I. M. Gelfand. The article on Anosov flows provides an example of ergodic flows on SL(2, **R**) and on Riemann surfaces of negative curvature. Much of the development described there generalizes to hyperbolic manifolds, since they can be viewed as quotients of the hyperbolic space by the action of a lattice in the semisimple Lie group SO(n,1). Ergodicity of the geodesic flow on Riemannian symmetric spaces was demonstrated by F. I. Mautner in 1957. In 1967 D. V. Anosov and Ya. G. Sinai proved ergodicity of the geodesic flow on compact manifolds of variable negative sectional curvature. A simple criterion for the ergodicity of a homogeneous flow on a homogeneous space of a semisimple Lie group was given by Calvin C. Moore in 1966. Many of the theorems and results from this area of study are typical of rigidity theory.

In the 1930s G. A. Hedlund proved that the horocycle flow on a compact hyperbolic surface is minimal and ergodic. Unique ergodicity of the flow was established by Hillel Furstenberg in 1972. Ratner's theorems provide a major generalization of ergodicity for unipotent flows on the homogeneous spaces of the form $\Gamma \backslash G$, where G is a Lie group and Γ is a lattice in G.

In the last 20 years, there have been many works trying to find a measure-classification theorem similar to Ratner's theorems but for diagonalizable actions, motivated by conjectures of Furstenberg and Margulis. An important partial result (solving those conjectures with an extra assumption of positive entropy) was proved by Elon Lindenstrauss, and he was awarded the Fields medal in 2010 for this result.

14.9   See also

- Chaos theory

- Ergodic hypothesis

- Ergodic process

- Lyapunov time – the time limit to the predictability of the system

- Maximal ergodic theorem

- Ornstein isomorphism theorem

- Statistical mechanics

- Symbolic dynamics

- Lindy Effect

14.10   References

[1] I: Functional Analysis : Volume 1 by Michael Reed, Barry Simon,Academic Press; REV edition (1980)

[2] (Walters 1982)

14.11   Historical references

- Birkhoff, George David (1931), "Proof of the ergodic theorem", *Proc Natl Acad Sci USA* **17** (12): 656–660, Bibcode:1931PNAS...17..656B, doi:10.1073/pnas.17.12.656, PMC 1076138, PMID 16577406.

- Birkhoff, George David (1942), "What is the ergodic theorem?", *American Mathematical Monthly* (The American Mathematical Monthly, Vol. 49, No. 4) **49** (4): 222–226, doi:10.2307/2303229, JSTOR 2303229.

- von Neumann, John (1932), "Proof of the Quasi-ergodic Hypothesis", *Proc Natl Acad Sci USA* **18** (1): 70–82, Bibcode:1932PNAS...18...70N, doi:10.1073/pnas.18.1.70, PMC 1076162, PMID 16577432.

- von Neumann, John (1932), "Physical Applications of the Ergodic Hypothesis", *Proc Natl Acad Sci USA* **18** (3): 263–266, Bibcode:1932PNAS...18..263N, doi:10.1073/pnas.18.3.263, JSTOR 86260, PMC 1076204, PMID 16587674.

- Hopf, Eberhard (1939), "Statistik der geodätischen Linien in Mannigfaltigkeiten negativer Krümmung", *Leipzig Ber. Verhandl. Sächs. Akad. Wiss.* **91**: 261–304.

- Fomin, Sergei V.; Gelfand, I. M. (1952), "Geodesic flows on manifolds of constant negative curvature", *Uspehi Mat. Nauk* **7** (1): 118–137.

- Mautner, F. I. (1957), "Geodesic flows on symmetric Riemann spaces", *Ann. Math.* (The Annals of Mathematics, Vol. 65, No. 3) **65** (3): 416–431, doi:10.2307/1970054, JSTOR 1970054.

- Moore, C. C. (1966), "Ergodicity of flows on homogeneous spaces", *Amer. J. Math.* (American Journal of Mathematics, Vol. 88, No. 1) **88** (1): 154–178, doi:10.2307/2373052, JSTOR 2373052.

14.12 Modern references

- D.V. Anosov (2001), "Ergodic theory", in Hazewinkel, Michiel, *Encyclopedia of Mathematics*, Springer, ISBN 978-1-55608-010-4

- *This article incorporates material from ergodic theorem on PlanetMath, which is licensed under the Creative Commons Attribution/Share-Alike License.*

- Vladimir Igorevich Arnol'd and André Avez, *Ergodic Problems of Classical Mechanics*. New York: W.A. Benjamin. 1968.

- Leo Breiman, *Probability*. Original edition published by Addison–Wesley, 1968; reprinted by Society for Industrial and Applied Mathematics, 1992. ISBN 0-89871-296-3. *(See Chapter 6.)*

- Walters, Peter (1982), *An introduction to ergodic theory*, Graduate Texts in Mathematics **79**, Springer-Verlag, ISBN 0-387-95152-0, Zbl 0475.28009

- Tim Bedford, Michael Keane and Caroline Series, *eds.* (1991), *Ergodic theory, symbolic dynamics and hyperbolic spaces*, Oxford University Press, ISBN 0-19-853390-X *(A survey of topics in ergodic theory; with exercises.)*

- Karl Petersen. Ergodic Theory (Cambridge Studies in Advanced Mathematics). Cambridge: Cambridge University Press. 1990.

- Joseph M. Rosenblatt and Máté Weirdl, *Pointwise ergodic theorems via harmonic analysis*, (1993) appearing in *Ergodic Theory and its Connections with Harmonic Analysis, Proceedings of the 1993 Alexandria Conference*, (1995) Karl E. Petersen and Ibrahim A. Salama, *eds.*, Cambridge University Press, Cambridge, ISBN 0-521-45999-0. *(An extensive survey of the ergodic properties of generalizations of the equidistribution theorem of shift maps on the unit interval. Focuses on methods developed by Bourgain.)*

- A.N. Shiryaev, *Probability*, 2nd ed., Springer 1996, Sec. V.3. ISBN 0-387-94549-0.

14.13 External links

- Ergodic Theory (16 Jun 2015) Notes by Cosma Rohilla Shalizi

- Ergodic theorem passes the test From Physics World

Chapter 15

Bergman space

In complex analysis, functional analysis and operator theory, a **Bergman space** is a function space of holomorphic functions in a domain D of the complex plane that are sufficiently well-behaved at the boundary that they are absolutely integrable. Specifically, for $0 < p < \infty$, the Bergman space $A^p(D)$ is the space of all holomorphic functions f in D for which the p-norm is finite:

$$\|f\|_{A^p(D)} := \left(\int_D |f(x+iy)|^p \, dx\, dy \right)^{1/p} < \infty.$$

The quantity $\|f\|_{A^p(D)}$ is called the *norm* of the function f; it is a true norm if $p \geq 1$. Thus $A^p(D)$ is the subspace of holomorphic functions that are in the space $L^p(D)$. The Bergman spaces are Banach spaces, which is a consequence of the estimate, valid on compact subsets K of D:

Thus convergence of a sequence of holomorphic functions in $L^p(D)$ implies also compact convergence, and so the limit function is also holomorphic.

If $p = 2$, then $A^p(D)$ is a reproducing kernel Hilbert space, whose kernel is given by the Bergman kernel.

15.1 Special cases and generalisations

If the domain D is bounded, then the norm is often given by

$$\|f\|_{A^p(D)} := \left(\int_D |f(z)|^p \, dA \right)^{1/p} \qquad (f \in A^p(D)),$$

where A is a normalised Lebesgue measure of the complex plane, i.e. $dA = dz/\text{Area}(D)<l$. Alternatively $dA = dz/\pi$ is used, regardless of the area of D. The Bergman space

is usually defined on the open unit disk $\mathbb{D}$ of the complex plane, in which case $A^p(\mathbb{C}) := A^p$. In the Hilbert space case, given $f(z) = \sum_{n=0}^{\infty} a_n z^n \in A^2$, we have

$$\|f\|_{A^2}^2 := \frac{1}{\pi} \int_{\mathbb{D}} |f(z)|^2 \, dz = \sum_{n=0}^{\infty} \frac{|a_n|^2}{n+1},$$

that is, A^2 is isometrically isomorphic to the weighted $\ell^p(1/(n+1))$ space.[1] In particular the polynomials are dense in A^2. Similarly, if $D = \mathbb{C}_+$), the right (or the upper) complex half-plane, then

$$\|F\|_{A^2(\mathbb{C}_+)}^2 := \frac{1}{\pi} \int_{\mathbb{C}_+} |F(z)|^2 \, dz = \int_0^{\infty} \frac{|f(t)|^2}{t},$$

where $F(z) = \int_0^{\infty} f(t) e^{-tz} \, dt$, that is, $A^2(\mathbb{C}_+)$ is isometrically isomorphic to the weighted $L^p 1/t\,(0,\infty)$ space (via the Laplace transform).[2][3]

The weighted Bergman space $A^p(D)$ is defined in an analogous way,[1] i.e.

$$\|f\|_{A^p_w(D)} := \left(\int_D |f(x+iy)|^2 \, w(x+iy) \, dx\, dy \right)^{1/p},$$

provided that $w : D \to [0, \infty)$ is chosen in such way, that $A^p_w(D)$ is a Banach space (or a Hilbert space, if $p=2$). In case where $D = \mathbb{D}$, by a weighted Bergman space A^p_α [4] we mean the space of all analytic functions f such that

$$\|f\|_{A^p_\alpha} := \left(\frac{1}{\pi} \int_{\mathbb{D}} |f(z)|^p \, (1-|z|^p)^\alpha dz \right)^{1/p} < \infty,$$

and similarly on the right half-plane (i.e. $A^p_\alpha(\mathbb{C}_+)$) we have[5]

$$\|f\|_{A^p_\alpha(\mathbb{C}_+)} := \left(\frac{1}{\pi} \int_{\mathbb{C}_+} |f(x+iy)|^p x^\alpha \, dx\, dy \right)^{1/p},$$

126

and this space is isometrically isomorphic, via the Laplace transform, to the space $L^2(\mathbb{R}_+, d\mu_\alpha)$,[6][7] where

$$d\mu_\alpha := \frac{\Gamma(\alpha + 1)}{2^\alpha t^{\alpha+1}} \, dt$$

(here Γ denotes the Gamma function).

Further generalisations are sometimes considered, for example A_ν^2 denotes a weighted Bergman space (often called a Zen space[3]) with respect to a translation-invariant positive regular Borel measure ν on the closed right complex half-plane $\overline{\mathbb{C}_+}$, that is

$$A_\nu^p := \left\{ f : \mathbb{C}_+ \longrightarrow \mathbb{C} \text{ analytic } : \|f\|_{A_\nu^p} := \left(\sup_{\epsilon>0} \int_{\mathbb{C}_+} |f(z+\epsilon)|^p \, d\nu(z) \right)^{1/p} < \infty \right\}.$$

15.2 Reproducing kernels

The reproducing kernel $k_z^{A^2}$ of A^2 at point $z \in \mathbb{D}$ is given by[1]

$$k_z^{A^2}(\zeta) = \frac{1}{(1 - \overline{z}\zeta)^2} \qquad (\zeta \in \mathbb{D}),$$

and similarly for $A^2(\mathbb{C}_+)$ we have[5]

$$k_z^{A^2(\mathbb{C}_+)}(\zeta) = \frac{1}{(\overline{z} + \zeta)^2} \qquad (\zeta \in \mathbb{C}_+),$$

In general, if φ maps a domain Ω conformally onto a domain D, then[1]

$$k_z^{A^2(\Omega)}(\zeta) = k_{\varphi(z)}^{\mathcal{A}^2(D)}(\varphi(\zeta)) \overline{\varphi'(z)} \varphi'(\zeta) \qquad (z, \zeta \in \Omega).$$

In weighted case we have[4]

$$k_z^{A_\alpha^2}(\zeta) = \frac{\alpha + 1}{(1 - \overline{z}\zeta)^{\alpha+2}} \qquad (z, \zeta \in \mathbb{D}),$$

and[5]

$$k_z^{A_\alpha^2(\mathbb{C}_+)}(\zeta) = \frac{2^\alpha(\alpha + 1)}{(\overline{z} + \zeta)^{\alpha+2}} \qquad (z, \zeta \in \mathbb{C}_+).$$

15.3 References

[1] Duren, Peter L.; Schuster, Alexander (2004), *Bergman spaces*, Mathematical Series and Monographs, American Mathematical Society, ISBN 978-0-8218-0810-8

[2] Duren, Peter L. (1969), *Extension of a theorem of Carleson* (PDF) **75**, Bulletin of the American Mathematical Society, pp. 143–146

[3] Jacob, Brigit; Partington, Jonathan R.; Pott, Sandra (2013-02-01), *On Laplace-Carleson embedding theorems* **264** (3), Journal of Functional Analysis, pp. 783–814

[4] Cowen, Carl; MacCluer, Barbara (1995-04-27), *Composition Operators on Spaces of Analytic Functions*, Studies in Advanced Mathematics, CRC Press, p. 27, ISBN 9780849384929

[5] Elliott, Sam J.; Wynn, Andrew (2011), *Composition Operators on the Weighted Bergman Spaces of the Half-Plane* **54** (2), Proceedings of the Edinburgh Mathematical Society, pp. 374–379

[6] Duren, Peter L.; Gallardo-Gutiérez, Eva A.; Montes-Rodríguez, Alfonso (2007-06-03), *A Paley-Wiener theorem for Bergman spaces with application to invariant subspaces* (PDF) **39** (3), Bulletin of the London Mathematical Society, pp. 459–466

[7] Gallrado-Gutiérez, Eva A.; Partington, Jonathan R.; Segura, Dolores (2009), *Cyclic vectors and invariant subspaces for Bergman and Dirichlet shifts* (PDF) **62** (1), Journal of Operator Theory, pp. 199–214

15.4 Further reading

- Bergman, Stefan (1970), *The kernel function and conformal mapping*, Mathematical Surveys **5** (2nd ed.), American Mathematical Society

- Hedenmalm, H.; Korenblum, B.; Zhu, K. (2000), *Theory of Bergman Spaces*, Springer, ISBN 978-0-387-98791-0

- Richter, Stefan (2001), "Bergman spaces", in Hazewinkel, Michiel, *Encyclopedia of Mathematics*, Springer, ISBN 978-1-55608-010-4.

15.5 See also

- Bergman kernel

- Banach space

- Hilbert space

- Reproducing kernel Hilbert space

- Hardy space

- Dirichlet space

Chapter 16

Reproducing kernel Hilbert space

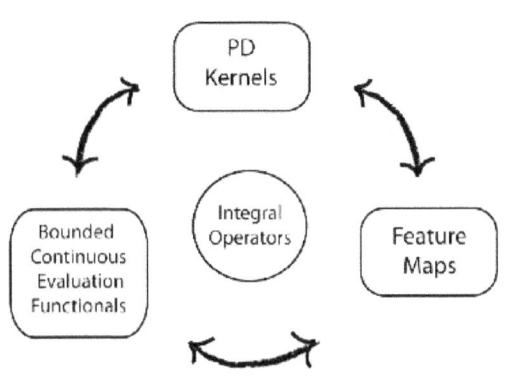

Figure illustrates related but varying approaches to viewing RKHS

In functional analysis (a branch of mathematics), a **reproducing kernel Hilbert space** (RKHS) is a Hilbert space associated with a kernel that reproduces every function in the space or, equivalently, where every evaluation functional is bounded. The reproducing kernel was first introduced in the 1907 work of Stanisław Zaremba concerning boundary value problems for harmonic and biharmonic functions. James Mercer simultaneously examined functions which satisfy the reproducing property in the theory of integral equations. The idea of the reproducing kernel remained untouched for nearly twenty years until it appeared in the dissertations of Gábor Szegő, Stefan Bergman, and Salomon Bochner. The subject was eventually systematically developed in the early 1950s by Nachman Aronszajn and Stefan Bergman. [1]

These spaces have wide applications, including complex analysis, harmonic analysis, and quantum mechanics. Reproducing kernel Hilbert spaces are particularly important in the field of statistical learning theory because of the celebrated Representer theorem which states that every function in an RKHS can be written as a linear combination of the kernel function evaluated at the training points. This is a practically useful result as it effectively simplifies the empirical risk minimization problem from an infinite dimensional to a finite dimensional optimization problem.

For ease of understanding, we provide the framework for real-valued Hilbert spaces. The theory can be easily extended to spaces of complex-valued functions and hence include the many important examples of reproducing kernel Hilbert spaces that are spaces of analytic functions. [2]

16.1 Definition

Let X be an arbitrary set and H a Hilbert space of real-valued functions on X. The evaluation functional over the Hilbert space of functions H is a linear functional that evaluates each function at a point x,

$$L_x : f \mapsto f(x) \; \forall f \in H.$$

We say that H is a **reproducing kernel Hilbert space** if L_x is a continuous function for any f in H or, equivalently, if for all x in X, L_x is a bounded operator on H, i.e. there exists some $M > 0$ such that

While property (**1**) is the weakest condition that ensures both the existence of an inner product and the evaluation of every function in H at every point in the domain, it does not lend itself to easy application in practice. A more intuitive definition of the RKHS can be obtained by observing that this property guarantees that the evaluation functional can be represented by taking the inner product of f with a function K_x in H. This function is the so-called **reproducing kernel** for the Hilbert space H from which the RKHS takes its name. More formally, the Riesz representation theorem implies that for all x in X there exists a unique element K_x of H with the reproducing property,

Since K_x is itself a function in H we have that for each x in X

$$K_y(x) = \langle K_y, \ K_x \rangle_H.$$

This allows us to define the reproducing kernel of H as a function $K : X \times X \to \mathbb{R}$ by

$$K(x, y) = \langle K_x, \ K_y \rangle_H.$$

From this definition it is easy to see that a function $K : X \times X \to \mathbb{R}$ is a reproducing kernel if it is both symmetric and positive definite, i.e.

$$\sum_{i,j=1}^{n} c_i c_j K(x_i, x_j) \geq 0$$

for any $n \in \mathbb{N}, x_1, \ldots, x_n \in X$, and $c_1, \ldots, c_n \in \mathbb{R}$. [3]

16.2   Example

The space of bandlimited functions H is a RKHS. Let

$$H = \{f \in L_2(\mathbb{R}) | \, \mathrm{supp}(\phi) \subset [-a, a], a < \infty\}$$

where $\phi(\omega) = \int f(x)e^{-i\omega x}dx$ is the Fourier transform of f. One can show that if $f \in H$ then

$$f(x) = \frac{1}{2\pi} \int_{-a}^{a} \phi(\omega)e^{ix\omega}d\omega$$

for some $\phi \in L_2[-a, a]$. It then follows by the Cauchy-Schwarz inequality and Plancherel's Theorem that

$$|f(x)| \leq \sqrt{\frac{1}{2\pi} \int_{-a}^{a} 1 d\omega} \sqrt{\frac{1}{2\pi} \int_{-a}^{a} |\phi(\omega)|^2 d\omega} = \sqrt{\frac{a}{\pi}}\|f\|.$$

As this inequality shows that the evaluation functional is bounded and H is also a Hilbert space, H is indeed a RKHS.

The kernel function K_x in this case is given by

$$K_x(y) = \frac{a}{\pi} \mathrm{sinc}(a(y - x)) = \frac{\sin(a(y - x))}{\pi(y - x)}$$

Note, that K_x in this case is the "bandlimited version" of the Dirac delta distribution and that K_x converges to $\delta(\cdot - x)$ in the weak sense, as explained in the entry for the sinc function.

16.3   Moore-Aronszajn Theorem

We have seen how a reproducing kernel Hilbert space defines a reproducing kernel function that is both symmetric and positive definite. The Moore-Aronszajn theorem goes in the other direction; it states that every symmetric, positive definite kernel defines a unique reproducing kernel Hilbert space. The theorem first appeared in Aronszajn's *Theory of Reproducing Kernels*, although he attributes it to E. H. Moore.

Theorem. Suppose K is a symmetric, positive definite kernel on a set X. Then there is a unique Hilbert space of functions on X for which K is a reproducing kernel.

Proof. For all x in X, define $Kx = K(x, \cdot)$. Let H_0 be the linear span of $\{Kx : x \in X\}$. Define an inner product on H_0 by

$$\left\langle \sum_{j=1}^{n} b_j K_{y_j}, \sum_{i=1}^{m} a_i K_{x_i} \right\rangle = \sum_{i=1}^{m}\sum_{j=1}^{n} a_i b_j K(y_j, x_i).$$

The symmetry of this inner product follows from the symmetry of K and the non-degeneracy follows from the fact that K is positive definite.

Let H be the completion of H_0 with respect to this inner product. Then H consists of functions of the form

$$f(x) = \sum_{i=1}^{\infty} a_i K_{x_i}(x)$$

where $\sum_{i=1}^{\infty} a_i^2 K(x_i, x_i) < \infty$. The fact that the above sum converges for every x follows from the Cauchy-Schwarz inequality.

Now we can check the reproducing property (**2**):

$$\langle f, K_x \rangle = \left\langle \sum_{i=1}^{\infty} a_i K_{x_i}, K_x \right\rangle = \sum_{i=1}^{\infty} a_i K(x_i, x) = f(x).$$

To prove uniqueness, let G be another Hilbert space of functions for which K is a reproducing kernel. For any x and y in X, (**2**) implies that

$$\langle K_x, K_y \rangle_H = K(x, y) = \langle K_x, K_y \rangle_G.$$

By linearity, $\langle \cdot, \cdot \rangle_H = \langle \cdot, \cdot \rangle_G$ on the span of $\{Kx : x \in X\}$. Then $G = H$ by the uniqueness of the completion.

16.4 Integral Operators and Mercer's Theorem

We may characterize a symmetric positive definite kernel K via the integral operator using Mercer's theorem and obtain an additional view of the RKHS. Let X be a compact space equipped with a strictly positive finite Borel measure μ and $K : X \times X \to \mathbb{R}$ a continuous, symmetric, and positive definite function. Define the integral operator $T_K : L_2(X) \to L_2(X)$ as

$$[T_K f](\cdot) = \int_X K(\cdot, t) f(t) \, d\mu(t)$$

where $L_2(X)$ is the space of square integrable functions with respect to μ.

Mercer's theorem states that the spectral decomposition of the integral operator T_K of K yields a series representation of K in terms of the eigenvalues and eigenfunctions of T_K. This then implies that K is a reproducing kernel so that the corresponding RKHS can be defined in terms of these eigenvalues and eigenfunctions. We provide the details below.

Under these assumptions T_k is a compact, continuous, self-adjoint, and positive operator. The spectral theorem for self-adjoint operators implies that there is an at most countable decreasing sequence $(\sigma_i)_i \geq 0$ such that $\lim_{i \to \infty} \sigma_i = 0$ and $T_K \phi_i(x) = \sigma_i \phi_i(x)$, where the $\{\phi_i\}$ form an orthonormal basis of $L_2(X)$. By the positivity T_k, $\sigma_i > 0 \ \forall i$. One can also show that T_k maps continuously into the space of continuous functions $C(X)$ and therefore we may choose continuous functions as the eigenvectors, that is, $\phi_i \in C(X) \ \forall i$. Then by Mercer's theorem K may be written in terms of the eigenvalues and continuous eigenfunctions as

$$K(x, y) = \sum_{j=1}^{\infty} \sigma_j \, \phi_j(x) \, \phi_j(y)$$

for all x, y in X such that $\lim_{n \to \infty} \sup_{u,v} |K(u, v) - \sum_{j=1}^{n} \sigma_j \, \phi_j(u) \, \phi_j(v)| = 0$. This above series representation is referred to as a Mercer kernel or Mercer representation of K.

Furthermore, it can be shown that the RKHS H of K is given by

$$H = \left\{ f \in L_2(X) \left| \sum_{i=1}^{\infty} \frac{\langle f, \phi_i \rangle^2}{\sigma_i} < \infty \right. \right\}$$

where the inner product of H given by $\langle f, g \rangle_H = \sum_{i=1}^{\infty} \frac{\langle f, \phi_i \rangle_{L_2} \langle g, \phi_i \rangle_{L_2}}{\sigma_i}$. This representation of the RKHS has application in probability and statistics, for example to the Karhunen-Loeve representation for stochastic processes and kernel PCA.

16.5 Feature Maps

A **feature map** is a map $\varphi : X \to F$, where F is a Hilbert space which we will call the feature space. The first sections presented the connection between bounded/continuous evaluation functions, positive definite functions, and integral operators and in this section we provide another representation of the RKHS in terms of feature maps.

We first note that every feature map defines a kernel via

Clearly K is symmetric and positive definiteness follows from the properties of inner product in F. Conversely, every positive definite function and corresponding reproducing kernel Hilbert space has infinitely many associated feature maps such that (**3**) holds.

For example, we can trivially take $F = H$ and $\varphi(x) = K_x$ for all $x \in X$. Then (**3**) is satisfied by the reproducing property. Another classical example of a feature map relates to the previous section regarding integral operators by taking $F = \ell^2$ and $\varphi(x) = (\sqrt{\sigma_i} \phi_i(x))_i$.

This connection between kernels and feature maps provides us with a new way to understand positive definite functions and hence reproducing kernels as inner products in H. Moreover, every feature map can naturally define a RKHS by means of the definition of a positive definite function.

Lastly, feature maps allow us to construct function spaces that reveal another perspective on the RKHS. Consider the linear space

$$H_\varphi = \{ f : X \to \mathbb{R} | \exists w \in F, f(x) = \langle w, \varphi(x) \rangle_F, \forall \, x \in X \}.$$

We can define a norm on H_φ by

$$\|f\|_\varphi = \inf\{ \|w\|_F : w \in F, f(x) = \langle w, \varphi(x) \rangle_F, \forall \, x \in X \}.$$

It can be shown that H_φ is a RKHS with kernel defined by $K(x, y) = \langle \varphi(x), \varphi(y) \rangle$. This representation implies that the elements of the RKHS are inner products of elements in the feature space and can accordingly be seen as hyperplanes. This view of the RKHS is related to the kernel trick in machine learning. [4]

16.6   Properties

The following properties of RKHSs may be useful to readers.

- Let $(X_i)_{i=1}^p$ be a sequence of sets and $(K_i)_{i=1}^p$ be a collection of corresponding positive definite functions on $(X_i)_{i=1}^p$. It then follows that

$$K((x_1,\ldots,x_p),(y_1,\ldots,y_p)) = K_1(x_1,y_1)\ldots K_p(x_p,y_p)$$

is a kernel on $X = X_1 \times \cdots \times X_p$.

- Let $X_0 \subset X$, then the restriction of K to $X_0 \times X_0$ is also a reproducing kernel.

- Consider a normalized kernel K such that $K(x,x) = 1$ for all $x \in X$. Define a pseudo-metric on X as

$$d_K(x,y) = \|K_x - K_y\|_H^2 = 2(1 - K(x,y)) \,\forall x \in X$$

By the Cauchy–Schwarz inequality,

$$K(x,y)^2 \leq K(x,x)K(y,y) \,\forall x,y \in X.$$

This inequality allows us to view K as a measure of similarity between inputs. If $x,y \in X$ are similar then $K(x,y)$ will be closer to 1 while if $x,y \in X$ are dissimilar then $K(x,y)$ will be closer to 0.

- The closure of the span of $\{K_x | x \in X\}$ coincides with H. [5]

16.7   Examples

Common examples of kernels include:

- **Linear Kernel**:

$$K(x,y) = \langle x,y \rangle$$

- **Polynomial Kernel**:

$$K(x,y) = (\alpha\langle x,y \rangle + 1)^d, \alpha \in \mathbb{R}, d \in \mathbb{N}$$

Other common examples are kernels which satisfy $K(x,y) = K(\|x - y\|)$. These are the radial basis function kernels.

- **Radial Basis Function Kernels**:

- **Gaussian Kernel**:

  Sometimes referred to as the Radial basis function kernel, or squared exponential kernel

  $$K(x,y) = e^{-\frac{\|x-y\|^2}{2\sigma^2}}, \sigma > 0$$

- **Laplacian Kernel**:

  $$K(x,y) = e^{-\frac{\|x-y\|}{\sigma}}, \sigma > 0$$

We also provide examples of Bergman kernels. Let X be finite and let H consist of all complex-valued functions on X. Then an element of H can be represented as an array of complex numbers. If the usual inner product is used, then Kx is the function whose value is 1 at x and 0 everywhere else, and $K(x,y)$ can be thought of as an identity matrix since $K(x,y)=1$ when $x=y$ and $K(x,y)=0$ otherwise. In this case, H is isomorphic to $\mathbf{C}^n$.

The case of $X = \mathbf{D}$ is more sophisticated, here the Bergman space $H^2(\mathbf{D})$ is the space of square-integrable holomorphic functions on $\mathbf{D}$. It can be shown that the reproducing kernel for $H^2(\mathbf{D})$ is

$$K(x,y) = \frac{1}{\pi}\frac{1}{(1 - x\overline{y})^2}.$$

Lastly, the space of band limited functions f in $L^2(\mathbb{R})$ with bandwidth π are a RKHS with reproducing kernel

$$K(x,y) = \frac{\sin \pi(x-y)}{\pi(x-y)}.$$

16.8   Extension to Vector-Valued Functions

In this section we extend the definition of the RKHS to spaces of vector-valued functions as this extension is particularly important in multi-task learning and manifold regularization. The main difference is that the reproducing kernel Γ is a symmetric function that is now a positive semi-definite *matrix* for any x,y in X. More formally, we define a vector-valued RKHS (vvRKHS) as a Hilbert space of functions $f : X \to \mathbb{R}^T$ such that for all $c \in \mathbb{R}^T$ and $x \in X$

$$\Gamma_x c(y) = \Gamma(x,y)c \in H \text{ for } y \in X$$

and

$\langle f, \Gamma_x c \rangle_H = f(x)^\mathsf{T} c.$

This second property parallels the reproducing property for the scalar-valued case. We note that this definition can also be connected to integral operators, bounded evaluation functions, and feature maps as we saw for the scalar-valued RKHS. We can equivalently define the vvRKHS as a vector-valued Hilbert space with a bounded evaluation functional and show that this implies the existence of a unique reproducing kernel by the Riesz Representation theorem. Mercer's theorem can also be extended to address the vector-valued setting and we can therefore obtain a feature map view of the vvRKHS. Lastly, it can also be shown that the closure of the span of $\{\Gamma_x c : x \in X, c \in \mathbb{R}^T\}$ coincides with H, another property similar to the scalar-valued case.

We can gain intuition for the vvRKHS by taking a component-wise perspective on these spaces. In particular, we find that every vvRKHS is isometrically isomorphic to a scalar-valued RKHS on a particular input space. Let $\Lambda = \{1, \dots, T\}$. Consider the space $X \times \Lambda$ and the corresponding reproducing kernel

As noted above, the RKHS associated to this reproducing kernel is given by the closure of the span of $\{\gamma_{(x,t)} : x \in X, t \in \Lambda\}$ where $\gamma_{(x,t)}(y, s) = \gamma((x, t), (y, s))$ for every set of pairs $(x, t), (y, s) \in X \times \Lambda$.

The connection to the scalar-valued RKHS can then be made by the fact that every matrix-valued kernel can be identified with a kernel of the form of (**4**) via

$\Gamma(x, y)_{(t,s)} = \gamma((x, t), (y, s)).$

Moreover, every kernel with the form of (**4**) defines a matrix-valued kernel with the above expression. Now letting the map $D : H_\Gamma \to H_\gamma$ be defined as

$(Df)(x, t) = \langle f(x), e_t \rangle_{\mathbb{R}^T}$

where e_t is the tth component of the canonical basis for $\mathbb{R}^T$, one can show that D is bijective and an isometry between H_Γ and H_γ.

While this view of the vvRKHS can be quite useful in multi-task learning, it should be noted that this isometry does not reduce the study of the vector-valued case to that of the scalar-valued case. In fact, this isometry procedure can make both the scalar-valued kernel and the input space too difficult to work with in practice as properties of the original kernels are often lost. [6] [7] [8]

An important class of matrix-valued reproducing kernels are *separable* kernels which can factorized as the product of a scalar valued kernel and a T-dimensional symmetric positive semi-definite matrix. In light of our previous discussion these kernels are of the form

$\gamma((x, t), (y, s)) = K(x, y) K_T(t, s)$

for all x, y in X and t, s in T. As the scalar-valued kernel encodes dependencies between the inputs, we can observe that the matrix-valued kernel encodes dependencies among both the inputs and the outputs.

We lastly remark that the above theory can be further extended to spaces of functions with values in function spaces but obtaining kernels for these spaces is a more difficult task. [9]

16.9   See also

- Positive definite kernel

- Mercer's theorem

- Kernel trick

- Kernel embedding of distributions

- Representer theorem

16.10   Notes

[1] Okutmustur

[2] Paulson

[3] Durrett

[4] Rosasco

[5] Rosasco

[6] De Vito

[7] Zhang

[8] Alvarez

[9] Rosasco

16.11   References

- Alvarez, Mauricio, Rosasco, Lorenzo and Lawrence, Neil, "Kernels for Vector-Valued Functions: a Review," http://arxiv.org/abs/1106.6251, June 2011.

- Aronszajn, Nachman (1950). "Theory of Reproducing Kernels". *Transactions of the American Mathematical Society* **68** (3): 337–404. doi:10.1090/S0002-9947-1950-0051437-7.   JSTOR 1990404.   MR 51437.

- Berlinet, Alain and Thomas, Christine. *Reproducing kernel Hilbert spaces in Probability and Statistics*, Kluwer Academic Publishers, 2004.

- Cucker, Felipe; Smale, Steve (2002).   "On the Mathematical Foundations of Learning". *Bulletin of the American Mathematical Society* **39** (1): 1–49. doi:10.1090/S0273-0979-01-00923-5. MR 1864085.

- De Vito, Ernest, Umanita, Veronica, and Villa, Silvia. "An extension of Mercer theorem to vector-valued measurable kernels," http://arxiv.org/pdf/1110.4017.pdf, June 2013.

- Durrett, Greg. 9.520 Course Notes, Massachusetts Institute of Technology, http://www.mit.edu/~{}9.520/scribe-notes/class03_gdurett.pdf, February 2010.

- Kimeldorf, George; Wahba, Grace (1971). "Some results on Tchebycheffian Spline Functions" (PDF). *Journal of Mathematical Analysis and Applications* **33** (1): 82–95. doi:10.1016/0022-247X(71)90184-3. MR 290013.

- Okutmustur, Baver.   "Reproducing Kernel Hilbert Spaces," Ph.D. dissertation, Bilkent University, http://www.thesis.bilkent.edu.tr/0002953.pdf,   August 2005.

- Paulsen, Vern. "An introduction to the theory of reproducing kernel Hilbert spaces," http://www.math.uh.edu/~{}vern/rkhs.pdf.

- Steinwart, Ingo; Scovel, Clint (2012). "Mercer's theorem on general domains: On the interaction between measures, kernels, and RKHSs". *Consty. Approx.* **35** (3): 363–417. MR 2914365.

- Rosasco, Lorenzo and Poggio, Thomas. "A Regularization Tour of Machine Learning - MIT 9.520 Lecture Notes" Manuscript, Dec. 2014.

- Wahba, Grace, *Spline Models for Observational Data*, SIAM, 1990.

- Zhang, Haizhang, Xu, Yuesheng, and Zhang, Qinghui (2012). "Refinement of Operator-valued Reproducing Kernels." Journal of Machine Learning Research 13 91-136.

Chapter 17

Fourier analysis

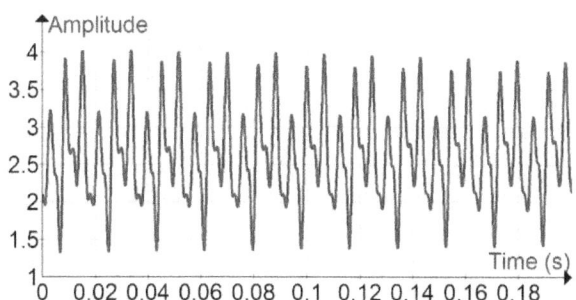

Bass guitar time signal of open string A note (55 Hz).

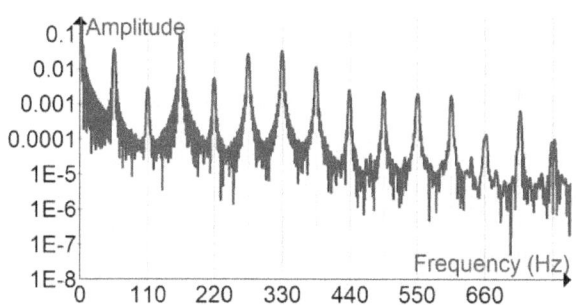

Fourier transform of bass guitar time signal of open string A note (55 Hz), computed with https://sourceforge.net/projects/ amoreaccuratefouriertransform/ . Fourier analysis reveals the oscillatory components of signals and functions.

In mathematics, **Fourier analysis** (English pronunciation: /ˈfɔərieɪ/) is the study of the way general functions may be represented or approximated by sums of simpler trigonometric functions. Fourier analysis grew from the study of Fourier series, and is named after Joseph Fourier, who showed that representing a function as a sum of trigonometric functions greatly simplifies the study of heat transfer.

Today, the subject of Fourier analysis encompasses a vast spectrum of mathematics. In the sciences and engineering, the process of decomposing a function into oscillatory components is often called Fourier analysis, while the operation of rebuilding the function from these pieces is known as

Fourier synthesis. For example, determining what component frequencies are present in a musical note would involve computing the Fourier transform of a sampled musical note. One could then re-synthesize the same sound by including the frequency components as revealed in the Fourier analysis. In mathematics, the term *Fourier analysis* often refers to the study of both operations.

The decomposition process itself is called a Fourier transformation. Its output, the Fourier transform, is often given a more specific name, which depends upon the domain and other properties of the function being transformed. Moreover, the original concept of Fourier analysis has been extended over time to apply to more and more abstract and general situations, and the general field is often known as harmonic analysis. Each transform used for analysis (see list of Fourier-related transforms) has a corresponding inverse transform that can be used for synthesis.

17.1 Applications

Fourier analysis has many scientific applications – in physics, partial differential equations, number theory, combinatorics, signal processing, imaging, probability theory, statistics, option pricing, cryptography, numerical analysis, acoustics, oceanography, sonar, optics, diffraction, geometry, protein structure analysis, and other areas.

This wide applicability stems from many useful properties of the transforms:

- The transforms are linear operators and, with proper normalization, are unitary as well (a property known as Parseval's theorem or, more generally, as the Plancherel theorem, and most generally via Pontryagin duality) (Rudin 1990).

- The transforms are usually invertible.

- The exponential functions are eigenfunctions of differentiation, which means that this representation transforms linear differential equations with constant coefficients into ordinary algebraic ones (Evans 1998). Therefore, the behavior of a linear time-invariant system can be analyzed at each frequency independently.

- By the convolution theorem, Fourier transforms turn the complicated convolution operation into simple multiplication, which means that they provide an efficient way to compute convolution-based operations such as polynomial multiplication and multiplying large numbers (Knuth 1997).

- The discrete version of the Fourier transform (see below) can be evaluated quickly on computers using Fast Fourier Transform (FFT) algorithms. (Conte & de Boor 1980)

Fourier transformation is also useful as a compact representation of a signal. For example, JPEG compression uses a variant of the Fourier transformation (discrete cosine transform) of small square pieces of a digital image. The Fourier components of each square are rounded to lower arithmetic precision, and weak components are eliminated entirely, so that the remaining components can be stored very compactly. In image reconstruction, each image square is reassembled from the preserved approximate Fourier-transformed components, which are then inverse-transformed to produce an approximation of the original image.

17.1.1   Applications in signal processing

When processing signals, such as audio, radio waves, light waves, seismic waves, and even images, Fourier analysis can isolate individual components of a compound waveform, concentrating them for easier detection and/or removal. A large family of signal processing techniques consist of Fourier-transforming a signal, manipulating the Fourier-transformed data in a simple way, and reversing the transformation. (Rabiner and Gold, 1975)

Some examples include:

- Equalization of audio recordings with a series of bandpass filters;

- Digital radio reception with no superheterodyne circuit, as in a modern cell phone or radio scanner;

- Image processing to remove periodic or anisotropic artifacts such as jaggies from interlaced video, stripe artifacts from strip aerial photography, or wave patterns from radio frequency interference in a digital camera;

- Cross correlation of similar images for co-alignment;

- X-ray crystallography to reconstruct a crystal structure from its diffraction pattern;

- Fourier transform ion cyclotron resonance mass spectrometry to determine the mass of ions from the frequency of cyclotron motion in a magnetic field.

- Many other forms of spectroscopy also rely upon Fourier Transforms to determine the three-dimensional structure and/or identity of the sample being analyzed, including Infrared and Nuclear Magnetic Resonance spectroscopies.

- Generation of sound spectrograms used to analyze sounds.

- Passive sonar used to classify targets based on machinery noise.

17.2   Variants of Fourier analysis

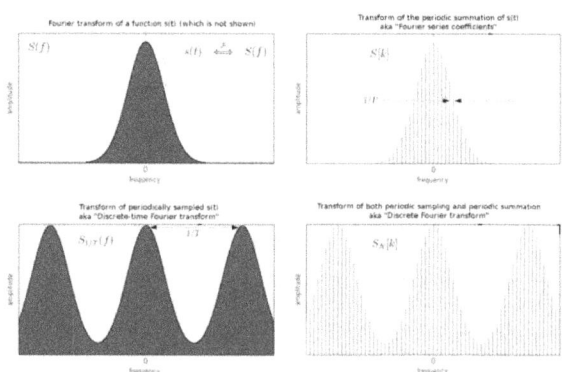

A Fourier transform and 3 variations caused by periodic sampling (at interval T) and/or periodic summation (at interval P) of the underlying time-domain function. The relative computational ease of the DFT sequence and the insight it gives into S(f) make it a popular analysis tool.

17.2.1   (Continuous) Fourier transform

Main article: Fourier transform

Most often, the unqualified term **Fourier transform** refers to the transform of functions of a continuous real argument, and it produces a continuous function of frequency, known as a *frequency distribution*. One function is transformed into another, and the operation is reversible. When the domain of the input (initial) function is time (t), and the domain of the output (final) function is ordinary frequency, the transform of function $s(t)$ at frequency f is given by the complex number:

$$S(f) = \int_{-\infty}^{\infty} s(t) \cdot e^{-i2\pi ft} dt.$$

Evaluating this quantity for all values of f produces the *frequency-domain* function. Then $s(t)$ can be represented as a recombination of complex exponentials of all possible frequencies:

$$s(t) = \int_{-\infty}^{\infty} S(f) \cdot e^{i2\pi ft} df,$$

which is the inverse transform formula. The complex number, $S(f)$, conveys both amplitude and phase of frequency f.

See Fourier transform for much more information, including:

- conventions for amplitude normalization and frequency scaling/units

- transform properties

- tabulated transforms of specific functions

- an extension/generalization for functions of multiple dimensions, such as images.

17.2.2 Fourier series

Main article: Fourier series

The Fourier transform of a periodic function, $sP(t)$, with period P, becomes a Dirac comb function, modulated by a sequence of complex coefficients:

$$S[k] = \tfrac{1}{P} \int_P sP(t) \cdot e^{-i2\pi \frac{k}{P} t} \, dt \text{ for all integer}$$
values of k,

and where $\int_P$ is the integral over any interval of length P.

The inverse transform, known as **Fourier series**, is a representation of $sP(t)$ in terms of a summation of a potentially infinite number of harmonically related sinusoids or complex exponential functions, each with an amplitude and phase specified by one of the coefficients:

$$s_\Gamma(t) = \sum_{k=-\infty}^{\infty} S[k] \cdot e^{i2\pi \frac{k}{P} t} \overset{\mathcal{F}}{\longleftrightarrow} \sum_{k=-\infty}^{+\infty} S[k] \, \delta\left(f - \frac{k}{P}\right).$$

When $sP(t)$, is expressed as a periodic summation of another function, $s(t)$:

$$s_P(t) \overset{\text{def}}{=} \sum_{k=-\infty}^{\infty} s(t - kP),$$

the coefficients are proportional to samples of $S(f)$ at discrete intervals of **1/P**:

$$S[k] = \tfrac{1}{P} \cdot S\left(\tfrac{k}{P}\right). \quad \text{[note 1]}$$

A sufficient condition for recovering $s(t)$ (and therefore $S(f)$) from just these samples is that the non-zero portion of $s(t)$ be confined to a known interval of duration P, which is the frequency domain dual of the Nyquist–Shannon sampling theorem.

See Fourier series for more information, including the historical development.

17.2.3 Discrete-time Fourier transform (DTFT)

The DTFT is the mathematical dual of the time-domain Fourier series. Thus, a convergent periodic summation in the frequency domain can be represented by a Fourier series, whose coefficients are samples of a related continuous time function:

$$S_{1/T}(f) \overset{\text{def}}{=} \overbrace{\underbrace{\sum_{k=-\infty}^{\infty} S\left(f - \frac{k}{T}\right) \equiv \sum_{n=-\infty}^{\infty} s[n] \cdot e^{-i2\pi fnT}}_{\text{Poisson summation formula}}}^{\text{(DTFT) series Fourier}}$$

$$= \mathcal{F}\left\{ \sum_{n=-\infty}^{\infty} s[n] \, \delta(t - nT) \right\},$$

which is known as the DTFT. Thus the **DTFT** of the $s[n]$ sequence is also the **Fourier transform** of the modulated Dirac comb function.[note 2]

The Fourier series coefficients (and inverse transform), are defined by:

$$s[n] \overset{\text{def}}{=} T \int_{1/T} S_{1/T}(f) \cdot e^{i2\pi fnT} \, df$$

$$= T \int_{-\infty}^{\infty} \underbrace{S(f) \cdot e^{i2\pi fnT} df}_{\overset{\text{def}}{=} \, s(nT)}$$

Parameter T corresponds to the sampling interval, and this Fourier series can now be recognized as a form of the Poisson summation formula. Thus we have the important result that when a discrete data sequence, $s[n]$, is proportional to samples of an underlying continuous function, $s(t)$, one can observe a periodic summation of the continuous Fourier transform, $S(f)$. That is a cornerstone in the foundation of digital signal processing. Furthermore, under certain idealized conditions one can theoretically recover $S(f)$ and $s(t)$ exactly. A sufficient condition for perfect recovery is that the non-zero portion of $S(f)$ be confined to a known frequency interval of width $1/T$. When that interval is $[-0.5/T, 0.5/T]$, the applicable reconstruction formula is the Whittaker–Shannon interpolation formula.

Another reason to be interested in $S1/T(f)$ is that it often provides insight into the amount of aliasing caused by the sampling process.

Applications of the DTFT are not limited to sampled functions. See Discrete-time Fourier transform for more information on this and other topics, including:

- normalized frequency units

- windowing (finite-length sequences)

- transform properties

- tabulated transforms of specific functions

17.2.4   Discrete Fourier transform (DFT)

Main article: Discrete Fourier transform

The DTFT of a periodic sequence, $sN[n]$, with period N, becomes another Dirac comb function, modulated by the coefficients of a **Fourier series**. And the integral formula for the coefficients simplifies to a summation (see DTFT/Periodic data):

$$S_N[k] = \frac{1}{NT} \underbrace{\sum_N s_N[n] \cdot e^{-i2\pi \frac{k}{N} n}}_{S_k} \text{ , where } \sum_N$$
is the sum over any n-sequence of length **N**.

The Sk sequence is what's customarily known as the **DFT** of sN. It is also N-periodic, so it is never necessary to compute more than N coefficients. In terms of Sk, the inverse transform is given by:

$$s_N[n] = \frac{1}{N} \sum_N S_k \cdot e^{i2\pi \frac{n}{N} k}, \text{ where } \sum_N \text{ is the}$$
sum over any k-sequence of length **N**.

When $sN[n]$ is expressed as a periodic summation of another function: $s_N[n] \stackrel{\text{def}}{=} \sum_{k=-\infty}^{\infty} s[n - kN]$, and $s[n] \stackrel{\text{def}}{=} T \cdot s(nT)$,

the coefficients are equivalent to samples of $S_1/T(f)$ at discrete intervals of $\mathbf{1/P = 1/NT}$: $S_k = S_{1/T}(k/P)$. (see DTFT/Sampling the DTFT)

Conversely, when one wants to compute an arbitrary number (N) of discrete samples of one cycle of a continuous DTFT, $S_{1/T}(f)$, it can be done by computing the relatively simple DFT of $sN[n]$, as defined above. In most cases, N is chosen equal to the length of non-zero portion of $s[n]$. Increasing N, known as *zero-padding* or *interpolation*, results in more closely spaced samples of one cycle of $S1/T(f)$. Decreasing N, causes overlap (adding) in the time-domain (analogous to aliasing), which corresponds to decimation in the frequency domain. (see Sampling the DTFT) In most cases of practical interest, the $s[n]$ sequence represents a longer sequence that was truncated by the application of a finite-length window function or FIR filter array.

The DFT can be computed using a fast Fourier transform (FFT) algorithm, which makes it a practical and important transformation on computers.

See Discrete Fourier transform for much more information, including:

- transform properties

- applications

- tabulated transforms of specific functions

17.2.5   Summary

For periodic functions, both the Fourier transform and the DTFT comprise only a discrete set of frequency components (Fourier series), and the transforms diverge at those frequencies. One common practice (not discussed above) is to handle that divergence via Dirac delta and Dirac comb functions. But the same spectral information can be discerned from just one cycle of the periodic function, since all the other cycles are identical. Similarly, finite-duration functions can be represented as a Fourier series, with no actual loss of information except that the periodicity of the inverse transform is a mere artifact. We also note that none of the formulas here require the duration of s to be limited to the period, **P** or **N**. But that is a common situation, in practice.

In the table below, associating the $\frac{1}{T}$ scale factor with function $S_{1/T}(f)$ results in some notational simplification without loss of generality.

17.2.6 Fourier transforms on arbitrary locally compact abelian topological groups

The Fourier variants can also be generalized to Fourier transforms on arbitrary locally compact abelian topological groups, which are studied in harmonic analysis; there, the Fourier transform takes functions on a group to functions on the dual group. This treatment also allows a general formulation of the convolution theorem, which relates Fourier transforms and convolutions. See also the Pontryagin duality for the generalized underpinnings of the Fourier transform.

17.2.7 Time–frequency transforms

For more details on this topic, see Time–frequency analysis.

In signal processing terms, a function (of time) is a representation of a signal with perfect *time resolution,* but no frequency information, while the Fourier transform has perfect *frequency resolution,* but no time information.

As alternatives to the Fourier transform, in time–frequency analysis, one uses time–frequency transforms to represent signals in a form that has some time information and some frequency information – by the uncertainty principle, there is a trade-off between these. These can be generalizations of the Fourier transform, such as the short-time Fourier transform, the Gabor transform or fractional Fourier transform (FRFT), or can use different functions to represent signals, as in wavelet transforms and chirplet transforms, with the wavelet analog of the (continuous) Fourier transform being the continuous wavelet transform.

17.3 History

See also: Fourier series § Historical development

A primitive form of harmonic series dates back to ancient Babylonian mathematics, where they were used to compute ephemerides (tables of astronomical positions).[1] The classical Greek concepts of deferent and epicycle in the Ptolemaic system of astronomy were related to Fourier series (see Deferent and epicycle: Mathematical formalism).

In modern times, variants of the discrete Fourier transform were used by Alexis Clairaut in 1754 to compute an orbit,[2] which has been described as the first formula for the DFT,[3] and in 1759 by Joseph Louis Lagrange, in computing the coefficients of a trigonometric series for a vibrating string.[4] Technically, Clairaut's work was a cosine-only se-

ries (a form of discrete cosine transform), while Lagrange's work was a sine-only series (a form of discrete sine transform); a true cosine+sine DFT was used by Gauss in 1805 for trigonometric interpolation of asteroid orbits.[5] Euler and Lagrange both discretized the vibrating string problem, using what would today be called samples.[4]

An early modern development toward Fourier analysis was the 1770 paper *Réflexions sur la résolution algébrique des équations* by Lagrange, which in the method of Lagrange resolvents used a complex Fourier decomposition to study the solution of a cubic:[6] Lagrange transformed the roots x_1, x_2, x_3 into the resolvents:

$$r_1 = x_1 + x_2 + x_3$$
$$r_2 = x_1 + \zeta x_2 + \zeta^2 x_3$$
$$r_3 = x_1 + \zeta^2 x_2 + \zeta x_3$$

where ζ is a cubic root of unity, which is the DFT of order 3.

A number of authors, notably Jean le Rond d'Alembert, and Carl Friedrich Gauss used trigonometric series to study the heat equation, but the breakthrough development was the 1807 paper *Mémoire sur la propagation de la chaleur dans les corps solides* by Joseph Fourier, whose crucial insight was to model *all* functions by trigonometric series, introducing the Fourier series.

Historians are divided as to how much to credit Lagrange and others for the development of Fourier theory: Daniel Bernoulli and Leonhard Euler had introduced trigonometric representations of functions,[3] and Lagrange had given the Fourier series solution to the wave equation,[3] so Fourier's contribution was mainly the bold claim that an arbitrary function could be represented by a Fourier series.[3]

The subsequent development of the field is known as harmonic analysis, and is also an early instance of representation theory.

The first fast Fourier transform (FFT) algorithm for the DFT was discovered around 1805 by Carl Friedrich Gauss when interpolating measurements of the orbit of the asteroids Juno and Pallas, although that particular FFT algorithm is more often attributed to its modern rediscoverers Cooley and Tukey.[5][7]

17.4 Interpretation in terms of time and frequency

In signal processing, the Fourier transform often takes a time series or a function of continuous time, and maps it into a frequency spectrum. That is, it takes a function from the time domain into the frequency domain; it is a decomposition of a function into sinusoids of different frequencies; in the case of a Fourier series or discrete Fourier transform, the sinusoids are harmonics of the fundamental frequency of the function being analyzed.

When the function f is a function of time and represents a physical signal, the transform has a standard interpretation as the frequency spectrum of the signal. The magnitude of the resulting complex-valued function F at frequency ω represents the amplitude of a frequency component whose initial phase is given by the phase of F.

Fourier transforms are not limited to functions of time, and temporal frequencies. They can equally be applied to analyze *spatial* frequencies, and indeed for nearly any function domain. This justifies their use in such diverse branches as image processing, heat conduction, and automatic control.

17.5 Notes

[1]

$$\int_P \left[\sum_{k=-\infty}^{\infty} s(t-kP) \right] \cdot e^{-i2\pi \frac{k}{P} t} dt = \underbrace{\int_{-\infty}^{\infty} s(t) \cdot e^{-i2\pi \frac{k}{P} t} dt}_{\overset{\text{def}}{=} S(k/P)}$$

[2] We may also note that: $\sum_{n=-\infty}^{+\infty} T \ s(nT) \ \delta(t-nT) = \sum_{n=-\infty}^{+\infty} T \ s(t) \ \delta(t-nT) = s(t) \cdot T \sum_{n=-\infty}^{+\infty} \delta(t-nT)$.
Consequently, a common practice is to model "sampling" as a multiplication by the Dirac comb function, which of course is only "possible" in a purely mathematical sense.

17.6 See also

- Generalized Fourier series
- Fourier-Bessel series
- Fourier-related transforms
- Laplace transform (LT)
- Two-sided Laplace transform
- Mellin transform
- Non-uniform discrete Fourier transform (NDFT)
- Quantum Fourier transform (QFT)
- Number-theoretic transform
- Least-squares spectral analysis
- Basis vectors

- Bispectrum
- Characteristic function (probability theory)
- Orthogonal functions
- Schwartz space
- Spectral density
- Spectral density estimation
- Spectral music
- Wavelet

17.7 Citations

[1] Prestini, Elena (2004), *The evolution of applied harmonic analysis: models of the real world*, Birkhäuser, ISBN 978-0-8176-4125-2, p. 62
Rota, Gian-Carlo; Palombi, Fabrizio (1997), *Indiscrete thoughts*, Birkhäuser, ISBN 978-0-8176-3866-5, p. 11
Neugebauer, Otto (1969) [1957], *The Exact Sciences in Antiquity* (2 ed.), Dover Publications, ISBN 978-0-486-22332-2
Brack-Bernsen, Lis; Brack, Matthias, *Analyzing shell structure from Babylonian and modern times*, arXiv:physics/0310126

[2] Terras, Audrey (1999), *Fourier analysis on finite groups and applications*, Cambridge University Press, ISBN 978-0-521-45718-7, p. 30

[3] Briggs, William L.; Henson, Van Emden (1995), *The DFT : an owner's manual for the discrete Fourier transform*, SIAM, ISBN 978-0-89871-342-8, p. 4

[4] Briggs, William L.; Henson, Van Emden (1995), *The DFT: an owner's manual for the discrete Fourier transform*, SIAM, ISBN 978-0-89871-342-8, p. 2

[5] Heideman, M. T., D. H. Johnson, and C. S. Burrus, "Gauss and the history of the fast Fourier transform," IEEE ASSP Magazine, 1, (4), 14–21 (1984)

[6] Knapp, Anthony W. (2006), *Basic algebra*, Springer, ISBN 978-0-8176-3248-9, p. 501

[7] Terras, Audrey (1999), *Fourier analysis on finite groups and applications*, Cambridge University Press, ISBN 978-0-521-45718-7, p. 31

17.8 References

- Conte, S. D.; de Boor, Carl (1980), *Elementary Numerical Analysis* (Third ed.), New York: McGraw Hill, Inc., ISBN 0-07-066228-2

- Evans, L. (1998), *Partial Differential Equations*, American Mathematical Society, ISBN 3-540-76124-1

- Howell, Kenneth B. (2001). *Principles of Fourier Analysis*, CRC Press. ISBN 978-0-8493-8275-8

- Kamen, E.W., and B.S. Heck. "Fundamentals of Signals and Systems Using the Web and Matlab". ISBN 0-13-017293-6

- Knuth, Donald E. (1997), *The Art of Computer Programming Volume 2: Seminumerical Algorithms* (3rd ed.), Section 4.3.3.C: Discrete Fourier transforms, pg.305: Addison-Wesley Professional, ISBN 0-201-89684-2

- Polyanin, A.D., and A.V. Manzhirov (1998). *Handbook of Integral Equations*, CRC Press, Boca Raton. ISBN 0-8493-2876-4

- Rabiner, Lawrence R., and Bernard Gold. "Theory and application of digital signal processing." Englewood Cliffs, NJ, Prentice-Hall, Inc., 1975. 777 p. 1 (1975).

- Rudin, Walter (1990), *Fourier Analysis on Groups*, Wiley-Interscience, ISBN 0-471-52364-X

- Smith, Steven W. (1999), *The Scientist and Engineer's Guide to Digital Signal Processing* (Second ed.), San Diego, Calif.: California Technical Publishing, ISBN 0-9660176-3-3

- Stein, E.M., and G. Weiss (1971). *Introduction to Fourier Analysis on Euclidean Spaces*. Princeton University Press. ISBN 0-691-08078-X

17.9 External links

- Tables of Integral Transforms at EqWorld: The World of Mathematical Equations.

- An Intuitive Explanation of Fourier Theory by Steven Lehar.

- Lectures on Image Processing: A collection of 18 lectures in pdf format from Vanderbilt University. Lecture 6 is on the 1- and 2-D Fourier Transform. Lectures 7–15 make use of it., by Alan Peters

- Moriarty, Philip; Bowley, Roger (2009). "Σ Summation (and Fourier Analysis)". *Sixty Symbols*. Brady Haran for the University of Nottingham.

Chapter 18

Gibbs phenomenon

In mathematics, the **Gibbs phenomenon,** discovered by Henry Wilbraham (1848)[1] and rediscovered by J. Willard Gibbs (1899),[2] is the peculiar manner in which the Fourier series of a piecewise continuously differentiable periodic function behaves at a jump discontinuity. The nth partial sum of the Fourier series has large oscillations near the jump, which might increase the maximum of the partial sum above that of the function itself. The overshoot does not die out as the frequency increases, but approaches a finite limit.[3] This sort of behavior was also observed by experimental physicists, but was believed to be due to imperfections in the measuring apparatuses.[4]

These are one cause of **ringing artifacts** in signal processing.

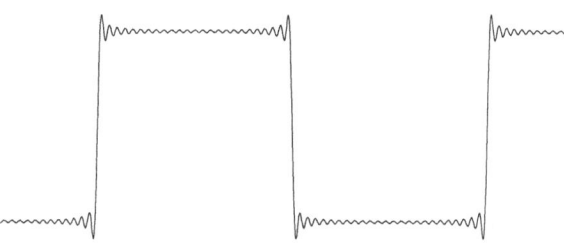

Functional approximation of square wave using 25 harmonics

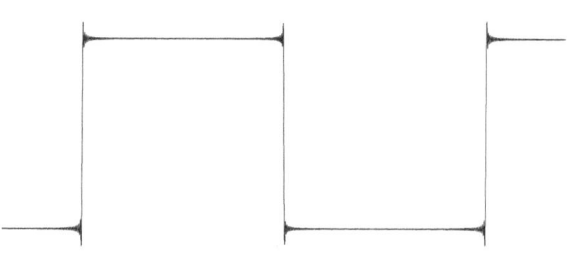

Functional approximation of square wave using 125 harmonics

18.1 Description

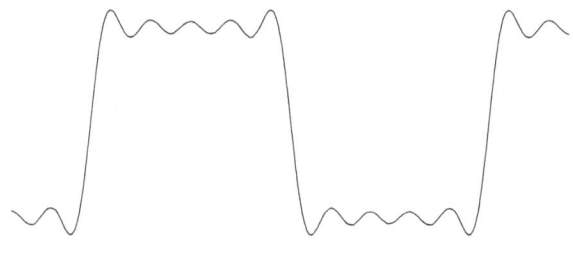

Functional approximation of square wave using 5 harmonics

The Gibbs phenomenon involves both the fact that Fourier sums overshoot at a jump discontinuity, and that this overshoot does not die out as the frequency increases.

The three pictures on the right demonstrate the phenomenon for a square wave (of height $\pi/4$) whose Fourier expansion is

More precisely, this is the function f which equals $\pi/4$ between $2n\pi$ and $(2n + 1)\pi$ and $-\pi/4$ between $(2n + 1)\pi$ and $(2n+2)\pi$ for every integer n; thus this square wave has a jump discontinuity of height $\pi/2$ at every integer multiple of π .

As can be seen, as the number of terms rises, the error of the approximation is reduced in width and energy, but converges to a fixed height. A calculation for the square wave (see Zygmund, chap. 8.5., or the computations at the end of this article) gives an explicit formula for the limit of the height of the error. It turns out that the Fourier series exceeds the height $\pi/4$ of the square wave by

$$\sin(x) + \frac{1}{3}\sin(3x) + \frac{1}{5}\sin(5x) + \cdots.$$

$$\frac{1}{2}\int_0^\pi \frac{\sin t}{t}\, dt - \frac{\pi}{4} = \frac{\pi}{2}\cdot(0.089490\ldots)$$

142

or about 9 percent. More generally, at any jump point of a piecewise continuously differentiable function with a jump of a, the nth partial Fourier series will (for n very large) overshoot this jump by approximately $a \cdot (0.089392\ldots)$ at one end and undershoot it by the same amount at the other end; thus the "jump" in the partial Fourier series will also be about 9% larger than the jump in the original function. At the location of the discontinuity itself, the partial Fourier series will converge to the midpoint of the jump (regardless of what the actual value of the original function is at this point). The quantity

$$\int_0^\pi \frac{\sin t}{t}\, dt = (1.851937052\ldots) = \frac{\pi}{2} + \pi \cdot (0.089392\ldots)$$

is sometimes known as the *Wilbraham–Gibbs constant*.

18.1.1   History

The Gibbs phenomenon was first noticed and analyzed by the obscure Henry Wilbraham.[1] He published a paper on it in 1848 that went unnoticed by the mathematical world.[5] Albert A. Michelson developed a device in 1898 that could compute and re-synthesize the Fourier series. A widespread myth says that when the Fourier coefficients for a square wave were input to the machine, the graph would oscillate at the discontinuities, and that because it was a physical device subject to manufacturing flaws, Michelson was convinced that the overshoot was caused by errors in the machine. In fact the graphs produced by the machine were not good enough to exhibit the Gibbs phenomenon clearly, and Michelson may not have noticed it as he made no mention of this effect in his paper (Michelson & Stratton 1898) about his machine or his later letters to *Nature*. Inspired by some correspondence in *Nature* between Michelson and Love about the convergence of the Fourier series of the square wave function, in 1898 J. Willard Gibbs published a short note in which he considered what today would be called a sawtooth wave and pointed out the important distinction between the limit of the graphs of the partial sums of the Fourier series, and the graph of the function that is the limit of those partial sums. In his first letter Gibbs failed to notice the Gibbs phenomenon, and the limit that he described for the graphs of the partial sums was inaccurate. In 1899 he published a correction in which he described the overshoot at the point of discontinuity (*Nature*: April 27, 1899, p. 606). In 1906, Maxime Bôcher gave a detailed mathematical analysis of that overshoot, which he called the "Gibbs phenomenon".[6]

18.1.2   Explanation

Informally, it reflects the difficulty inherent in approximating a discontinuous function by a *finite* series of continuous sine and cosine waves. It is important to put emphasis on the word *finite* because even though every partial sum of the Fourier series overshoots the function it is approximating, the limit of the partial sums does not. The value of x where the maximum overshoot is achieved moves closer and closer to the discontinuity as the number of terms summed increases so, again informally, once the overshoot has passed by a particular x, convergence at the value of x is possible.

There is no contradiction in the overshoot converging to a non-zero amount, but the limit of the partial sums having no overshoot, because where that overshoot happens moves. We have pointwise convergence, but not uniform convergence. For a piecewise C^1 function the Fourier series converges to the function at *every point* except at the jump discontinuities. At the jump discontinuities themselves the limit will converge to the average of the values of the function on either side of the jump. This is a consequence of the Dirichlet theorem.[7]

The Gibbs phenomenon is also closely related to the principle that the decay of the Fourier coefficients of a function at infinity is controlled by the smoothness of that function; very smooth functions will have very rapidly decaying Fourier coefficients (resulting in the rapid convergence of the Fourier series), whereas discontinuous functions will have very slowly decaying Fourier coefficients (causing the Fourier series to converge very slowly). Note for instance that the Fourier coefficients 1, −1/3, 1/5, ... of the discontinuous square wave described above decay only as fast as the harmonic series, which is not absolutely convergent; indeed, the above Fourier series turns out to be only conditionally convergent for almost every value of *x*. This provides a partial explanation of the Gibbs phenomenon, since Fourier series with absolutely convergent Fourier coefficients would be uniformly convergent by the Weierstrass M-test and would thus be unable to exhibit the above oscillatory behavior. By the same token, it is impossible for a discontinuous function to have absolutely convergent Fourier coefficients, since the function would thus be the uniform limit of continuous functions and therefore be continuous, a contradiction. See more about absolute convergence of Fourier series.

18.1.3   Solutions

In practice, the difficulties associated with the Gibbs phenomenon can be ameliorated by using a smoother method of Fourier series summation, such as Fejér summation or Riesz summation, or by using sigma-approximation. Using a wavelet transform with Haar basis functions, the Gibbs phenomenon does not occur in the case of continuous data at jump discontinuities,[8] and is minimal in the discrete case at large change points. In wavelet analysis, this is commonly referred to as the Longo phenomenon.

18.2 Formal mathematical description of the phenomenon

Let $f : \mathbb{R} \to \mathbb{R}$ be a piecewise continuously differentiable function which is periodic with some period $L > 0$. Suppose that at some point x_0, the left limit $f(x_0^-)$ and right limit $f(x_0^+)$ of the function f differ by a non-zero gap a:

$$f(x_0^+) - f(x_0^-) = a \neq 0.$$

For each positive integer $N \geq 1$, let $SN f$ be the Nth partial Fourier series

$$S_N f(x) := \sum_{-N \leq n \leq N} \hat{f}(n) e^{\frac{2i\pi n x}{L}} = \frac{1}{2} a_0 + \sum_{n=1}^{N} \left(a_n \cos\left(\frac{2\pi n x}{L}\right) + b_n \sin\left(\frac{2\pi n x}{L}\right) \right),$$

where the Fourier coefficients $\hat{f}(n), a_n, b_n$ are given by the usual formulae

$$\hat{f}(n) := \frac{1}{L} \int_0^L f(x) e^{-2i\pi n x/L} \, dx$$

$$a_n := \frac{2}{L} \int_0^L f(x) \cos\left(\frac{2\pi n x}{L}\right) dx$$

$$b_n := \frac{2}{L} \int_0^L f(x) \sin\left(\frac{2\pi n x}{L}\right) dx.$$

Then we have

$$\lim_{N \to \infty} S_N f\left(x_0 + \frac{L}{2N}\right) = f(x_0^+) + a \cdot (0.089392\ldots)$$

and

$$\lim_{N \to \infty} S_N f\left(x_0 - \frac{L}{2N}\right) = f(x_0^-) - a \cdot (0.089392\ldots)$$

but

$$\lim_{N \to \infty} S_N f(x_0) = \frac{f(x_0^-) + f(x_0^+)}{2}.$$

More generally, if x_N is any sequence of real numbers which converges to x_0 as $N \to \infty$, and if the gap a is positive then

$$\limsup_{N \to \infty} S_N f(x_N) \leq f(x_0^+) + a \cdot (0.089392\ldots)$$

and

$$\liminf_{N \to \infty} S_N f(x_N) \geq f(x_0^-) - a \cdot (0.089392\ldots).$$

If instead the gap a is negative, one needs to interchange limit superior with limit inferior, and also interchange the $\leq$ and $\geq$ signs, in the above two inequalities.

18.3 Signal processing explanation

For more details on this topic, see Ringing artifacts.

From the point of view of signal processing, the Gibbs

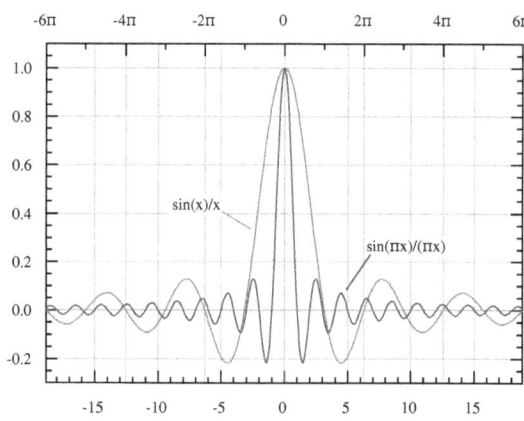

The sinc function, the impulse response of an ideal low-pass filter. Scaling narrows the function, and correspondingly increases magnitude (which is not shown here), but does not reduce the magnitude of the undershoot, which is the integral of the tail.

phenomenon is the step response of a low-pass filter, and the oscillations are called ringing or ringing artifacts. Truncating the Fourier transform of a signal on the real line, or the Fourier series of a periodic signal (equivalently, a signal on the circle) corresponds to filtering out the higher frequencies by an ideal (brick-wall) low-pass/high-cut filter. This can be represented as convolution of the original signal with the impulse response of the filter (also known as the kernel), which is the sinc function. Thus the Gibbs phenomenon can be seen as the result of convolving a Heaviside step function (if periodicity is not required) or a square wave (if periodic)

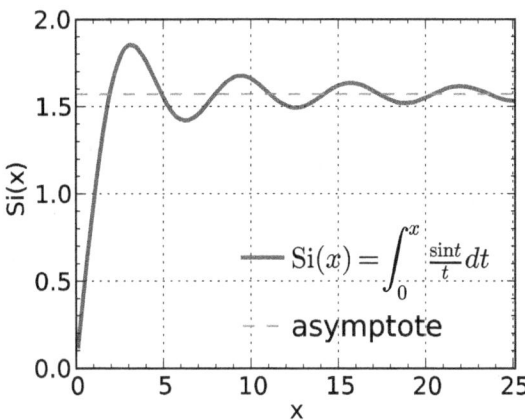

The sine integral, exhibiting the Gibbs phenomenon for a step function on the real line.

tor, leaving the integrals between corresponding points unchanged. This is a general feature of the Fourier transform: widening in one domain corresponds to narrowing and increasing height in the other. This results in the oscillations in sinc being narrower and taller and, in the filtered function (after convolution), yields oscillations that are narrower and thus have less *area*, but does *not* reduce the *magnitude*: cutting off at any finite frequency results in a sinc function, however narrow, with the same tail integrals. This explains the persistence of the overshoot and undershoot.

- Oscillations can be interpreted as convolution with a sinc.

- Higher cutoff makes the sinc narrower but taller, with the same magnitude tail integrals, yielding higher frequency oscillations, but whose magnitude does not vanish.

Thus the features of the Gibbs phenomenon are interpreted as follows:

- the undershoot is due to the impulse response having a negative tail integral, which is possible because the function takes negative values;

- the overshoot offsets this, by symmetry (the overall integral does not change under filtering);

- the persistence of the oscillations is because increasing the cutoff narrows the impulse response, but does not reduce its integral – the oscillations thus move towards the discontinuity, but do not decrease in magnitude.

with a sinc function: the oscillations in the sinc function cause the ripples in the output.

In the case of convolving with a Heaviside step function, the resulting function is exactly the integral of the sinc function, the sine integral; for a square wave the description is not as simply stated. For the step function, the magnitude of the undershoot is thus exactly the integral of the (left) tail, integrating to the first negative zero: for the normalized sinc of unit sampling period, this is $\int_{-\infty}^{-1} \frac{\sin(\pi x)}{\pi x} \, dx$. The overshoot is accordingly of the same magnitude: the integral of the right tail, or, which amounts to the same thing, the difference between the integral from negative infinity to the first positive zero, minus 1 (the non-overshooting value).

The overshoot and undershoot can be understood thus: kernels are generally normalized to have integral 1, so they result in a mapping of constant functions to constant functions – otherwise they have gain. The value of a convolution at a point is a linear combination of the input signal, with coefficients (weights) the values of the kernel. If a kernel is non-negative, such as for a Gaussian kernel, then the value of the filtered signal will be a convex combination of the input values (the coefficients (the kernel) integrate to 1, and are non-negative), and will thus fall between the minimum and maximum of the input signal – it will not undershoot or overshoot. If, on the other hand, the kernel assumes negative values, such as the sinc function, then the value of the filtered signal will instead be an affine combination of the input values, and may fall outside of the minimum and maximum of the input signal, resulting in undershoot and overshoot, as in the Gibbs phenomenon.

Taking a longer expansion – cutting at a higher frequency – corresponds in the frequency domain to widening the brickwall, which in the time domain corresponds to narrowing the sinc function and increasing its height by the same fac-

18.4   The square wave example

We now illustrate the above Gibbs phenomenon in the case of the square wave described earlier. In this case the period L is 2π, the discontinuity x_0 is at zero, and the jump a is equal to $\pi/2$. For simplicity let us just deal with the case when N is even (the case of odd N is very similar). Then we have

$$S_N f(x) = \sin(x) + \frac{1}{3}\sin(3x) + \cdots + \frac{1}{N-1}\sin((N-1)x).$$

Substituting $x = 0$, we obtain

$$S_N f(0) = 0 = \frac{-\frac{\pi}{4} + \frac{\pi}{4}}{2} = \frac{f(0^-) + f(0^+)}{2}$$

as claimed above. Next, we compute

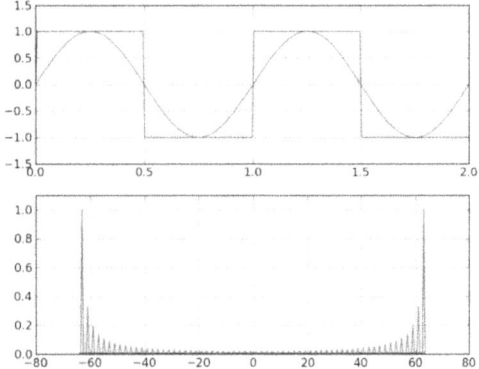

Animation of the additive synthesis of a square wave with an increasing number of harmonics. The Gibbs phenomenon is visible especially when the number of harmonics is large.

$$S_N f\left(\frac{2\pi}{2N}\right) = \sin\left(\frac{\pi}{N}\right) + \frac{1}{3}\sin\left(\frac{3\pi}{N}\right) + \cdots + \frac{1}{N-1}\sin\left(\frac{(N-1)\pi}{N}\right).$$

If we introduce the normalized sinc function, $\mathrm{sinc}(x)$, we can rewrite this as

$$S_N f\left(\frac{2\pi}{2N}\right) = \frac{\pi}{2}\left[\frac{2}{N}\mathrm{sinc}\left(\frac{1}{N}\right) + \frac{2}{N}\mathrm{sinc}\left(\frac{3}{N}\right) + ..\right.$$

$$\left. ..+ \frac{2}{N}\mathrm{sinc}\left(\frac{(N-1)}{N}\right)\right].$$

But the expression in square brackets is a numerical integration approximation to the integral $\int_0^1 \mathrm{sinc}(x)\,dx$ (more precisely, it is a midpoint rule approximation with spacing $2/N$). Since the sinc function is continuous, this approximation converges to the actual integral as $N \to \infty$. Thus we have

$$\lim_{N \to \infty} S_N f\left(\frac{2\pi}{2N}\right) = \frac{\pi}{2}\int_0^1 \mathrm{sinc}(x)\,dx$$

$$= \frac{1}{2}\int_{x=0}^1 \frac{\sin(\pi x)}{\pi x}\,d(\pi x)$$

$$= \frac{1}{2}\int_0^\pi \frac{\sin(t)}{t}\,dt \quad = \quad \frac{\pi}{4} + \frac{\pi}{2}\cdot(0.089490\ldots),$$

which was what was claimed in the previous section. A similar computation shows

$$\lim_{N \to \infty} S_N f\left(-\frac{2\pi}{2N}\right) = -\frac{\pi}{2}\int_0^1 \mathrm{sinc}(x)\,dx = -\frac{\pi}{4} - \frac{\pi}{2}$$

$$\cdot(0.089490\ldots).$$

18.5   Consequences

In signal processing, the Gibbs phenomenon is undesirable because it causes artifacts, namely clipping from the overshoot and undershoot, and ringing artifacts from the oscillations. In the case of low-pass filtering, these can be reduced or eliminated by using different low-pass filters.

In MRI, the Gibbs phenomenon causes artifacts in the presence of adjacent regions of markedly differing signal intensity. This is most commonly encountered in spinal MR imaging, where the Gibbs phenomenon may simulate the appearance of syringomyelia.

18.6   See also

- Pinsky phenomenon

- Compare with Runge's phenomenon for polynomial approximations

- Sigma approximation

- Sine integral

18.7   Notes

[1] Hewitt, Edwin; Hewitt, Robert E. (1979). "The Gibbs-Wilbraham phenomenon: An episode in Fourier analysis". *Archive for History of Exact Sciences* **21** (2): 129–160. doi:10.1007/BF00330404. Retrieved 16 September 2011. Available on-line at: National Chiao Tung University: Open Course Ware: Hewitt & Hewitt, 1979.

[2] Andrew Dimarogonas. *Vibration for engineers.* ISBN 0-13-462938-8.

[3] H. S. Carslaw (1930). *Introduction to the theory of Fourier's series and integrals* (Third ed.). New York: Dover Publications Inc. Chapter IX.

[4] Vretblad 2000 Section 4.7.

[5] Wilbraham, Henry (1848) "On a certain periodic function," *The Cambridge and Dublin Mathematical Journal*, **3** : 198-201.

[6] Bôcher, Maxime (April 1906) "Introduction to the theory of Fourier's series," *Annals of Mathethematics*, second series, **7** (3) : 81-152. The Gibbs phenomenon is discussed on pages 123-132; Gibbs' role is mentioned on page 129.

[7] M. Pinsky (2002). *Introduction to Fourier Analysis and Wavelets.* United states of America: Brooks/Cole. p. 27.

[8] Kelly, Susan E. "Gibbs Phenomenon for Wavelets." Applied and Computational Harmonic Analysis 3, 1995. http://www.uwlax.edu/faculty/kelly/Publications/GibbsJan.pdf

18.8 References

- Gibbs, J. Willard (1898), "Fourier's Series", *Nature* **59** (1522): 200, doi:10.1038/059200b0, ISSN 0028-0836

- Gibbs, J. Willard (1899), "Fourier's Series", *Nature* **59** (1539): 606, doi:10.1038/059606a0, ISSN 0028-0836

- Michelson, A. A.; Stratton, S. W. (1898), "A new harmonic analyser", *Philosophical Magazine* **5** (45): 85–91

- Antoni Zygmund, *Trigonometrical series*, Dover publications, 1955.

- Wilbraham, Henry (1848), "On a certain periodic function", *The Cambridge and Dublin Mathematical Journal* **3**: 198–201

- Paul J. Nahin, *Dr. Euler's Fabulous Formula*, Princeton University Press, 2006. Ch. 4, Sect. 4.

- Vretblad, Anders (2000), *Fourier Analysis and its Applications*, Graduate Texts in Mathematics **223**, New York: Springer Publishing, ISBN 0-387-00836-5

18.9 External links

- Hazewinkel, Michiel, ed. (2001), "Gibbs phenomenon", *Encyclopedia of Mathematics*, Springer, ISBN 978-1-55608-010-4

- Weisstein, Eric W., "*Gibbs Phenomenon*". From MathWorld—A Wolfram Web Resource.

- Prandoni, Paolo, "*Gibbs Phenomenon*".

- Radaelli-Sanchez, Ricardo, and Richard Baraniuk, "*Gibbs Phenomenon*". The Connexions Project. (Creative Commons Attribution License)

- Horatio S Carslaw : Introduction to the theory of Fourier's series and integrals.pdf (introductiontot00unkngoog.pdf) at archive.org

Chapter 19

Weak convergence (Hilbert space)

In mathematics, **weak convergence** in a Hilbert space is convergence of a sequence of points in the weak topology.

19.1   Definition

A sequence of points (x_n) in a Hilbert space H is said to **converge weakly** to a point x in H if

$$\langle x_n, y \rangle \to \langle x, y \rangle$$

for all y in H. Here, $\langle \cdot, \cdot \rangle$ is understood to be the inner product on the Hilbert space. The notation

$$x_n \rightharpoonup x$$

is sometimes used to denote this kind of convergence.

19.2   Properties

- If a sequence converges strongly, then it converges weakly as well.

- Since every closed and bounded set is weakly relatively compact (its closure in the weak topology is compact), every bounded sequence x_n in a Hilbert space H contains a weakly convergent subsequence. Note that closed and bounded sets are not in general weakly compact in Hilbert spaces (consider the set consisting of an orthonormal basis in an infinitely dimensional Hilbert space which is closed and bounded but not weakly compact since it doesn't contain 0). However, bounded and weakly closed sets are weakly compact so as a consequence every convex bounded closed set is weakly compact.

- As a consequence of the principle of uniform boundedness, every weakly convergent sequence is bounded.

- The norm is (sequentially) weakly lower-semicontinuous: if x_n converges weakly to x, then

$$\|x\| \le \liminf_{n \to \infty} \|x_n\|,$$

and this inequality is strict whenever the convergence is not strong. For example, infinite orthonormal sequences converge weakly to zero, as demonstrated below.

- If x_n converges weakly to x and we have the additional assumption that $\|x_n\| \to \|x\|$, then x_n converges to x strongly:

$$\langle x - x_n, x - x_n \rangle = \langle x, x \rangle + \langle x_n, x_n \rangle - \langle x_n, x \rangle - \langle x, x_n \rangle \to 0.$$

- If the Hilbert space is finite-dimensional, i.e. a Euclidean space, then the concepts of weak convergence and strong convergence are the same.

19.2.1   Example

The Hilbert space $L^2[0, 2\pi]$ is the space of the square-integrable functions on the interval $[0, 2\pi]$ equipped with the inner product defined by

$$\langle f, g \rangle = \int_0^{2\pi} f(x) \cdot g(x)\, dx,$$

(see L^p space). The sequence of functions $f_1, f_2, \ldots$ defined by

$$f_n(x) = \sin(nx)$$

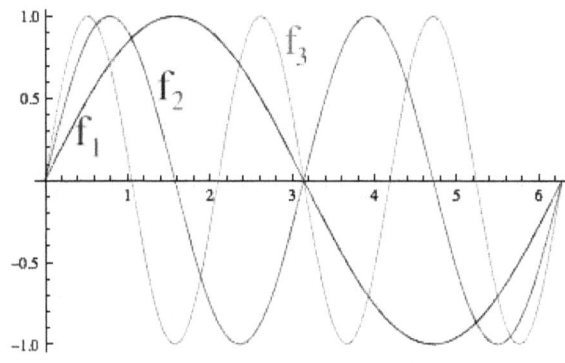

The first 3 functions in the sequence $f_n(x) = \sin(nx)$ on $[0, 2\pi]$. As $n \to \infty$ f_n converges weakly to $f = 0$.

converges weakly to the zero function in $L^2[0, 2\pi]$, as the integral

$$\int_0^{2\pi} \sin(nx) \cdot g(x) \, dx.$$

tends to zero for any square-integrable function g on $[0, 2\pi]$ when n goes to infinity, i.e.

$$\langle f_n, g \rangle \to \langle 0, g \rangle = 0.$$

Although f_n has an increasing number of 0's in $[0, 2\pi]$ as n goes to infinity, it is of course not equal to the zero function for any n. Note that f_n does not converge to 0 in the L_∞ or L_2 norms. This dissimilarity is one of the reasons why this type of convergence is considered to be "weak."

19.2.2 Weak convergence of orthonormal sequences

Consider a sequence e_n which was constructed to be orthonormal, that is,

$$\langle e_n, e_m \rangle = \delta_{mn}$$

where δ_{mn} equals one if $m = n$ and zero otherwise. We claim that if the sequence is infinite, then it converges weakly to zero. A simple proof is as follows. For $x \in H$, we have

$$\sum_n |\langle e_n, x \rangle|^2 \le \|x\|^2 \quad \text{(Bessel's inequality)}$$

where equality holds when $\{en\}$ is a Hilbert space basis. Therefore

$$|\langle e_n, x \rangle|^2 \to 0$$

i.e.

$$\langle e_n, x \rangle \to 0.$$

19.3 Banach–Saks theorem

The **Banach–Saks theorem** states that every bounded sequence x_n contains a subsequence x_{n_k} and a point x such that

$$\frac{1}{N} \sum_{k=1}^{N} x_{n_k}$$

converges strongly to x as N goes to infinity.

19.4 Generalizations

See also: Weak topology and Weak topology (polar topology)

The definition of weak convergence can be extended to Banach spaces. A sequence of points (x_n) in a Banach space B is said to **converge weakly** to a point x in B if

$$f(x_n) \to f(x)$$

for any bounded linear functional f defined on B, that is, for any f in the dual space B'. If B is a Hilbert space, then, by the Riesz representation theorem, any such f has the form

$$f(\cdot) = \langle \cdot, y \rangle$$

for some y in B, so one obtains the Hilbert space definition of weak convergence.

Chapter 20

Banach space

In mathematics, more specifically in functional analysis, a **Banach space** (pronounced [ˈbanax]) is a complete normed vector space. Thus, a Banach space is a vector space with a metric that allows the computation of vector length and distance between vectors and is complete in the sense that a Cauchy sequence of vectors always converges to a well defined limit that is within the space.

Banach spaces are named after the Polish mathematician Stefan Banach, who introduced and made a systematic study of them in 1920–1922 along with Hans Hahn and Eduard Helly.[1] Banach spaces originally grew out of the study of function spaces by Hilbert, Fréchet, and Riesz earlier in the century. Banach spaces play a central role in functional analysis. In other areas of analysis, the spaces under study are often Banach spaces.

20.1 Definition

A Banach space is a vector space X over the field **R** of real numbers, or over the field **C** of complex numbers, which is equipped with a norm and which is complete with respect to that norm, that is to say, for every Cauchy sequence $\{xn\}$ in X, there exists an element x in X such that

$$\lim_{n \to \infty} x_n = x,$$

or equivalently:

$$\lim_{n \to \infty} \|x_n - x\|_X = 0.$$

The vector space structure allows one to relate the behavior of Cauchy sequences to that of converging series of vectors. A normed space X is a Banach space if and only if each absolutely convergent series in X converges,[2]

$$\sum_{n=1}^{\infty} \|v_n\|_X < \infty \quad \text{that implies} \quad \sum_{n=1}^{\infty} v_n \text{ in converges } X.$$

Completeness of a normed space is preserved if the given norm is replaced by an equivalent one.

All norms on a finite-dimensional vector space are equivalent. Every finite-dimensional normed space over **R** or **C** is a Banach space.[3]

20.2 General theory

20.2.1 Linear operators, isomorphisms

Main article: Bounded operator

If X and Y are normed spaces over the same ground field **K**, the set of all continuous **K**-linear maps $T : X \to Y$ is denoted by $B(X, Y)$. In infinite-dimensional spaces, not all linear maps are continuous. A linear mapping from a normed space X to another normed space is continuous if and only if it is bounded on the closed unit ball of X. Thus, the vector space $B(X, Y)$ can be given the operator norm

$$\|T\| = \sup \{\|Tx\|_Y \mid x \in X, \|x\|_X \le 1\}.$$

For Y a Banach space, the space $B(X, Y)$ is a Banach space with respect to this norm.

If X is a Banach space, the space $B(X) = B(X, X)$ forms a unital Banach algebra; the multiplication operation is given by the composition of linear maps.

If X and Y are normed spaces, they are **isomorphic normed spaces** if there exists a linear bijection $T : X \to Y$ such that T and its inverse T^{-1} are continuous. If one of the two spaces X or Y is complete (or reflexive, separable, etc.) then so is the other space. Two normed spaces X and Y are **isometrically isomorphic** if in addition, T is an isometry, i.e., ||T(x)|| = ||x|| for every x in X. The Banach-Mazur distance $d(X, Y)$ between two isomorphic but not isometric spaces X and Y gives a measure of how much the two spaces X and Y differ.

20.2.2 Basic notions

Every normed space X can be isometrically embedded in a Banach space. More precisely, there is a Banach space Y and an isometric mapping $T : X \to Y$ such that $T(X)$ is dense in Y. If Z is another Banach space such that there is an isometric isomorphism from X onto a dense subset of Z, then Z is isometrically isomorphic to Y.

This Banach space Y is the **completion** of the normed space X. The underlying metric space for Y is the same as the metric completion of X, with the vector space operations extended from X to Y. The completion of X is often denoted by $\widehat{X}$.

The cartesian product $X \times Y$ of two normed spaces is not canonically equipped with a norm. However, several equivalent norms are commonly used,[4] such as

$$\|(x, y)\|_1 = \|x\| + \|y\|, \qquad \|(x, y)\|_\infty = \max(\|x\|, \|y\|)$$

and give rise to isomorphic normed spaces. In this sense, the product $X \times Y$ (or the direct sum $X \oplus Y$) is complete if and only if the two factors are complete.

If M is a closed linear subspace of a normed space X, there is a natural norm on the **quotient space** X / M,

$$\|x + M\| = \inf_{m \in M} \|x + m\|.$$

The quotient X / M is a Banach space when X is complete.[5] The **quotient map** from X onto X / M, sending x in X to its class $x + M$, is linear, onto and has norm 1, except when $M = X$, in which case the quotient is the null space.

The closed linear subspace M of X is said to be a **complemented subspace** of X if M is the range of a bounded linear projection P from X onto M. In this case, the space X is isomorphic to the direct sum of M and Ker(P), the kernel of the projection P.

Suppose that X and Y are Banach spaces and that $T \in B(X, Y)$. There exists a **canonical factorization** of T as[5]

$$T = T_1 \circ \pi, \quad T : X \xrightarrow{\pi} X / \operatorname{Ker}(T) \xrightarrow{T_1} Y$$

where the first map π is the quotient map, and the second map T_1 sends every class $x + \operatorname{Ker}(T)$ in the quotient to the image $T(x)$ in Y. This is well defined because all elements in the same class have the same image. The mapping T_1 is a linear bijection from $X / \operatorname{Ker}(T)$ onto the range $T(X)$, whose inverse need not be bounded.

20.2.3 Classical spaces

Basic examples[6] of Banach spaces include: the L^p spaces and their special cases, the sequence spaces ℓ^p that consist of scalar sequences indexed by **N**; among them, the space ℓ^1 of absolutely summable sequences and the space ℓ^2 of square summable sequences; the space c_0 of sequences tending to zero and the space ℓ^∞ of bounded sequences; the space $C(K)$ of continuous scalar functions on a compact Hausdorff space K, equipped with the max norm,

$$\|f\|_{C(K)} = \max\{|f(x)| : x \in K\}, \quad f \in C(K).$$

According to the Banach–Mazur theorem, every Banach space is isometrically isomorphic to a subspace of some $C(K)$.[7] For every separable Banach space X, there is a closed subspace M of ℓ^1 such that $X \cong \ell^1/M$.[8]

Any Hilbert space serves as an example of a Banach space. A Hilbert space H on **K** = **R**, **C** is complete for a norm of the form

$$\|x\|_H = \sqrt{\langle x, x \rangle},$$

where

$$\langle \cdot, \cdot \rangle : H \times H \to \mathbf{K}$$

is the inner product, linear in its first argument that satisfies the following:

$$\forall x, y \in H : \quad \langle y, x \rangle = \overline{\langle x, y \rangle},$$
$$\forall x \in H : \quad \langle x, x \rangle \geq 0,$$
$$\langle x, x \rangle = 0 \Leftrightarrow x = 0.$$

For example, the space L^2 is a Hilbert space.

The Hardy spaces, the Sobolev spaces are examples of Banach spaces that are related to L^p spaces and have additional structure. They are important in different branches of analysis, Harmonic analysis and Partial differential equations among others.

20.2.4 Banach algebras

A **Banach algebra** is a Banach space Λ over **K** = **R** or **C**, together with a structure of algebra over **K**, such that the product map $(a, b) \in A \times A \to ab \in A$ is continuous. An equivalent norm on A can be found so that $\|ab\| \leq \|a\| \, \|b\|$ for all $a, b \in A$.

Examples

- The Banach space $C(K)$, with the pointwise product, is a Banach algebra.

- The disk algebra $A(\mathbf{D})$ consists of functions holomorphic in the open unit disk $\mathbf{D} \subset \mathbf{C}$ and continuous on its closure: $\mathbf{D}$. Equipped with the max norm on $\mathbf{D}$, the disk algebra $A(\mathbf{D})$ is a closed subalgebra of $C(\mathbf{D})$.

- The Wiener algebra $A(\mathbf{T})$ is the algebra of functions on the unit circle $\mathbf{T}$ with absolutely convergent Fourier series. Via the map associating a function on $\mathbf{T}$ to the sequence of its Fourier coefficients, this algebra is isomorphic to the Banach algebra $\ell^1(\mathbf{Z})$, where the product is the convolution of sequences.

- For every Banach space X, the space $B(X)$ of bounded linear operators on X, with the composition of maps as product, is a Banach algebra.

- A C*-algebra is a complex Banach algebra A with an antilinear involution $a \to a^*$ such that $\|a^*a\| = \|a\|^2$. The space $B(H)$ of bounded linear operators on a Hilbert space H is a fundamental example of C*-algebra. The Gelfand–Naimark theorem states that every C*-algebra is isometrically isomorphic to a C*-subalgebra of some $B(H)$. The space $C(K)$ of complex continuous functions on a compact Hausdorff space K is an example of commutative C*-algebra, where the involution associates to every function f its complex conjugate f .

20.2.5 Dual space

Main article: Dual space

If X is a normed space and $\mathbf{K}$ the underlying field (either the real or the complex numbers), the **continuous dual space** is the space of continuous linear maps from X into $\mathbf{K}$, or **continuous linear functionals**. The notation for the continuous dual is $X' = B(X, \mathbf{K})$ in this article.[9] Since $\mathbf{K}$ is a Banach space (using the absolute value as norm), the dual X' is a Banach space, for every normed space X.

The main tool for proving the existence of continuous linear functionals is the Hahn–Banach theorem.

> **Hahn–Banach theorem.** Let X be a vector space over the field $\mathbf{K} = \mathbf{R}, \mathbf{C}$. Let further
>
> - $Y \subseteq X$ be a linear subspace,
> - $p : X \to \mathbf{R}$ be a sublinear function and

> - $f : Y \to \mathbf{K}$ be a linear functional so that Re($f(y)) \le p(y)$ for all y in Y.
>
> Then, there exists a linear functional $F : X \to \mathbf{K}$ so that
>
> $$F|_Y = f, \quad \text{and} \quad \forall x \in X, \ \mathrm{Re}(F(x)) \le p(x).$$

In particular, every continuous linear functional on a subspace of a normed space can be continuously extended to the whole space, without increasing the norm of the functional.[10] An important special case is the following: for every vector x in a normed space X, there exists a continuous linear functional f on X such that

$$f(x) = \|x\|_X, \quad \|f\|_{X'} \le 1.$$

When x is not equal to the $\mathbf{0}$ vector, the functional f must have norm one, and is called a **norming functional** for x.

The Hahn–Banach separation theorem states that two disjoint non-empty convex sets in a real Banach space, one of them open, can be separated by a closed affine hyperplane. The open convex set lies strictly on one side of the hyperplane, the second convex set lies on the other side but may touch the hyperplane.[11]

A subset S in a Banach space X is **total** if the linear span of S is dense in X. The subset S is total in X if and only if the only continuous linear functional that vanishes on S is the $\mathbf{0}$ functional: this equivalence follows from the Hahn–Banach theorem.

If X is the direct sum of two closed linear subspaces M and N, then the dual X' of X is isomorphic to the direct sum of the duals of M and N.[12] If M is a closed linear subspace in X, one can associate the *orthogonal of* M in the dual,

$$M^\perp = \{x' \in X' : x'(m) = 0, \ \forall m \in M\}.$$

The orthogonal $M^\perp$ is a closed linear subspace of the dual. The dual of M is isometrically isomorphic to $X'/M^\perp$. The dual of X/M is isometrically isomorphic to $M^\perp$.[13]

The dual of a separable Banach space need not be separable, but:

> **Theorem.**[14] Let X be a normed space. If X' is separable, then X is separable.

When X' is separable, the above criterion for totality can be used for proving the existence of a countable total subset in X.

Weak topologies

The **weak topology** on a Banach space X is the coarsest topology on X for which all elements x' in the continuous dual space X' are continuous. The norm topology is therefore finer than the weak topology. It follows from the Hahn–Banach separation theorem that the weak topology is Hausdorff, and that a norm-closed convex subset of a Banach space is also weakly closed.[15] A norm-continuous linear map between two Banach spaces X and Y is also **weakly continuous**, i.e., continuous from the weak topology of X to that of Y.[16]

If X is infinite-dimensional, there exist linear maps which are not continuous. The space X^* of all linear maps from X to the underlying field **K** (this space X^* is called the algebraic dual space, to distinguish it from X') also induces a topology on X which is finer than the weak topology, and much less used in functional analysis.

On a dual space X', there is a topology weaker than the weak topology of X', called **weak* topology**. It is the coarsest topology on X' for which all evaluation maps $x' \in X' \to x'(x)$, $x \in X$, are continuous. Its importance comes from the Banach–Alaoglu theorem.

> **Banach–Alaoglu Theorem.** Let X be a normed vector space. Then the closed unit ball $B' = \{x' \in X' : \|x'\| \le 1\}$ of the dual space is compact in the weak* topology.

The Banach–Alaoglu theorem depends on Tychonoff's theorem about infinite products of compact spaces. When X is separable, the unit ball B' of the dual is a metrizable compact in the weak* topology.[17]

Examples of dual spaces

The dual of c_0 is isometrically isomorphic to ℓ^1: for every bounded linear functional f on c_0, there is a unique element $y = \{y_n\} \in \ell^1$ such that

$$f(x) = \sum_{n \in \mathbf{N}} x_n y_n, \qquad x = \{x_n\} \in c_0, \text{ and } \|f\|_{(c_0)'} = \|y\|_{\ell_1}.$$

The dual of ℓ^1 is isometrically isomorphic to ℓ^∞. The dual of $L^p([0, 1])$ is isometrically isomorphic to $L^q([0, 1])$ when $1 \le p < \infty$ and $1/p + 1/q = 1$.

For every vector y in a Hilbert space H, the mapping

$$x \in H \to f_y(x) = \langle x, y \rangle$$

defines a continuous linear functional fy on H. The Riesz representation theorem states that every continuous linear functional on H is of the form fy for a uniquely defined vector y in H. The mapping $y \in H \to fy$ is an antilinear isometric bijection from H onto its dual H'. When the scalars are real, this map is an isometric isomorphism.

When K is a compact Hausdorff topological space, the dual $M(K)$ of $C(K)$ is the space of Radon measures in the sense of Bourbaki.[18] The subset $P(K)$ of $M(K)$ consisting of non-negative measures of mass 1 (probability measures) is a convex w*-closed subset of the unit ball of $M(K)$. The extreme points of $P(K)$ are the Dirac measures on K. The set of Dirac measures on K, equipped with the w*-topology, is homeomorphic to K.

> **Banach-Stone Theorem.** If K and L are compact Hausdorff spaces and if $C(K)$ and $C(L)$ are isometrically isomorphic, then the topological spaces K and L are homeomorphic.[19][20]

The result has been extended by Amir[21] and Cambern[22] to the case when the multiplicative Banach–Mazur distance between $C(K)$ and $C(L)$ is < 2. The theorem is no longer true when the distance is = 2.[23]

In the commutative Banach algebra $C(K)$, the maximal ideals are precisely kernels of Dirac mesures on K,

$$I_x = \ker \delta_x = \{f \in C(K) : f(x) = 0\}, \quad x \in K.$$

More generally, by the Gelfand-Mazur theorem, the maximal ideals of a unital commutative Banach algebra can be identified with its characters---not merely as sets but as topological spaces: the former with the hull-kernel topology and the latter with the w*-topology. In this identification, the maximal ideal space can be viewed as a w*-compact subset of the unit ball in the dual A'.

> **Theorem.** If K is a compact Hausdorff space, then the maximal ideal space Ξ of the Banach algebra $C(K)$ is homeomorphic to K.[19]

Not every unital commutative Banach algebra is of the form $C(K)$ for some compact Hausdorff space K. However, this statement holds if one places $C(K)$ in the smaller category of commutative C*-algebras. Gelfand's representation theorem for commutative C* algebras states that every commutative unital C*-algebra A is isometrically isomorphic to a $C(K)$ space.[24] The Hausdorff compact space K here is again the maximal ideal space, also called the spectrum of A in the C*-algebra context.

Bidual

If X is a normed space, the (continuous) dual X'' of the dual X' is called **bidual**, or **second dual** of X. For every normed space X, there is a natural map,

$$\begin{cases} F_X : X \to X'' \\ F_X(x)(f) = f(x) \quad \forall x \in X, \forall f \in X' \end{cases}$$

This defines $FX(x)$ as a continuous linear functional on X', i.e., an element of X''. The map $FX : x \to FX(x)$ is a linear map from X to X''. As a consequence of the existence of a norming functional f for every x in X, this map FX is isometric, thus injective.

For example, the dual of $X = c_0$ is identified with ℓ^1, and the dual of ℓ^1 is identified with ℓ^∞, the space of bounded scalar sequences. Under these identifications, FX is the inclusion map from c_0 to ℓ^∞. It is indeed isometric, but not onto.

If FX is surjective, then the normed space X is called **reflexive** (see below). Being the dual of a normed space, the bidual X'' is complete, therefore, every reflexive normed space is a Banach space.

Using the isometric embedding FX, it is customary to consider a normed space X as a subset of its bidual. When X is a Banach space, it is viewed as a closed linear subspace of X''. If X is not reflexive, the unit ball of X is a proper subset of the unit ball of X''. The Goldstine theorem states that the unit ball of a normed space is weakly*-dense in the unit ball of the bidual. In other words, for every x'' in the bidual, there exists a net $\{xj\}$ in X so that

$$\sup_j \|x_j\| \le \|x''\|, \quad x''(f) = \lim_j f(x_j), \quad f \in X'.$$

The net may be replaced by a weakly*-convergent sequence when the dual X' is separable. On the other hand, elements of the bidual of ℓ^1 that are not in ℓ^1 cannot be weak*-limit of *sequences* in ℓ^1, since ℓ^1 is weakly sequentially complete.

20.2.6 Banach's theorems

Here are the main general results about Banach spaces that go back to the time of Banach's book (Banach (1932)) and are related to the Baire category theorem. According to this theorem, a complete metric space (such as a Banach space, a Fréchet space or an F-space) cannot be equal to a union of countably many closed subsets with empty interiors. Therefore, a Banach space cannot be the union of countably many closed subspaces, unless it is already equal to one of them; a Banach space with a countable Hamel basis is finite-dimensional.

Banach–Steinhaus Theorem. Let X be a Banach space and Y be a normed vector space. Suppose that F is a collection of continuous linear operators from X to Y. The uniform boundedness principle states that if for all x in X we have $\sup T \in F \, \|T(x)\| Y < \infty$, then $\sup T \in F \, \|T\| Y < \infty$.

The Banach–Steinhaus theorem is not limited to Banach spaces. It can be extended for example to the case where X is a Fréchet space, provided the conclusion is modified as follows: under the same hypothesis, there exists a neighborhood U of **0** in X such that all T in F are uniformly bounded on U,

$$\sup_{T \in F} \sup_{x \in U} \|T(x)\|_Y < \infty.$$

The Open Mapping Theorem. Let X and Y be Banach spaces and $T : X \to Y$ be a continuous linear operator. Then T is surjective if and only if T is an open map.

Corollary. Every one-to-one bounded linear operator from a Banach space onto a Banach space is an isomorphism.

The First Isomorphism Theorem for Banach spaces. Suppose that X and Y are Banach spaces and that $T \in B(X, Y)$. Suppose further that the range of T is closed in Y. Then $X/\operatorname{Ker}(T)$ is isomorphic to $T(X)$.

This result is a direct consequence of the preceding *Banach isomorphism theorem* and of the canonical factorization of bounded linear maps.

Corollary. If a Banach space X is the internal direct sum of closed subspaces $M_1, ..., Mn$, then X is isomorphic to $M_1 \oplus ... \oplus Mn$.

This is another consequence of Banach's isomorphism theorem, applied to the continuous bijection from $M_1 \oplus ... \oplus Mn$ onto X sending $(m_1, ..., mn)$ to the sum $m_1 + ... + mn$.

The Closed Graph Theorem. Let $T : X \to Y$ be a linear mapping between Banach spaces. The graph of T is closed in $X \times Y$ if and only if T is continuous.

20.2.7 Reflexivity

Main article: Reflexive space

The normed space X is called **reflexive** when the natural map

$$\begin{cases} F_X : X \to X'' \\ F_X(x)(f) = f(x) \quad \forall x \in X, \forall f \in X' \end{cases}$$

is surjective. Reflexive normed spaces are Banach spaces.

Theorem. If X is a reflexive Banach space, every closed subspace of X and every quotient space of X are reflexive.

This is a consequence of the Hahn–Banach theorem. Further, by the open mapping theorem, if there is a bounded linear operator from the Banach space X onto the Banach space Y, then Y is reflexive.

Theorem. If X is a Banach space, then X is reflexive if and only if X' is reflexive.

Corollary. Let X be a reflexive Banach space. Then X is separable if and only if X' is separable.

Indeed, if the dual Y' of a Banach space Y is separable, then Y is separable. If X is reflexive and separable, then the dual of X' is separable, so X' is separable.

Theorem. Suppose that $X_1, ..., Xn$ are normed spaces and that $X = X_1 \oplus ... \oplus Xn$. Then X is reflexive if and only if each Xj is reflexive.

Hilbert spaces are reflexive. The L^p spaces are reflexive when $1 < p < \infty$. More generally, uniformly convex spaces are reflexive, by the Milman–Pettis theorem. The spaces c_0, ℓ^1, $L^1([0, 1])$, $C([0, 1])$ are not reflexive. In these examples of non-reflexive spaces X, the bidual X'' is "much larger" than X. Namely, under the natural isometric embedding of X into X'' given by the Hahn–Banach theorem, the quotient X'' / X is infinite-dimensional, and even nonseparable. However, Robert C. James has constructed an example[25] of a non-reflexive space, usually called "*the James space*" and denoted by J,[26] such that the quotient J'' / J is one-dimensional. Furthermore, this space J is isometrically isomorphic to its bidual.

Theorem. A Banach space X is reflexive if and only if its unit ball is compact in the weak topology.

When X is reflexive, it follows that all closed and bounded convex subsets of X are weakly compact. In a Hilbert space H, the weak compactness of the unit ball is very often used in the following way: every bounded sequence in H has weakly convergent subsequences.

Weak compactness of the unit ball provides a tool for finding solutions in reflexive spaces to certain optimization problems. For example, every convex continuous function on the unit ball B of a reflexive space attains its minimum at some point in B.

As a special case of the preceding result, when X is a reflexive space over **R**, every continuous linear functional f in X' attains its maximum $\| f \|$ on the unit ball of X. The following theorem of Robert C. James provides a converse statement.

James' Theorem. For a Banach space the following two properties are equivalent:

- X is reflexive.
- for all f in X' there exists x in X with $\|x\| \leq 1$, so that $f(x) = \| f \|$.

The theorem can be extended to give a characterization of weakly compact convex sets.

On every non-reflexive Banach space X, there exist continuous linear functionals that are not *norm-attaining*. However, the Bishop–Phelps theorem[27] states that norm-attaining functionals are norm dense in the dual X' of X.

20.2.8 Weak convergences of sequences

A sequence $\{xn\}$ in a Banach space X is **weakly convergent** to a vector $x \in X$ if $f(xn)$ converges to $f(x)$ for every continuous linear functional f in the dual X'. The sequence $\{xn\}$ is a **weakly Cauchy sequence** if $f(xn)$ converges to a scalar limit $L(f)$, for every f in X'. A sequence $\{fn\}$ in the dual X' is **weakly* convergent** to a functional $f \in X'$ if $fn(x)$ converges to $f(x)$ for every x in X. Weakly Cauchy sequences, weakly convergent and weakly* convergent sequences are norm bounded, as a consequence of the Banach–Steinhaus theorem.

When the sequence $\{xn\}$ in X is a weakly Cauchy sequence, the limit L above defines a bounded linear functional on the dual X', i.e., an element L of the bidual of X, and L is the limit of $\{xn\}$ in the weak*-topology of the bidual. The Banach space X is **weakly sequentially complete** if every weakly Cauchy sequence is weakly convergent in X. It follows from the preceding discussion that reflexive spaces are weakly sequentially complete.

Theorem. [28] For every measure μ, the space $L^1(\mu)$ is weakly sequentially complete.

An orthonormal sequence in a Hilbert space is a simple example of a weakly convergent sequence, with limit equal to the **0** vector. The unit vector basis of ℓ^p, $1 < p < \infty$, or of c_0, is another example of a **weakly null sequence**, i.e., a sequence that converges weakly to **0**. For every weakly null sequence in a Banach space, there exists a sequence of convex combinations of vectors from the given sequence that is norm-converging to **0**.[29]

The unit vector basis of ℓ^1 is not weakly Cauchy. Weakly Cauchy sequences in ℓ^1 are weakly convergent, since L^1-spaces are weakly sequentially complete. Actually, weakly convergent sequences in ℓ^1 are norm convergent.[30] This means that ℓ^1 satisfies Schur's property.

Results involving the ℓ^1 basis

Weakly Cauchy sequences and the ℓ^1 basis are the opposite cases of the dichotomy established in the following deep result of H. P. Rosenthal.[31]

> **Theorem.**[32] Let $\{xn\}$ be a bounded sequence in a Banach space. Either $\{xn\}$ has a weakly Cauchy subsequence, or it admits a subsequence equivalent to the standard unit vector basis of ℓ^1.

A complement to this result is due to Odell and Rosenthal (1975).

> **Theorem.**[33] Let X be a separable Banach space. The following are equivalent:
>
> - The space X does not contain a closed subspace isomorphic to ℓ^1.
> - Every element of the bidual X'' is the weak*-limit of a sequence $\{xn\}$ in X.

By the Goldstine theorem, every element of the unit ball B'' of X'' is weak*-limit of a net in the unit ball of X. When X does not contain ℓ^1, every element of B'' is weak*-limit of a *sequence* in the unit ball of X.[34]

When the Banach space X is separable, the unit ball of the dual X', equipped with the weak*-topology, is a metrizable compact space K,[17] and every element x'' in the bidual X'' defines a bounded function on K:

$$x' \in K \mapsto x''(x'), \quad |x''(x')| \leq \|x''\|.$$

This function is continuous for the compact topology of K if and only if x'' is actually in X, considered as subset of X''. Assume in addition for the rest of the paragraph that X does not contain ℓ^1. By the preceding result of Odell and Rosenthal, the function x'' is the pointwise limit on K of a sequence $\{xn\} \subset X$ of continuous functions on K, it is therefore a first Baire class function on K. The unit ball of the bidual is a pointwise compact subset of the first Baire class on K.[35]

Sequences, weak and weak* compactness

When X is separable, the unit ball of the dual is weak*-compact by Banach–Alaoglu and metrizable for the weak* topology,[17] hence every bounded sequence in the dual has weakly* convergent subsequences. This applies to separable reflexive spaces, but more is true in this case, as stated below.

The weak topology of a Banach space X is metrizable if and only if X is finite-dimensional.[36] If the dual X' is separable, the weak topology of the unit ball of X is metrizable. This applies in particular to separable reflexive Banach spaces. Although the weak topology of the unit ball is not metrizable in general, one can characterize weak compactness using sequences.

> **Eberlein–Šmulian theorem.**[37] A set A in a Banach space is relatively weakly compact if and only if every sequence $\{an\}$ in A has a weakly convergent subsequence.

A Banach space X is reflexive if and only if each bounded sequence in X has a weakly convergent subsequence.[38]

A weakly compact subset A in ℓ^1 is norm-compact. Indeed, every sequence in A has weakly convergent subsequences by Eberlein–Šmulian, that are norm convergent by the Schur property of ℓ^1.

20.3   Schauder bases

Main article: Schauder basis

A **Schauder basis** in a Banach space X is a sequence $\{en\}n \geq 0$ of vectors in X with the property that for every vector x in X, there exist *uniquely* defined scalars $\{xn\}n \geq 0$ depending on x, such that

$$x = \sum_{n=0}^{\infty} x_n e_n, \quad i.e., \quad x = \lim_n P_n(x), \ P_n(x) := \sum_{k=0}^{n} x_k e_k.$$

Banach spaces with a Schauder basis are necessarily separable, because the countable set of finite linear combinations with rational coefficients (say) is dense.

It follows from the Banach–Steinhaus theorem that the linear mappings {Pn} are uniformly bounded by some constant C. Let {e_*^n} denote the coordinate functionals which assign to every x in X the coordinate xn of x in the above expansion. They are called **biorthogonal functionals**. When the basis vectors have norm 1, the coordinate functionals {e_*^n} have norm ≤ $2C$ in the dual of X.

Most classical separable spaces have explicit bases. The Haar system {hn} is a basis for $L^p([0, 1])$, $1 \leq p < \infty$. The trigonometric system is a basis in $L^p(\mathbf{T})$ when $1 < p < \infty$. The Schauder system is a basis in the space $C([0, 1])$.[39] The question of whether the disk algebra $A(\mathbf{D})$ has a basis[40] remained open for more than forty years, until Bočkarev showed in 1974 that $A(\mathbf{D})$ admits a basis constructed from the Franklin system.[41]

Since every vector x in a Banach space X with a basis is the limit of $Pn(x)$, with Pn of finite rank and uniformly bounded, the space X satisfies the bounded approximation property. The first example[42] by Enflo of a space failing the approximation property was at the same time the first example of a separable Banach space without a Schauder basis.

Robert C. James characterized reflexivity in Banach spaces with a basis: the space X with a Schauder basis is reflexive if and only if the basis is both shrinking and boundedly complete.[43] In this case, the biorthogonal functionals form a basis of the dual of X.

20.4 Tensor product

Main article: Tensor product
Let X and Y be two **K**-vector spaces. The tensor product

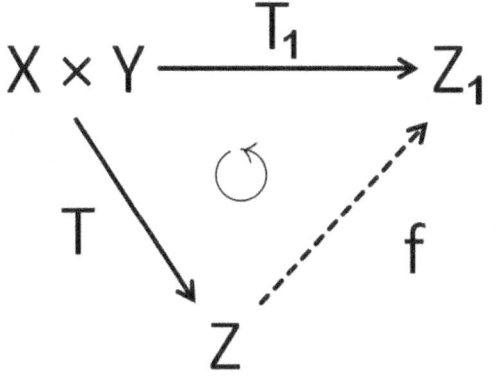

$X \otimes Y$ of X and Y is a **K**-vector space Z with a bilinear mapping $T : X \times Y \to Z$ which has the following universal property:

> If $T_1 : X \times Y \to Z_1$ is any bilinear mapping into a **K**-vector space Z_1, then there exists a unique linear mapping $f : Z \to Z_1$ such that $T_1 = f \circ T$.

The image under T of a couple (x, y) in $X \times Y$ is denoted by $x \otimes y$, and called a **simple tensor**. Every element z in $X \otimes Y$ is a finite sum of such simple tensors.

There are various norms that can be placed on the tensor product of the underlying vector spaces, amongst others the projective cross norm and injective cross norm introduced by A. Grothendieck in 1955.[44]

In general, the tensor product of complete spaces is not complete again. When working with Banach spaces, it is customary to call **projective tensor product**[45] of two Banach spaces X and Y the *completion* $X \widehat{\otimes}_\pi Y$ of the algebraic tensor product $X \otimes Y$ equipped with the projective tensor norm, and similarly for the **injective tensor product**[46] $X \widehat{\otimes}_\varepsilon Y$. Grothendieck proved in particular that[47]

$$C(K) \widehat{\otimes}_\varepsilon Y \simeq C(K, Y),$$
$$L^1([0,1]) \widehat{\otimes}_\pi Y \simeq L^1([0,1], Y),$$

where K is a compact Hausdorff space, $C(K, Y)$ the Banach space of continuous functions from K to Y and $L^1([0, 1], Y)$ the space of Bochner-measurable and integrable functions from [0, 1] to Y, and where the isomorphisms are isometric. The two isomorphisms above are the respective extensions of the map sending the tensor $f \otimes y$ to the vector-valued function $s \in K \to f(s)y \in Y$.

20.4.1 Tensor products and the approximation property

Let X be a Banach space. The tensor product $X' \widehat{\otimes}_\varepsilon X$ is identified isometrically with the closure in $B(X)$ of the set of finite rank operators. When X has the approximation property, this closure coincides with the space of compact operators on X.

For every Banach space Y, there is a natural norm 1 linear map

$$Y \widehat{\otimes}_\pi X \to Y \widehat{\otimes}_\varepsilon X$$

obtained by extending the identity map of the algebraic tensor product. Grothendieck related the approximation problem to the question of whether this map is one-to-one when Y is the dual of X. Precisely, for every Banach space X, the map

$$X' \widehat{\otimes}_\pi X \longrightarrow X' \widehat{\otimes}_\varepsilon X$$

is one-to-one if and only if X has the approximation property.[48]

Grothendieck conjectured that $X \widehat{\otimes}_\pi Y$ and $X \widehat{\otimes}_\varepsilon Y$ must be different whenever X and Y are infinite-dimensional Banach spaces. This was disproved by Gilles Pisier in 1983.[49] Pisier constructed an infinite-dimensional Banach space X such that $X \widehat{\otimes}_\pi X$ and $X \widehat{\otimes}_\varepsilon X$ are equal. Furthermore, just as Enflo's example, this space X is a "hand-made" space that fails to have the approximation property. On the other hand, Szankowski proved that the classical space $B(\ell^2)$ does not have the approximation property.[50]

20.5 Some classification results

20.5.1 Characterizations of Hilbert space among Banach spaces

A necessary and sufficient condition for the norm of a Banach space X to be associated to an inner product is the parallelogram identity:

$$\forall x, y \in X : \qquad \|x+y\|^2 + \|x-y\|^2 = 2\left(\|x\|^2 + \|y\|^2\right).$$

It follows, for example, that the Lebesgue space $L^p([0, 1])$ is a Hilbert space only when $p = 2$. If this identity is satisfied, the associated inner product is given by the polarization identity. In the case of real scalars, this gives:

$$\langle x, y \rangle = \tfrac{1}{4}\left(\|x+y\|^2 - \|x-y\|^2\right).$$

For complex scalars, defining the inner product so as to be **C**-linear in x, antilinear in y, the polarization identity gives:

$$\langle x, y \rangle = \tfrac{1}{4}\left(\|x+y\|^2 - \|x-y\|^2 + i\left(\|x+iy\|^2 - \|x\right.\right.$$
$$\left.\left. - iy\|^2\right)\right).$$

To see that the parallelogram law is sufficient, one observes in the real case that $<x, y>$ is symmetric, and in the complex case, that it satisfies the Hermitian symmetry property and $<ix, y> = i<x, y>$. The parallelogram law implies that $<x, y>$ is additive in x. It follows that it is linear over the rationals, thus linear by continuity.

Several characterizations of spaces isomorphic (rather than isometric) to Hilbert spaces are available. The parallelogram law can be extended to more than two vectors, and weakened by the introduction of a two-sided inequality with a constant $c \geq 1$: Kwapień proved that if

$$c^{-2} \sum_{k=1}^{n} \|x_k\|^2 \leq \mathrm{Ave}_\pm \left\| \sum_{k=1}^{n} \pm x_k \right\|^2 \leq c^2 \sum_{k=1}^{n} \|x_k\|^2$$

for every integer n and all families of vectors $\{x_1, ..., x_n\} \subset X$, then the Banach space X is isomorphic to a Hilbert space.[51] Here, Ave± denotes the average over the 2^n possible choices of signs ±1. In the same article, Kwapień proved that the validity of a Banach-valued Parseval's theorem for the Fourier transform characterizes Banach spaces isomorphic to Hilbert spaces.

Lindenstrauss and Tzafriri proved that a Banach space in which every closed linear subspace is complemented (that is, is the range of a bounded linear projection) is isomorphic to a Hilbert space.[52] The proof rests upon Dvoretzky's theorem about Euclidean sections of high-dimensional centrally symmetric convex bodies. In other words, Dvoretzky's theorem states that for every integer n, any finite-dimensional normed space, with dimension sufficiently large compared to n, contains subspaces nearly isometric to the n-dimensional Euclidean space.

The next result gives the solution of the so-called *homogeneous space problem*. An infinite-dimensional Banach space X is said to be **homogeneous** if it is isomorphic to all its infinite-dimensional closed subspaces. A Banach space isomorphic to ℓ^2 is homogeneous, and Banach asked for the converse.[53]

> **Theorem.**[54] A Banach space isomorphic to all its infinite-dimensional closed subspaces is isomorphic to a separable Hilbert space.

An infinite-dimensional Banach space is **hereditarily indecomposable** when no subspace of it can be isomorphic to the direct sum of two infinite-dimensional Banach spaces. The Gowers dichotomy theorem[54] asserts that every infinite-dimensional Banach space X contains, either a subspace Y with unconditional basis, or a hereditarily indecomposable subspace Z, and in particular, Z is not isomorphic to its closed hyperplanes.[55] If X is homogeneous, it must therefore have an unconditional basis. It follows then from the partial solution obtained by Komorowski and Tomczak–Jaegermann, for spaces with an unconditional basis,[56] that X is isomorphic to ℓ^2.

20.5.2 Spaces of continuous functions

When two compact Hausdorff spaces K_1 and K_2 are homeomorphic, the Banach spaces $C(K_1)$ and $C(K_2)$ are isometric. Conversely, when K_1 is not homeomorphic to K_2, the (multiplicative) Banach–Mazur distance between

$C(K_1)$ and $C(K_2)$ must be greater than or equal to 2, see above the results by Amir and Cambern. Although uncountable compact metric spaces can have different homeomorphy types, one has the following result due to Milutin:[57]

> **Theorem.**[58] Let K be an uncountable compact metric space. Then $C(K)$ is isomorphic to $C([0, 1])$.

The situation is different for countably infinite compact Hausdorff spaces. Every countably infinite compact K is homeomorphic to some closed interval of ordinal numbers

$$\langle 1, \alpha \rangle = \{\gamma \; : \; 1 \le \gamma \le \alpha\}$$

equipped with the order topology, where α is a countably infinite ordinal.[59] The Banach space $C(K)$ is then isometric to $C(<1, \alpha>)$. When α, β are two countably infinite ordinals, and assuming $\alpha \le \beta$, the spaces $C(<1, \alpha>)$ and $C(<1, \beta>)$ are isomorphic if and only if $\beta < \alpha^\omega$.[60] For example, the Banach spaces

$$C(\langle 1, \omega \rangle), \; C(\langle 1, \omega^\omega \rangle), \; C(\langle 1, \omega^{\omega^2} \rangle), \; C(\langle 1, \omega^{\omega^3}$$

$$\rangle), \cdots, C(\langle 1, \omega^{\omega^\omega} \rangle), \cdots$$

are mutually non-isomorphic.

20.6 Examples

Main article: List of Banach spaces

A glossary of symbols:

- **K = R, C**;

- X is a compact Hausdorff space;

- I is a closed and bounded interval $[a, b]$;

- p, q are real numbers with $1 < p, q < \infty$ so that $1/p + 1/q = 1$.

- Σ is a σ-algebra of sets;

- Ξ is an algebra of sets (for spaces only requiring finite additivity, such as the ba space);

- μ is a measure with variation $|\mu|$.

20.7 Derivatives

Several concepts of a derivative may be defined on a Banach space. See the articles on the Fréchet derivative and the Gâteaux derivative for details. The Fréchet derivative allows for an extension of the concept of a directional derivative to Banach spaces. The Gâteaux derivative allows for an extension of a directional derivative to locally convex topological vector spaces. Fréchet differentiability is a stronger condition than Gâteaux differentiability. The quasi-derivative is another generalization of directional derivative that implies a stronger condition than Gâteaux differentiability, but a weaker condition than Fréchet differentiability.

20.8 Generalizations

Several important spaces in functional analysis, for instance the space of all infinitely often differentiable functions **R** → **R**, or the space of all distributions on **R**, are complete but are not normed vector spaces and hence not Banach spaces. In Fréchet spaces one still has a complete metric, while LF-spaces are complete uniform vector spaces arising as limits of Fréchet spaces.

20.9 See also

- Space (mathematics)
  - Hilbert space
  - Lp space
  - Sobolev space
  - Hardy space
- Interpolation space
- Distortion problem

20.10 Notes

[1] Bourbaki 1987, V.86

[2] see Theorem 1.3.9, p. 20 in Megginson (1998).

[3] see Corollary 1.4.18, p. 32 in Megginson (1998).

[4] see Banach (1932), p. 182.

[5] see pp. 17–19 in Carothers (2005).

[6] see Banach (1932), pp. 11-12.

[7] see Banach (1932), Th. 9 p. 185.

[8] see Theorem 6.1, p. 55 in Carothers (2005)

[9] Several books about functional analysis use the notation X^* for the continuous dual, for example Carothers (2005), Lindenstrauss & Tzafriri (1977), Megginson (1998), Ryan (2002), Wojtaszczyk (1991).

[10] Theorem 1.9.6, p. 75 in Megginson (1998)

[11] see also Theorem 2.2.26, p. 179 in Megginson (1998)

[12] see p. 19 in Carothers (2005).

[13] Theorems 1.10.16, 1.10.17 pp.94–95 in Megginson (1998)

[14] Theorem 1.12.11, p. 112 in Megginson (1998)

[15] Theorem 2.5.16, p. 216 in Megginson (1998).

[16] see II.A.8, p. 29 in Wojtaszczyk (1991)

[17] see Theorem 2.6.23, p. 231 in Megginson (1998).

[18] see N. Bourbaki, (2004), "Integration I", Springer Verlag, ISBN 3-540-41129-1.

[19] Eilenberg, Samuel (Jan 22, 1942). "Banach Space Methods in Topology". *Annals of Mathematics* **43** (3): 568. doi:10.2307/1968812. Check date values in: |access-date= (help);

[20] see also Banach (1932), p. 170 for metrizable K and L.

[21] see D. Amir, "On isomorphisms of continuous function spaces". Israel J. Math. **3** (1965), 205–210.

[22] M. Cambern, "A generalized Banach-Stone theorem". Proc. Amer. Math. Soc. **17** (1966), 396–400, and "On isomorphisms with small bound". Proc. Amer. Math. Soc. **18** (1967), 1062–1066.

[23] H. B. Cohen, "A bound-two isomorphism between $C(X)$ Banach spaces". Proc. Amer. Math. Soc. **50** (1975), 215–217.

[24] see for example W. Arveson, (1976), "An Invitation to C*-Algebra", Springer-Verlag, ISBN 0-387-90176-0.

[25] R. C. James (1951). "A non-reflexive Banach space isometric with its second conjugate space". *Proc. Natl. Acad. Sci. U.S.A.* **37**: 174–177. doi:10.1073/pnas.37.3.174.

[26] see Lindenstrauss & Tzafriri (1977), p. 25.

[27] see E. Bishop and R. Phelps, "A proof that every Banach space is subreflexive". Bull. Amer. Math. Soc. **67** (1961), 97–98.

[28] see III.C.14, p. 140 in Wojtaszczyk (1991).

[29] see Corollary 2, p. 11 in Diestel (1984).

[30] see p. 85 in Diestel (1984).

[31] Rosenthal, Haskell P. (1974), "A characterization of Banach spaces containing ℓ^1", Proc. Nat. Acad. Sci. U.S.A. **71**:2411–2413. Rosenthal's proof is for real scalars. The complex version of the result is due to L. Dor, in Dor, Leonard E. (1975), "On sequences spanning a complex ℓ^1 space", Proc. Amer. Math. Soc. **47**:515–516.

[32] see p. 201 in Diestel (1984).

[33] Odell, Edward W.; Rosenthal, Haskell P. (1975), "A double-dual characterization of separable Banach spaces containing ℓ^1", *Israel J. Math.* **20**: 375–384, doi:10.1007/bf02760341.

[34] Odell and Rosenthal, Sublemma p. 378 and Remark p. 379.

[35] for more on pointwise compact subsets of the Baire class, see Bourgain, Jean; Fremlin, D. H.; Talagrand, Michel (1978), "Pointwise Compact Sets of Baire-Measurable Functions", *American J. of Math.* **100**: 845–886, doi:10.2307/2373913.

[36] see Proposition 2.5.14, p. 215 in Megginson (1998).

[37] see for example p. 49, II.C.3 in Wojtaszczyk (1991).

[38] see Corollary 2.8.9, p. 251 in Megginson (1998).

[39] see Lindenstrauss & Tzafriri (1977) p. 3.

[40] the question appears p. 238, §3 in Banach's book, Banach (1932).

[41] see S. V. Bočkarev, "Existence of a basis in the space of functions analytic in the disc, and some properties of Franklin's system". (Russian) Mat. Sb. (N.S.) 95(137) (1974), 3–18, 159.

[42] see P. Enflo, "A counterexample to the approximation property in Banach spaces". Acta Math. 130, 309–317(1973).

[43] see R.C. James, "Bases and reflexivity of Banach spaces". Ann. of Math. (2) 52, (1950). 518–527. See also Lindenstrauss & Tzafriri (1977) p. 9.

[44] see A. Grothendieck, "Produits tensoriels topologiques et espaces nucléaires". Mem. Amer. Math. Soc. 1955 (1955), no. 16, 140 pp., and A. Grothendieck, "Résumé de la théorie métrique des produits tensoriels topologiques". Bol. Soc. Mat. São Paulo 8 1953 1–79.

[45] see chap. 2, p. 15 in Ryan (2002).

[46] see chap. 3, p. 45 in Ryan (2002).

[47] see Example. 2.19, p. 29, and pp. 49–50 in Ryan (2002).

[48] see Proposition 4.6, p. 74 in Ryan (2002).

[49] see Pisier, Gilles (1983), "Counterexamples to a conjecture of Grothendieck", Acta Math. **151**:181–208.

[50] see Szankowski, Andrzej (1981), "$B(H)$ does not have the approximation property", Acta Math. **147**: 89–108. Ryan claims that this result is due to Per Enflo, p. 74 in Ryan (2002).

[51] see Kwapień, S. (1970), "A linear topological characterization of inner-product spaces", Studia Math. **38**:277–278.

[52] see Lindenstrauss, J. and Tzafriri, L. (1971), "On the complemented subspaces problem", Israel J. Math. **9**:263–269.

[53] see p. 245 in Banach (1932). The homogeneity property is called "propriété (15)" there. Banach writes: "on ne connaît aucun exemple d'espace à une infinité de dimensions qui, sans être isomorphe avec (L^2), possède la propriété (15)".

[54] Gowers, W. T. (1996), "A new dichotomy for Banach spaces", Geom. Funct. Anal. **6**:1083–1093.

[55] see Gowers, W. T. (1994), "A solution to Banach's hyperplane problem", Bull. London Math. Soc. **26**:523–530.

[56] see Komorowski, Ryszard A. and Tomczak–Jaegermann, Nicole (1995), "Banach spaces without local unconditional structure", Israel J. Math. **89**:205–226 and also (1998), "Erratum to: Banach spaces without local unconditional structure", Israel J. Math. **105**:85–92.

[57] Milyutin, Alekseĭ A. (1966), "Isomorphism of the spaces of continuous functions over compact sets of the cardinality of the continuum". (Russian) Teor. Funkciĭ Funkcional. Anal. i Priložen. Vyp. **2**:150–156.

[58] Milutin. See also Rosenthal, Haskell P., "The Banach spaces C(K)" in Handbook of the geometry of Banach spaces, Vol. 2, 1547–1602, North-Holland, Amsterdam, 2003.

[59] One can take $\alpha = \omega^{\beta n}$, where $\beta + 1$ is the Cantor–Bendixson rank of K, and $n > 0$ is the finite number of points in the β-th derived set $K^{(\beta)}$ of K. See Mazurkiewicz, Stefan; Sierpiński, Wacław (1920), "Contribution à la topologie des ensembles dénombrables", Fundamenta Mathematicae 1: 17–27.

[60] Bessaga, Czesław; Pełczyński, Aleksander (1960), "Spaces of continuous functions. IV. On isomorphical classification of spaces of continuous functions", Studia Math. **19**:53–62.

20.11 References

- Banach, Stefan (1932), *Théorie des opérations linéaires*, Monografie Matematyczne **1**, Warszawa: Subwencji Funduszu Kultury Narodowej, Zbl 0005.20901.

- Beauzamy, Bernard (1985) [1982], *Introduction to Banach Spaces and their Geometry* (Second revised ed.), North-Holland.

- Bourbaki, Nicolas (1987), *Topological vector spaces*, Elements of mathematics, Berlin: Springer-Verlag, ISBN 978-3-540-13627-9.

- Carothers, Neal L. (2005), *A short course on Banach space theory*, London Mathematical Society Student Texts **64**, Cambridge: Cambridge University Press, pp. xii+184, ISBN 0-521-84283-2.

- Diestel, Joseph (1984), *Sequences and series in Banach spaces*, Graduate Texts in Mathematics **92**, New York: Springer-Verlag, pp. xii+261, ISBN 0-387-90859-5.

- Dunford, Nelson; Schwartz, Jacob T. with the assistance of W. G. Bade and R. G. Bartle (1958), *Linear Operators. I. General Theory*, Pure and Applied Mathematics **7**, New York: Interscience Publishers, Inc., MR 0117523

- Lindenstrauss, Joram; Tzafriri, Lior (1977), *Classical Banach Spaces I, Sequence Spaces*, Ergebnisse der Mathematik und ihrer Grenzgebiete **92**, Berlin: Springer-Verlag, ISBN 3-540-08072-4.

- Megginson, Robert E. (1998), *An introduction to Banach space theory*, Graduate Texts in Mathematics **183**, New York: Springer-Verlag, pp. xx+596, ISBN 0-387-98431-3.

- Ryan, Raymond A. (2002), *Introduction to Tensor Products of Banach Spaces*, Springer Monographs in Mathematics, London: Springer-Verlag, pp. xiv+225, ISBN 1-85233-437-1.

- Wojtaszczyk, Przemysław (1991), *Banach spaces for analysts*, Cambridge Studies in Advanced Mathematics **25**, Cambridge: Cambridge University Press, pp. xiv+382, ISBN 0-521-35618-0.

20.12 External links

- Hazewinkel, Michiel, ed. (2001), "Banach space", *Encyclopedia of Mathematics*, Springer, ISBN 978-1-55608-010-4

- Weisstein, Eric W., "Banach Space", *MathWorld*.

Chapter 21

Tensor product of Hilbert spaces

In mathematics, and in particular functional analysis, the **tensor product of Hilbert spaces** is a way to extend the tensor product construction so that the result of taking a tensor product of two Hilbert spaces is another Hilbert space. Roughly speaking, the tensor product is the metric space completion of the ordinary tensor product. This is a special case of a topological tensor product. The tensor product allows the Hilbert space to be described by a symmetric monoidal category.[1]

21.1 Definition

Since Hilbert spaces have inner products, one would like to introduce an inner product, and therefore a topology, on the tensor product that arise naturally from those of the factors. Let H_1 and H_2 be two Hilbert spaces with inner products $\langle \cdot, \cdot \rangle_1$ and $\langle \cdot, \cdot \rangle_2$, respectively. Construct the tensor product of H_1 and H_2 as vector spaces as explained in the article on tensor products. We can turn this vector space tensor product into an inner product space by defining

$$\langle \phi_1 \otimes \phi_2, \psi_1 \otimes \psi_2 \rangle = \langle \phi_1, \psi_1 \rangle_1 \langle \phi_2, \psi_2 \rangle_2 \quad \text{for all } \phi_1$$

$$, \psi_1 \in H_1 \text{ and } \phi_2, \psi_2 \in H_2$$

and extending by linearity. That this inner product is the natural one is justified by the identification of scalar-valued bilinear maps on $H_1 \times H_2$ and linear functionals on their vector space tensor product. Finally, take the completion under this inner product. The resulting Hilbert space is the tensor product of H_1 and H_2.

21.1.1 Explicit construction

The tensor product can also be defined without appealing to the metric space completion. If H_1 and H_2 are two Hilbert spaces, one associates to every simple tensor product $x_1 \otimes x_2$ the rank one operator from H_1^* to H_2 that maps a given $x^* \in H_1^*$ as

$$x^* \mapsto x^*(x_1) x_2$$

This extends to a linear identification between $H_1 \otimes H_2$ and the space of finite rank operators from H_1^* to H_2. The finite rank operators are embedded in the Hilbert space $HS(H_1^*, H_2)$ of Hilbert–Schmidt operators from H_1^* to H_2. The scalar product in $HS(H_1^*, H_2)$ is given by

$$\langle T_1, T_2 \rangle = \sum_n \langle T_1 e_n^*, T_2 e_n^* \rangle,$$

where (e_n^*) is an arbitrary orthonormal basis of H_1^*.

Under the preceding identification, one can define the Hilbertian tensor product of H_1 and H_2, that is isometrically and linearly isomorphic to $HS(H_1^*, H_2)$.

21.1.2 Universal property

The Hilbert tensor product $H = H_1 \otimes H_2$ is characterized by the following universal property (Kadison & Ringrose 1983, Theorem 2.6.4):

- There is a weakly Hilbert–Schmidt mapping $p : H_1 \times H_2 \to H$ such that, given any weakly Hilbert–Schmidt mapping $L : H_1 \times H_2 \to K$ to a Hilbert space K, there is a unique bounded operator $T : H \to K$ such that $L = Tp$.

A weakly Hilbert-Schmidt mapping $L : H_1 \times H_2 \to K$ is defined as a bilinear map for which a real number d exists, such that $\sum_{i,j=1}^{\infty} \left| \langle L(e_i, f_j), u \rangle \right|^2 \leq d^2 \|u\|^2$ for all $u \in K$ and one (hence all) orthonormal basis $e_1, e_2, \ldots$ of H_1 and $f_1, f_2, \ldots$ of H_2.

As with any universal property, this characterizes the tensor product H uniquely, up to isomorphism. The same universal property, with obvious modifications, also applies for the tensor product of any finite number of Hilbert spaces. It is essentially the same universal property shared by all definitions of tensor products, irrespective of the spaces being tensored: this implies that any space with a tensor product is a symmetric monoidal category, and Hilbert spaces are a particular example thereof.

21.1.3 Infinite tensor products

If H_n is a collection of Hilbert spaces and ξ_n is a collection of unit vectors in these Hilbert spaces then the incomplete tensor product (or Guichardet tensor product) is the L^2 completion of the set of all finite linear combinations of simple tensor vectors $\otimes_{n=1}^{\infty} \psi_n$ where all but finitely many of the ψ_n 's equal the corresponding ξ_n .[2]

21.1.4 Operator algebras

Let $\mathfrak{A}_i$ be the von Neumann algebra of bounded operators on H_i for $i = 1, 2$. Then the von Neumann tensor product of the von Neumann algebras is the strong completion of the set of all finite linear combinations of simple tensor products $A_1 \otimes A_2$ where $A_i \in \mathfrak{A}_i$ for $i = 1, 2$. This is exactly equal to the von Neumann algebra of bounded operators of $H_1 \otimes H_2$. Unlike for Hilbert spaces, one may take infinite tensor products of von Neumann algebras, and for that matter C*-algebras of operators, without defining reference states.[2] This is one advantage of the "algebraic" method in quantum statistical mechanics.

21.2 Properties

If H_1 and H_2 have orthonormal bases $\{\varphi k\}$ and $\{\psi l\}$, respectively, then $\{\varphi k \otimes \psi l\}$ is an orthonormal basis for $H_1 \otimes H_2$. In particular, the Hilbert dimension of the tensor product is the product (as cardinal numbers) of the Hilbert dimensions.

21.3 Examples and applications

The following examples show how tensor products arise naturally.

Given two measure spaces X and Y, with measures μ and ν respectively, one may look at $L^2(X \times Y)$, the space of functions on $X \times Y$ that are square integrable with respect to the product measure $\mu \times \nu$. If f is a square integrable function on X, and g is a square integrable function on Y, then we can define a function h on $X \times Y$ by $h(x,y) = f(x) g(y)$. The definition of the product measure ensures that all functions

of this form are square integrable, so this defines a bilinear mapping $L^2(X) \times L^2(Y) \to L^2(X \times Y)$. Linear combinations of functions of the form $f(x) g(y)$ are also in $L^2(X \times Y)$. It turns out that the set of linear combinations is in fact dense in $L^2(X \times Y)$, if $L^2(X)$ and $L^2(Y)$ are separable. This shows that $L^2(X) \otimes L^2(Y)$ is isomorphic to $L^2(X \times Y)$, and it also explains why we need to take the completion in the construction of the Hilbert space tensor product.

Similarly, we can show that $L^2(X; H)$, denoting the space of square integrable functions $X \to H$, is isomorphic to $L^2(X) \otimes H$ if this space is separable. The isomorphism maps $f(x) \otimes \varphi \in L^2(X) \otimes H$ to $f(x)\varphi \in L^2(X; H)$. We can combine this with the previous example and conclude that $L^2(X) \otimes L^2(Y)$ and $L^2(X \times Y)$ are both isomorphic to $L^2(X; L^2(Y))$.

Tensor products of Hilbert spaces arise often in quantum mechanics. If some particle is described by the Hilbert space H_1, and another particle is described by H_2, then the system consisting of both particles is described by the tensor product of H_1 and H_2. For example, the state space of a quantum harmonic oscillator is $L^2(\mathbf{R})$, so the state space of two oscillators is $L^2(\mathbf{R}) \otimes L^2(\mathbf{R})$, which is isomorphic to $L^2(\mathbf{R}^2)$. Therefore, the two-particle system is described by wave functions of the form $\varphi(x_1, x_2)$. A more intricate example is provided by the Fock spaces, which describe a variable number of particles.

21.4 References

[1] B. Coecke and E. O. Paquette, Categories for the practising physicist, in: New Structures for Physics, B. Coecke (ed.), Springer Lecture Notes in Physics, 2009. arXiv:0905.3010

[2] Bratteli, O. and Robinson, D: *Operator Algebras and Quantum Statistical Mechanics v.1, 2nd ed.*, page 144. Springer-Verlag, 2002.

• Kadison, Richard V.; Ringrose, John R. (1997), *Fundamentals of the theory of operator algebras. Vol. I*, Graduate Studies in Mathematics **15**, Providence, R.I.: American Mathematical Society, ISBN 978-0-8218-0819-1, MR 1468229.

• Weidmann, Joachim (1980), *Linear operators in Hilbert spaces*, Graduate Texts in Mathematics **68**, Berlin, New York: Springer-Verlag, ISBN 978-0-387-90427-6, MR 566954.

Chapter 22

Hilbert algebra

In mathematics, **Hilbert algebras** occur in the theory of von Neumann algebras in:

- Commutation theorem

- Tomita–Takesaki theory

Chapter 23

Hilbert manifold

In mathematics, a **Hilbert manifold** is a manifold modeled on Hilbert spaces. Thus it is a separable Hausdorff space in which each point has a neighbourhood homeomorphic to an infinite dimensional Hilbert space. The concept of a Hilbert manifold provides a possibility of extending the theory of manifolds to infinite-dimensional setting. Analogously to the finite-dimensional situation, one can define a *differentiable* Hilbert manifold by considering a maximal atlas in which the transition maps are differentiable.

23.1 Properties

Many basic constructions of the manifold theory, such as the tangent space of a manifold and a tubular neighbourhood of a submanifold (of finite codimension) carry over from the finite dimensional situation to the Hilbert setting with little change. However, in statements involving maps between manifolds, one often has to restrict consideration to *Fredholm maps*, i.e. maps whose differential at every point is Fredholm. The reason for this is that Sard's lemma holds for Fredholm maps, but not in general. Notwithstanding this difference, Hilbert manifolds have several very nice properties.

- **Kuiper's theorem**: If X is a compact topological space or has the homotopy type of a CW-Complex then every (real or complex) Hilbert space bundle over X is trivial. In particular, every Hilbert manifold is parallelizable.

- Every smooth Hilbert manifold can be smoothly embedded onto an open subset of the model Hilbert space.

- Every homotopy equivalence between two Hilbert manifolds is homotopic to a diffeomorphism. In particular every two homotopy equivalent Hilbert manifolds are already diffeomorphic. This stands in contrast to lens spaces and exotic spheres, which demonstrate that in the finite-dimensional situation, homo-topy equivalence, homeomorphism, and diffeomorphism of manifolds are distinct properties.

- Although Sard's Theorem does not hold in general, every continuous map $f : X \to \mathbf{R}^n$ from a Hilbert manifold can be arbitrary closely approximated by a smooth map $g : X \to \mathbf{R}^n$ which has no critical points

23.2 Examples

- Any Hilbert space H is a Hilbert manifold with a single global chart given by the identity function on H. Moreover, since H is a vector space, the tangent space $\mathrm{T}pH$ to H at any point $p \in H$ is canonically isomorphic to H itself, and so has a natural inner product, the "same" as the one on H. Thus, H can be given the structure of a Riemannian manifold with metric

$$g(v, w)(p) := \langle v, w \rangle_H \text{ for } v, w \in \mathrm{T}_p H,$$

where $\langle \cdot, \cdot \rangle H$ denotes the inner product in H.

- Similarly, any open subset of a Hilbert space is a Hilbert manifold and a Riemannian manifold under the same construction as for the whole space.

- There are several mapping spaces between manifolds which can be viewed as Hilbert spaces by only considering maps of suitable Sobolev class. For example we can consider the space LM of all H^1 maps from the unit circle $\mathbf{S}^1$ into a manifold M. This can be topologized via the compact open topology as a subspace of the space of all continuous mappings from the circle to M, i.e. the free loop space of M. The Sobolev kind mapping space LM described above is homotopy equivalent to the free loop space. This makes it suited to the study of algebraic topology of the free loop space, especially in the field of string topology. We

can do an analogous Sobolev construction for the loop
space, making it a codimension d Hilbert submanifold
of LM, where d is the dimension of M.

23.3 See also

- Banach manifold

23.4 References

- Klingenberg, Wilhelm (1982), *Riemannian Geometry*, Berlin: W. de Gruyter, ISBN 978-3-11-008673-7.
  Contains a general introduction to Hilbert manifolds
  and many details about the free loop space.

- Lang, Serge (1995), *Differential and Riemannian
  Manifolds*, New York: Springer, ISBN 978-
  0387943381.   Another introduction with more
  differential topology.

- N. Kuiper, The homotopy type of the unitary group of
  Hilbert spaces", Topology 3, 19-30

- J. Eells, K. D. Elworthy, "On the differential topology
  of Hilbert manifolds", Global analysis. Proceedings
  of Symposia in Pure Mathematics, Volume XV 1970,
  41-44.

- J. Eells, K. D. Elworthy, "Open embeddings of certain
  Banach manifolds", Annals of Mathematics 91 (1970),
  465-485

- D. Chataur, "A Bordism Approach to String Topology", preprint http://arxiv.org/abs/math.at/0306080

23.5 External links

- Hilbert manifold at the Manifold Atlas

Chapter 24

Rigged Hilbert space

In mathematics, a **rigged Hilbert space** (**Gelfand triple**, **nested Hilbert space**, **equipped Hilbert space**) is a construction designed to link the distribution and square-integrable aspects of functional analysis. Such spaces were introduced to study spectral theory in the broad sense. They bring together the 'bound state' (eigenvector) and 'continuous spectrum', in one place.

24.1 Motivation

A function such as the canonical homomorphism of the real line into the complex plane

$$x \mapsto e^{ix},$$

is an eigenfunction of the differential operator

$$-i\frac{d}{dx}$$

on the real line **R**, but isn't square-integrable for the usual Borel measure on **R**. To properly consider this function as an eigenfunction requires some way of stepping outside the strict confines of the Hilbert space theory. This was supplied by the apparatus of Schwartz distributions, and a *generalized eigenfunction* theory was developed in the years after 1950.

24.2 Functional analysis approach

The concept of rigged Hilbert space places this idea in an abstract functional-analytic framework. Formally, a rigged Hilbert space consists of a Hilbert space H, together with a subspace Φ which carries a finer topology, that is one for which the natural inclusion

$$\Phi \subseteq H$$

is continuous. It is no loss to assume that Φ is dense in H for the Hilbert norm. We consider the inclusion of dual spaces H^* in Φ^*. The latter, dual to Φ in its 'test function' topology, is realised as a space of distributions or generalised functions of some sort, and the linear functionals on the subspace Φ of type

$$\phi \mapsto \langle v, \phi \rangle$$

for v in H are faithfully represented as distributions (because we assume Φ dense).

Now by applying the Riesz representation theorem we can identify H^* with H. Therefore the definition of *rigged Hilbert space* is in terms of a sandwich:

$$\Phi \subseteq H \subseteq \Phi^*.$$

The most significant examples are those for which Φ is a nuclear space; this comment is an abstract expression of the idea that Φ consists of test functions and Φ^* of the corresponding distributions. Also, a simple example is given by Sobolev spaces: Here (in the simplest case of Sobolev spaces on $\mathbb{R}^n$)

$$H = L^2(\mathbb{R}^n), \ \Phi = H^s(\mathbb{R}^n), \ \Phi^* = H^{-s}(\mathbb{R}^n)$$

where $s > 0$.

24.3 Formal definition (Gelfand triple)

A **rigged Hilbert space** is a pair (H, Φ) with H a Hilbert space, Φ a dense subspace, such that Φ is given a topological vector space structure for which the inclusion map i is continuous.

Identifying H with its dual space H^*, the adjoint to i is the map

$$i^* : H = H^* \to \Phi^*.$$

The duality pairing between Φ and Φ^* is then compatible with the inner product on H, in the sense that:

$$\langle u, v \rangle_{\Phi \times \Phi^*} = (u, v)_H$$

whenever $u \in \Phi \subset H$ and $v \in H = H^* \subset \Phi^*$. In the case of complex Hilbert spaces one of u or v on the left should be complex conjugated, depending on whether one uses the physics or maths convention, respectively, of hermitian scalar product.

The triple $(\Phi, \ H, \ \Phi^*)$ is often named the "Gelfand triple" (after the mathematician Israel Gelfand).

Note that even though Φ is isomorphic to Φ^* if it happens that Φ is a Hilbert space in its own right, this isomorphism is *not* the same as the composition of the inclusion i with its adjoint $i*$

$$i^* i : \Phi \subset H = H^* \to \Phi^*.$$

24.4   References

- J.-P. Antoine, *Quantum Mechanics Beyond Hilbert Space* (1996), appearing in *Irreversibility and Causality, Semigroups and Rigged Hilbert Spaces*, Arno Bohm, Heinz-Dietrich Doebner, Piotr Kielanowski, eds., Springer-Verlag, ISBN 3-540-64305-2. *(Provides a survey overview.)*

- Jean Dieudonné, *Éléments d'analyse* VII (1978). *(See paragraphs 23.8 and 23.32)*

- I. M. Gelfand and N. J. Vilenkin. Generalized Functions, vol. 4: Some Applications of Harmonic Analysis. Rigged Hilbert Spaces. Academic Press, New York, 1964.

- R. de la Madrid, "The role of the rigged Hilbert space in Quantum Mechanics," Eur. J. Phys. 26, 287 (2005); quant-ph/0502053.

- K. Maurin, *Generalized Eigenfunction Expansions and Unitary Representations of Topological Groups*, Polish Scientific Publishers, Warsaw, 1968.

- Minlos, R.A. (2001), "Rigged_Hilbert_space", in Hazewinkel, Michiel, *Encyclopedia of Mathematics*, Springer, ISBN 978-1-55608-010-4

24.5 Text and image sources, contributors, and licenses

24.5.1 Text

- **Hilbert space** *Source:* https://en.wikipedia.org/wiki/Hilbert_space?oldid=685686974 *Contributors:* AxelBoldt, Bryan Derksen, Zundark, AstroNomer~enwiki, Taw, Magnus~enwiki, XJaM, Toby Bartels, Miguel~enwiki, Kurt Jansson, Maury Markowitz, Youandme, Edward, Michael Hardy, Wshun, MartinHarper, TakuyaMurata, GTBacchus, Looxix~enwiki, Rossami, Nikai, Ideyal, Charles Matthews, Dysprosia, Jitse Niesen, Doradus, Prumpf, Nickshanks, Jph, Chuunen Baka, Robbot, MathMartin, Rorro, Blainster, Timrollpickering, UtherSRG, Aetheling, Widsith, Diberri, Tobias Bergemann, Giftlite, Rs2, Barbara Shack, Mikez, Harp, BenFrantzDale, Lethe, Lupin, MathKnight, Fastfission, Peruvianllama, Anville, Shibboleth, CSTAR, Maximaximax, Troels Arvin, Hellisp, TheObtuseAngleOfDoom, PhotoBox, Chris Howard, Rich Farmbrough, Guanabot, Pj.de.bruin, Mani1, Paul August, Bender235, Tompw, MisterSheik, C S, Tsirel, LutzL, Passw0rd, Vilemiasma, Dirac1933, Cmprince, WilliamKF, Joriki, Kohtala, Linas, Igny, Joke137, Palica, RuM, Rjwilmsi, Koavf, Salix alba, R.e.b., GreenLocust, RexNL, Vonkje, Chobot, YurikBot, Wavelength, KSmrq, Archelon, Gaius Cornelius, NawlinWiki, Robertvan1, Buster79, Grafen, Jpowell, DAJF, Larsobrien, Hakeem.gadi, Tetracube, Arthur Rubin, SteveWitham, Scineram, Pred, Bo Jacoby, Zvika, Kgf0, SmackBot, Elonka, Nihonjoe, Nitelm, Septegram, Chris the speller, Autarch, Silly rabbit, Nbarth, DHN-bot~enwiki, Colonies Chris, Vanished User 0001, 1diot, Tesseran, Lambiam, Nishkid64, Terry Bollinger, Zarniwoot, Dll99, Mets501, Mathsci, Colonel Warden, CBM, NickW557, Mct mht, NormHardy, Mblumber, Endpoint, RelHistBuff, Thijs!bot, Headbomb, Rlupsa, Frank, Kev11sky, Eleuther, Dkemper, Ben pcc, Lakripun, Janton~enwiki, Salgueiro~enwiki, Steelpillow, Onkel Tuca~enwiki, MER-C, Matthew Fennell, Awilley, Thenub314, OhanaUnited, Yill577, LordFoom, Sojourner001, Usien6, Tercer, Pagw, R'n'B, Abecedare, Maurice Carbonaro, Salih, Brickc1, Daniele.tampieri, Aram33~enwiki, Policron, Sigmundur, Idioma-bot, Jmcdon10, Kyle the bot, Matumba, Hqb, Geometry guy, Pellerv, Jmath666, John189, Pierre-Alain Gouanvic, Arcfrk, SieBot, Paolo.dL, JackSchmidt, JL-Bot, StewartMH, Yasmar, Plastikspork, JuPitEer, Niceguyedc, Brews ohare, Cenarium, Cemalgencoglu, StevenDH, Vanished user tj23rpoij4tikkd, TimothyRias, Marc van Leeuwen, Charles Sturm, Ie1833, Addbot, Dmhowarth26, AkhtaBot, Topology Expert, Favonian, Zorrobot, Luckas-bot, Yobot, Ht686rg90, Kilom691, Tynpeddler, GateKeeper, 黑超, AnomieBOT, Palpher, Rubinbot, Citation bot, Qorilla, LilHelpa, Xqbot, Bdmy, ProtectionTaggingBot, Contraverse, FrescoBot, T8191, Sławomir Biały, Cs32en, Citation bot 1, Rc3002, DrilBot, Kiefer.Wolfowitz, Jonesey95, Gruntler, RedBot, Differenxe, Ybungalobill, Le Docteur, Katovatzschyn, EmausBot, Racerx11, Netheril96, Slawekb, Tocz0001, Berthold1954, Jacksccsi, Brandmeister, RockMagnetist, Liuyipei, ClueBot NG, Uber82, Helpful Pixie Bot, Bibcode Bot, Elferdo, Jeanpaulusa, Solomon7968, Dlituiev, Randomguess, George.ad.stamatiou, Uber84, Dbcooney, Wanze, Darvii, Jnhlnd, Mya3318, 2 Hertz, Airwoz, Austrartsua, Mgkrupa, VInit24, Soumilm, YeOldeGentleman, KasparBot, KPomorski and Anonymous: 175

- **Hilbert curve** *Source:* https://en.wikipedia.org/wiki/Hilbert_curve?oldid=678845126 *Contributors:* Eob, SGBailey, Lkesteloot, Ashwin, Giftlite, BenFrantzDale, Tomruen, Qef, Andrejj, Chbarts, Blotwell, Tromp, Krellis, Burn, Oleg Alexandrov, Japanese Searobin, WilliamKF, Linas, Lovingboth, Mathbot, Ysangkok, Vossman, WriterHound, Wavelength, Nol Aders, Arichnad, Merosonox, Johnpseudo, Robertd, Stepa, Iskunk, Chris the speller, Jtxx000, Atoll, Belizefan, Tó campos~enwiki, Hanche, Pciszek, Escarbot, IanOsgood, Raphman, Jarekt, David Eppstein, Gwern, Verdatum, Prokofiev2, Laurusnobilis, PMajer, TXiKiBoT, VanishedUserABC, Spinningspark, Svick, PerryTachett, Jwz, PixelBot, Watmough, Addbot, Sasepeev, Enormator, Zaspam.akaunt, Lightbot, Yobot, GrouchoBot, Pauladin, False vacuum, ZYV, Cleanback, 00Ragora00, Mgboulton, Xnn, WikitanvirBot, Atila rey, Derekleungtszhei, Mikhail Ryazanov, ClueBot NG, BattyBot, Stefano1952, Bwvdnbro, Dme26, CalmWords, Chessguy3 and Anonymous: 51

- **Lp space** *Source:* https://en.wikipedia.org/wiki/Lp_space?oldid=679926807 *Contributors:* AxelBoldt, Zundark, Tarquin, Miguel~enwiki, Michael Hardy, Menchi, GTBacchus, Loisel, Charles Matthews, Dcoetzee, Jitse Niesen, Prumpf, Maximus Rex, AndrewKepert, Josh Cherry, Guan, MathMartin, Bkell, Aetheling, Tobias Bergemann, Pdenapo, Giftlite, BenFrantzDale, Lethe, Lupin, JanHRichter, Pucicu, CSTAR, Gadykozma, Mat cross, Paul August, Gauge, Crisófilax, Msh210, Arcenciel, Eric Kvaalen, Diego Moya, Aitter, Oleg Alexandrov, Joriki, Linas, GregorB, Marudubshinki, Eslip17, SixWingedSeraph, Rjwilmsi, John Baez, Quuxplusone, Sodin, Krishnavedala, Bgwhite, Algebraist, Archelon, Buster79, LMSchmitt, Schmock, Crasshopper, Bota47, Pred, Das-g, SmackBot, InverseHypercube, Spireguy, KYN, Chuyelchulo~enwiki, Oli Filth, Silly rabbit, RayAYang, Nbarth, Tim81~enwiki, AdamSmithee, Tesseran, Lambiam, OlegSmirnov, Zero sharp, Ylloh, Biscay, Jackzhp, Mct mht, A876, Blaisorblade, 黑超, Headbomb, Cj67, Em3rguy, Thenub314, Swpb, Sullivan.t.j, David Eppstein, Martynas Patasius, TomyDuby, Daniele.tampieri, Policron, Remember the dot, HyDeckar, Matumba, Rei-bot, McM.bot, Gamesou, Jmath666, Antixt, GirasoleDE, MikeRumex, Quietbritishjim, SieBot, BotMultichill, DaBler, Lightmouse, ClueBot, Plastikspork, Nsk92, UKoch, Carolus m, SchreiberBike, Addbot, Favonian, Ozob, Zorrobot, Luckas-bot, Yobot, Ht686rg90, TaBOT-zerem, AnomieBOT, Erel Segal, Citation bot, DannyAsher, LilHelpa, Xqbot, Bdmy, Mr.gondolier, FrescoBot, 4bpp, Sławomir Biały, Citation bot 1, Kiefer.Wolfowitz, Foobarnix, Tcnuk, Xmuhcd, MidgleyC, DASHBot, KHamsun, Slawekb, ZéroBot, Nomen4Omen, Quondum, Parodi, Luke18:2-8, Dreambother, ClueBot NG, Helpful Pixie Bot, Koertefa, BG19bot, Apatch1265, Stausifr, Freeze S, Mathwu, Kephir, Heinzerich~enwiki, Gmkwo, Ludvonga, Mdjara, LHSPhantom, GeoffreyT2000, SoSivr and Anonymous: 127

- **Sequence space** *Source:* https://en.wikipedia.org/wiki/Sequence_space?oldid=679926706 *Contributors:* Patrick, Joshuabowman, Aleph4, MathMartin, Tobias Bergemann, Giftlite, Dmb000006, Oleg Alexandrov, BD2412, Rjwilmsi, Wavelength, Nbarth, Henning Makholm, JR-Spriggs, Cydebot, Nadav1, Futurebird, MER-C, StevenJohnston, Drmies, Alexbot, Addbot, Erel Segal, Bdmy, Erik9bot, FrescoBot, Sławomir Biały, Citation bot 1, Kiefer.Wolfowitz, RjwilmsiBot, EmausBot, Slawekb, BG19bot, Kephir, Evolution and evolvability, Noix07, LHSPhantom and Anonymous: 8

- **Sobolev space** *Source:* https://en.wikipedia.org/wiki/Sobolev_space?oldid=689346338 *Contributors:* Gareth Owen, Michael Hardy, GTBacchus, Loisel, Vargenau, Charles Matthews, Jitse Niesen, Tobias Bergemann, Giftlite, BenFrantzDale, Marcika, Stern~enwiki, Waltpohl, Sam Hocevar, Gadykozma, Mat cross, Gauge, Eric Kvaalen, Kusma, Oleg Alexandrov, Igny, MFH, Grammarbot, Rjwilmsi, Tammojan, R.e.b., RussBot, Tong~enwiki, Grafen, Brian Tvedt, Zvika, GrafZahl, SmackBot, Silly rabbit, RayAYang, DHN-bot~enwiki, Khazar, Comech, Lavaka, Makeemlighter, Yrodro, Nuwewsco, Thijs!bot, Bocgnr, Cj67, Sullivan.t.j, Mir76, CarlFeynman, R'n'B, TomyDuby, Daniele.tampieri, Haseldon, Yurakm, Temurjin, TXiKiBoT, Wikiisawesome, Jmath666, Picojeff, SieBot, 4wajzkd02, Perturbationist, RockabyeJames, Addbot, Hughett, LaaknorBot, Wikomidia, Yobot, Ptbotgourou, MichalKotowski, Semimartingale, MauritsBot, Xqbot, Bdmy, Omnipaedista, Sławomir Biały, Citation bot 1, Jonesey95, SepIHw, Trappist the monk, RjwilmsiBot, Gdzie, Dewritech, Bonnnnn2010, Joerg Bader, FunctionspaceInvader, Helpful Pixie Bot, Frank2910, Christian Jaeh, Katterjohn, David9550 and Anonymous: 62

- **Generalized function** *Source:* https://en.wikipedia.org/wiki/Generalized_function?oldid=684887829 *Contributors:* Michael Hardy, Charles Matthews, Dino, Hyacinth, Giftlite, Gauge, Alai, Oleg Alexandrov, MFH, Dpv, Rjwilmsi, Mathbot, Siddhant, Piet Delport, Tong~enwiki, Yonir, SmackBot, PouyaDT, Domitori, CRGreathouse, CmdrObot, Barticus88, WinBot, Myanw, Magioladitis, David Eppstein, Tparameter, SieBot, ClueBot, CohesionBot, Addbot, DOI bot, Lightbot, Yobot, Omnipaedista, Arid Zkwelty, Some standardized rigour, Citation bot 1, Tkuvho, Kishmakov, Filip Albert, Bogdanb (nou), Super-real dance, Bibcode Bot, BG19bot, ChrisGualtieri, PF1200, Limit-theorem, Rutgerjanlange, Nigellwh, Giardello, Reader634, Monkbot, NCC1701F, KasparBot and Anonymous: 17

- **Hardy space** *Source:* https://en.wikipedia.org/wiki/Hardy_space?oldid=676191601 *Contributors:* Michael Hardy, Charles Matthews, Dysprosia, The Anomebot, ComplexZeta, Giftlite, Lethe, Linas, Rjwilmsi, R.e.b., Wavelength, Jugander, Chris the speller, Silly rabbit, CRGreathouse, WISo, Martin Hogbin, Thenub314, Leyo, Henry Delforn (old), Alexbot, Addbot, Cuaxdon, Yobot, The Earwig, Citation bot, Bdmy, Citation bot 1, Kiefer.Wolfowitz, RedBot, Suslindisambiguator, Quondam, Sednodna, Coroner's jury, Anrnusna, K9re11, Stoverc and Anonymous: 20

- **Holomorphic function** *Source:* https://en.wikipedia.org/wiki/Holomorphic_function?oldid=689604303 *Contributors:* Damian Yerrick, AxelBoldt, Tobias Hoevekamp, Tarquin, XJaM, PierreAbbat, Patrick, Michael Hardy, Chinju, TakuyaMurata, Charles Matthews, Dcoetzee, Dysprosia, Jusjih, Robbot, Rvollmert, Altenmann, Naddy, Aetheling, Tobias Bergemann, Giftlite, Abiola Lapite, Jao, Lethe, Dratman, Frencheigh, Mike40033, Jason Quinn, Eequor, DragonflySixtyseven, Almit39, ObsessiveMathsFreak, EmilJ, Aaronbrick, Crust, Stephen Bain, Thorfinn~enwiki, Sligocki, Reaverdrop, Oleg Alexandrov, Linas, FuriousScribble, Dzordzm, Graham87, HannsEwald, Salix alba, Maxim Razin, Mathbot, Adoniscik, YurikBot, Archelon, NickBush24, Crasshopper, DomenicDenicola, Tetracube, Jobh, CrniBombarder!!!, Lunch, KnightRider~enwiki, SmackBot, MalafayaBot, Silly rabbit, Weierstraß, Vanished User 0001, Lhf, Acepectif, Duckbill, Unco, Jim.belk, Alf kadett, CRGreathouse, FilipeS, Thijs!bot, Irigi, Ben pcc, Dougher, JAnDbot, Saleemsan, JamesBWatson, Michael Goodyear, Email4mobile, Michael K. Edwards, Christian.Mercat, Robert Illes, Policron, DavidCBryant, LokiClock, Hesam7, Jmath666, Riwnodennyk, Fakhredinblog, NicoleTedesco, The Diagonal Prince, Addbot, Some jerk on the Internet, LinkFA-Bot, Ozob, Prim Ethics, PV=nRT, Luckas-bot, AnomieBOT, SvartMan, ArthurBot, Xqbot, Bdmy, ΑΠΧΙΒΑΡΙΥC, RibotBOT, Howard McCay, Tcnuk, TobeBot, Laurent MAYER, John of Reading, Slawekb, ZéroBot, Quondam, ChuispastonBot, Mittgaurav, Llightex, Isocliff, ClueBot NG, Alexander E Ross, Erick GR, Mokhtari34, Mike abc, Brad7777, YFdyh-bot, Dexbot, Brirush, ChrisLPhys12, Susan.grayeff, Moby122, Rottenshark, KasparBot and Anonymous: 72

- **Pythagorean theorem** *Source:* https://en.wikipedia.org/wiki/Pythagorean_theorem?oldid=687172313 *Contributors:* AxelBoldt, Magnus Manske, Mav, Zundark, The Anome, Tarquin, Grouse, Moly, Vignaux, XJaM, Hari, Toby~enwiki, Toby Bartels, PierreAbbat, William Avery, Roadrunner, FvdP, Infrogmation, D, Michael Hardy, Wshun, Booyabazooka, Llywrch, Wapcaplet, Ixfd64, Chinju, GTBacchus, Eric119, Minesweeper, Ahoerstemeier, John Webb, Stevenj, Snoyes, Susan Mason, Mark Foskey, Александър, Poor Yorick, Rotem Dan, Andres, Cimon Avaro, Samw, Cherkash, Mxn, Ideyal, Revolver, Trainspotter~enwiki, Charles Matthews, Wikiborg, Jitse Niesen, Gutza, Tpbradbury, E23~enwiki, ReciprocityProject, Bevo, AnonMoos, Olathe, Denelson83, Jni, Robbot, Sander123, 1984, Fredrik, Schutz, Gandalf61, Math-Martin, Wjhonson, Tualha, ZekeMacNeil, Meelar, Bkell, Hadal, Paul Murray, Komet, Wile E. Heresiarch, Cedars, Tosha, Giftlite, Smjg, Wolf-keeper, Nunh-huh, Lethe, Lupin, Herbee, Dissident, Tom Radulovich, Ds13, Everyking, No Guru, Gus Polly, Dratman, Curps, Cantus, Rick Block, Leonard G., Dav4is, Guanaco, Avsa, Eequor, Softssa, Bobblewik, Lucky 6.9, Ato, Knutux, Noe, Antandrus, OverlordQ, Kaldari, APH, Anythingyouwant, Gauss, Icairns, Wroscel, Starx, Ukexpat, ELApro, SYSS Mouse, Zro, L-H, David Sneek, Jayjg, ThreeE, Freakofnurture, Poccil, Wfaulk, Bornintheguz, Discospinster, Rich Farmbrough, Guanabot, Supercoop, FT2, Inkypaws, Kooo, Dbachmann, Paul August, Damaru, Danny B-), Mediocretes, Rick MILLER~enwiki, Rgdboer, Hayabusa future, Crisófilax, ́EmilJ, West London Dweller, Causa sui, Bobo192, BrokenSegue, Shenme, SpeedyGonsales, PeterisP, RichardNeill, Obradovic Goran, Aelscha, Merope, Ranveig, Jumbuck, Stephen G. Brown, Mrzaius, Anthony Appleyard, Arthena, Keenan Pepper, Bart133, Cburnett, Evil Monkey, Jheald, Bsadowski1, BlastOButter42, BDD, Agutie, Tintin1107, Oleg Alexandrov, Lssilva, Gmaxwell, Shreevatsa, LOL, Borb, Splintax, Ruud Koot, -Ril-, Wces423, Scm83x, SDC, Zzyzx11, Prashanthns, Gerbrant, King of Hearts (old account 2), Graham87, Pranathi, Mendaliv, Canderson7, Rjwilmsi, P3Pp3r, JVz, Koavf, Hack-Man, Tangotango, Salix alba, Nneonneo, Bhadani, Matt Deres, FlaBot, VKokielov, Alexb@cut-the-knot.com, RobertG, Mathbot, Nihiltres, RexNL, Gurch, DannyZ, Terrx, Goudzovski, Sodin, Hermajesty, Glenn L, King of Hearts, N8cantor, Chobot, DVdm, Gwernol, EamonnPKeane, Roboto de Ajvol, The Rambling Man, YurikBot, Wavelength, Karlscherer3, Sceptre, Deeptrivia, Dmharvey, Wolfmankurd, Adam1213, Anonymous editor, KSmrq, Stephenb, Zimbricchio, Alex Bakharev, Wimt, Gustavb, Finbarr Saunders, NawlinWiki, Hillcino368, Rick Norwood, DragonHawk, Wiki alf, Pringle, Astral, Grafen, MathMan64, Eighty~enwiki, JocK, The Obfuscator, Aaron Brenneman, Anetode, Mlouns, Mission9801, Bota47, Everyguy, David Underdown, Tonywalton, Caroline Sanford, Genjix, FF2010, 21655, Closedmouth, Spondoolicks, Red Jay, Djsmem, MathsIsFun, CWenger, Danny-w, HereToHelp, Willtron, Pred, Mikus, Katieh5584, Hagie, GrinBot~enwiki, Zvika, Ccerer, DVD R W, Finell, Choi9999, AndrewWTaylor, Luk, Fuzzyblob, Sardanaphalus, Attilios, Amalthea, SmackBot, RDBury, BeteNoir, Cubs Fan, Prodego, InverseHypercube, KnowledgeOfSelf, Argyll Lassie, David.Mestel, C.Fred, Msn26586, Bomac, Jbaldus~enwiki, Weatherman90, Jagged 85, Dreprince, Eskimbot, Alsandro, Ollieollieollie, Gilliam, Hmains, Chaojoker, Kurykh, Oli Filth, Miquonranger03, MalafayaBot, Silly rabbit, SchfiftyThree, SEIBasaurus, DarthInsinuate, Spellchecker, DHN-bot~enwiki, Nazgjunk, Saiswa, Zsinj, Dethme0w, Can't sleep, clown will eat me, MyNameIsVlad, TheGerm, Smallbones, AeroSpace, Writtenright, Awh, OrphanBot, Rrburke, Krich, Swainstonation, NoIdeaNick, Nakon, Jiddisch~enwiki, Dreadstar, Mini-Geek, Hgilbert, PatrickA, IMaRocketMan, Astroview120mm, Bryanmcdonald, Jbergquist, Ultraexactzz, Sammy1339, Richtcs, Copysan, Bidabadi~enwiki, Nin10do, Adsllc, Mental Blank, SashatoBot, Lambiam, Dono, Harryboyles, Mr415, Cronholm144, Alex Arnold, Sir Nicholas de Mimsy-Porpington, Edwy, Mgiganteus1, Javit, Aaronpaul, Jim.belk, Scetoaux, Aleenf1, Valepert, Stwalkerster, Kirbytime, Hargle, Dicklyon, Mets501, Michael Greiner, Limaner, Galactor213, Phantom C, BranStark, HisSpaceResearch, K, Spebudmak, Peter M Dodge, Wayfarers43, NativeForeigner, Sander Säde, Eestolano, Mrdthree, CapitalR, A. Pichler, Aque0us, Audiosmurf, Anger22, Zadil, Tawkerbot2, Conrad.Irwin, MightyWarrior, InvisibleK, Mebizzare, Tanthalas39, Geremia, Aherunar, CBM, Pr0t0type, JohnCD, Inferno 619, CWY2190, Basawala, Friendofthehose, Jokes Free4Me, 345Kai, WeggeBot, Gregbard, Nilfanion, Themightyquill, Herd of Swine, MC10, Steel, Mato, Rifleman 82, Alvesgaspar, Felinoel, Jack Phoenix, DumbBOT, Chrislk02, Eisenberg, Dogwhelk, Omicronpersei8, Wexcan, Sam may, Thijs!bot, Epbr123, Elfred, RohanDhruva, Robsinden, Pajz, Qwyrxian, Karadoc~enwiki, Mojo Hand, Chickenflicker, Headbomb, Pjvpjv, John254, James086, Tapuzi, Poe Joe, Big Bird, Dawnseeker2000, Natalie Erin, Northumbrian, TKLM, Stannered, Anti-VandalBot, Luna Santin, Seaphoto, John.d.page, Schellack, CobraWiki, Billscottbob, Fashionslide, Jj137, Exteray, Coyets, Dinferno, Danger, Farosdaughter, AmericanXplorer13, Braindrain0000, Spouima, Fireice, P.L.A.R., Ashleyy osaurus, Karadimos, Golgofrinchian, MikeLynch, .alyn.post., JAnDbot, GromXXVII, Mckee, Em3ryguy, MER-C, Skomorokh, Instinct, Jonemerson, Ricardo sandoval, Rearete, Db099221, Tengfred, Hut 8.5, Agol, TheEditrix2, Acroterion, Tarif Ezaz, Meeples, Animaly2k2, Magioladitis, Celithemis, Bongwarrior, VoABot II, Fusionmix, JNW, Peetvanschalkwyk, JamesBWatson, Rivertorch, BanRay, Avicennasis, Johnbibby, Blanko4, 28421u2232nfenfcenc, David Eppstein, Peterhi, Laur2ro, Matt Adore, DerHexer, JaGa, Philg88, Mabuhelwa, GuelphGryphon98, Falcor84, The Light6, Kayau, Rickterp,

Ssafarik, Srleffler, Kri, R160K, Chobot, Gwernol, Algebraist, YurikBot, Wavelength, Spacepotato, Hairy Dude, RussBot, Michael Slone, CambridgeBayWeather, Rick Norwood, Kinser, Guruparan, Trovatore, Vanished user 1029384756, Nick, Bota47, BraneJ, Martinwilke1980, Antiduh, Arthur Rubin, Lonerville, Netrapt, Curpsbot-unicodify, Cjfsyntropy, Paul D. Anderson, GrinBot~enwiki, SmackBot, RDBury, InverseHypercube, KocjoBot~enwiki, Davidsiegel, Chris the speller, SlimJim, SMP, Silly rabbit, Complexica, Nbarth, DHN-bot~enwiki, Colonies Chris, Chlewbot, Vanished User 0001, Cícero, Cybercobra, Daqu, Mattpat, James084, Lambiam, Tbjw, Breno, Terry Bollinger, Michael Kinyon, Lim Wei Quan, Rcowlagi, SandyGeorgia, Whackawhackawoo, Inquisitus, Rschwieb, Levineps, Madmath789, Markan~enwiki, Tawkerbot2, Igni, CRGreathouse, Mct mht, Cydebot, Danman3459, Guitardemon666, Mikewax, Thijs!bot, Headbomb, RobHar, CharlotteWebb, Urdutext, Escarbot, JAnDbot, Thenub314, Englebert, Magioladitis, Jakob.scholbach, Kookas, SwiftBot, WhatamIdoing, David Eppstein, Cpl Syx, Charitwo, Akhil999in, Infovarius, Frenchef, TechnoFaye, CommonsDelinker, Paranomia, Michaelp7, Mitsuruaoyama, Trumpet marietta 45750, Daniele.tampieri, Gombang, Policron, Fylwind, Cartiod, Camrn86, AlnoktaBOT, Hakankösem~enwiki, TXiKiBoT, Hlevkin, Gwib, Anonymous Dissident, Imasleepviking, Hrrr, Mechakucha, Geometry guy, Terabyte06, Tommyinla, Wikithesource, Staka, AlleborgoBot, Deconstructhis, Newbyguesses, YohanN7, SieBot, Ivan Štambuk, Portalian, ToePeu.bot, Lucasbfrbot, Tiptoety, Paolo.dL, Henry Delforn (old), Thehotelambush, JackSchmidt, Jorgen W, AlanUS, Randomblue, Jludwig, ClueBot, Alksentrs, Nsk92, JP.Martin-Flatin, FractalFusion, Niceguyedc, DifferCake, Auntof6, 0ladne, PixelBot, Brews ohare, Jotterbot, Hans Adler, SchreiberBike, Jasanas~enwiki, Humanengr, TimothyRias, BodhisattvaBot, SilvonenBot, Jaan Vajakas, Addbot, Gabriele ricci, AndrewHarvey4, Topology Expert, NjardarBot, Looie496, Uncia, ChenzwBot, Ozob, Wikomidia, TeH nOmInAtOr, Jarble, CountryBot, Yobot, Kan8eDie, THEN WHO WAS PHONE?, Edurazee, AnomieBOT, ^musaz, Götz, Citation bot, Xqbot, Txebixev, GeometryGirl, Point-set topologist, RibotBOT, Charvest, Quartl, Lisp21, FrescoBot, Nageh, Rckrone, Sławomir Biały, Citation bot 1, Kiefer.Wolfowitz, Jonesey95, MarcelB612, Stpasha, Mathstudent3000, Jujutacular, Dashed, TobeBot, Javierito92, January, Setitup, TjBot, EmausBot, WikitanvirBot, Brydustin, Fly by Night, Slawekb, Chricho, Ldboer, Quondum, D.Lazard, Milad pourrahmani, RaptureBot, Cloudmichael, ClueBot NG, Wcherowi, Chitransh Gaurav, Jiri 1984, Joel B. Lewis, Widr, Helpful Pixie Bot, Ma snx, David815, Alesak23, Probability0001, JOSmithIII, Duxwing, PsiEpsilon, IkamusumeFan, Uᵾᵾᵾ Ꮯuhᵾᵾᵾᵾᵾ, IPWAI, JYBot, Dexbot, Catclock, Tch3n93, Fycafterpro, CsDix, Hella.chillz, Jose Brox, François Robere, Loganfalco, Newestcastleman, UY Scuti, K9re11, Monkbot, AntiqueReader, Aditya8795, KurtHeckman, Isambard Kingdom, Shivakrishna .Srinivas. Dasari, NateLloydClark and Anonymous: 216

- **Lebesgue integration** *Source:* https://en.wikipedia.org/wiki/Lebesgue_integration?oldid=690090108 *Contributors:* AxelBoldt, Chenyu, Zundark, The Anome, Tarquin, AstroNomer~enwiki, Ap, Gareth Owen, Miguel~enwiki, Roadrunner, Patrick, Chas zzz brown, Michael Hardy, Dominus, Loisel, Stevenj, AugPi, Poor Yorick, Fthulin, Bjcairns, Charles Matthews, Dysprosia, Jitse Niesen, Birkett, MathMartin, Sverdrup, Henrygb, Rorro, Lesonyrra, Daniel Dickman, Lupo, Diberri, Wile E. Heresiarch, Pengo, Tobias Bergemann, Pdenapo, Tosha, Giftlite, Rs2, Lupin, MathKnight, Anville, Dratman, CSTAR, Icairns, Fintor, Guanabot, Mat cross, ArnoldReinhold, AlanBarrett, Paul August, Bender235, Srbauer, Kwamikagami, AnyFile, Rbj, Jumbuck, Denis.arnaud, Eric Kvaalen, BernardH, Aegis Maelstrom, Totalcynic, Oleg Alexandrov, Joriki, DealPete, Woohookitty, Igny, Guardian of Light, Graham87, Melesse, Grammarbot, MarSch, Salix alba, Jenny Harrison, YurikBot, Baccala@freesoft.org, KSmrq, Trovatore, Crasshopper, Curpsbot-unicodify, Bo Jacoby, SmackBot, Eskimbot, Chris the speller, Shawn M. O'Hare, Silly rabbit, RayAYang, Nbarth, Lox, DRLB, Dezmo~enwiki, Germandemat, Ncmathsadist, Ohconfucius, Lim Wei Quan, Makyen, Belizefan, Mets501, Inquisitus, Jakubukaj, Bernard the Varanid, DivideByZero14, King Bee, Headbomb, Cj67, Escarbot, Gossamers, Scintillatingstuffs, JAnDbot, Sanchom, .anacondabot, Sullivan.t.j, David Eppstein, R'n'B, Pomte, Peskydan, Daniele.tampieri, Plasticup, Policron, Sigmundur, LokiClock, TXiKiBoT, Aaron Rotenberg, PaulTanenbaum, Geometry guy, Wikiisawesome, Sylviaelse, Vanished user lkdfj39u3mfk4, Stigin, AlleborgoBot, Stca74, Benjaminveal, S2000magician, Alexbot, Marc van Leeuwen, Jaan Vajakas, Addbot, Jncraton, MrOllie, LinkFA-Bot, Jasper Deng, Lightbot, Matěj Grabovský, Zorrobot, Balabiot, Yobot, Ht686rg90, Ptbotgourou, Jordsan, Amirobot, AnomieBOT, JackieBot, 9258fahsflkh917fas, Kingpin13, Citation bot, Dustankb, LilHelpa, Bdmy, ManningBartlett, Erud, Duncan.hawthorne, Sleeping-Charles, Constructive editor, Dave Ordinary, Finn Diesel, Craig Pemberton, Citation bot 1, DrilBot, MastiBot, Kerrick Staley, John of Reading, KHamsun, Chaohuang, Slawekb, Chricho, Maschen, Ankanbansal, Demonsquirrel, Mesoderm, Solomon7968, ImJimHill, Brad7777, Randomguess, Freeze S, CsDix, Martin Ueding, Wkschwartz, Lily.r.s, Borderlands1920, SillyBunnies, KasparBot, Olazznog, Schidan and Anonymous: 139

- **Ergodic theory** *Source:* https://en.wikipedia.org/wiki/Ergodic_theory?oldid=680957278 *Contributors:* The Anome, Michael Hardy, Dominus, Dcljr, William M. Connolley, Den fjättrade ankan~enwiki, Charles Matthews, Dysprosia, Fredrik, Ojigiri~enwiki, Wile E. Heresiarch, Giftlite, Rs2, Klaus scheicher, D6, Mani1, Bender235, Tobacman, Mdd, Phils, Hammertime, Jheald, Oleg Alexandrov, Linas, Armando, Benbest, XaosBits, Nanite, Rjwilmsi, Joe Decker, Mathbot, YurikBot, RussBot, Tong~enwiki, Zwobot, Msuzen, Zvika, Diligent, That Guy, From That Show!, SmackBot, K-UNIT, Pokipsy76, Silly rabbit, SerialJaywalker, Mhym, Huon, Negrello, Rspanton, JorisvS, Jim.belk, Catquas, Yuide, Stotr~enwiki, Paul venter, CRGreathouse, Jackzhp, OOOOOOOO, CBM, Mamluk, Harej bot, MarkusQ, Myasuda, Headbomb, RobHar, Takwan, Avaya1, DirkOliverTheis, Vegasprof, Chiswick Chap, Lemur235, JohnBlackburne, PMajer, Jmath666, Joseph Grcar, Arcfrk, Melcombe, Neithan Agarwaen, Mild Bill Hiccup, Spmeyn, Torsten Nielsen, NoEdward, Luckas-bot, AdamSiska, Citation bot, DirlBot, Bdmy, LamaO, Sławomir Biały, Citation bot 1, Tinton5, Bowenthebeard, Hmasoom, 777sms, Tiled, Suslindisambiguator, Vramasub, JFB80, Vergilden, MerllwBot, Bibcode Bot, 李晓晗, Brad7777, ChrisGualtieri, Deltahedron, Galaas, Stefoo, LHSPhantom, Jishnu Bose, KasparBot, Volvox globator and Anonymous: 64

- **Bergman space** *Source:* https://en.wikipedia.org/wiki/Bergman_space?oldid=679163270 *Contributors:* Michael Hardy, Charles Matthews, Mathsci, RobHar, YohanN7, Denisarona, Addbot, Materialscientist, Thehelpfulbot, Sławomir Biały, Slawekb, BG19bot, Jansnieg and Anonymous: 7

- **Reproducing kernel Hilbert space** *Source:* https://en.wikipedia.org/wiki/Reproducing_kernel_Hilbert_space?oldid=690065654 *Contributors:* The Anome, Michael Hardy, SebastianHelm, A5, Charles Matthews, Aetheling, Giftlite, Lupin, CSTAR, 3mta3, Wiki-uk, Oleg Alexandrov, Linas, Salix alba, SmackBot, Bluebot, Silly rabbit, Kjetil1001, Memming, James pic, Myasuda, Mct mht, ClydeC, Partha lal, Feynman81, Jmath666, Ayyuru, Jneem, Niceguyedc, Addbot, Ronhjones, Lightbot, Yobot, Ziyuang, Citation bot, Bdmy, SW;xkyz, X7q, Sławomir Biały, Dewritech, ZéroBot, Quondum, Koertefa, Randomguess, Veil007, Tmfs10, Mark viking, Uixs23, LouisLChen, Ematsen, Gronsbellj, Pwensing, Nzer0, Zhanxiong and Anonymous: 50

- **Fourier analysis** *Source:* https://en.wikipedia.org/wiki/Fourier_analysis?oldid=688893823 *Contributors:* Damian Yerrick, AxelBoldt, Zundark, The Anome, Tbackstr, Gareth Owen, Rade Kutil, ChangChienFu, JohnOwens, Michael Hardy, Nixdorf, Theanthrope, SebastianHelm, Ejrh, Ahoerstemeier, Stevenj, Kevin Baas, Smack, Kat, Charles Matthews, Reddi, Dysprosia, Jitse Niesen, Furrykef, Saltine, AndrewKepert, Yardgnome, Omegatron, Samsara, Hankwang, BenBreen2003, Gandalf61, Cdnc, Sverdrup, Ojigiri~enwiki, Cyrius, Wile E. Heresiarch, Tobias Bergemann, Enochlau, Pdenapo, Giftlite, Rs2, Jyril, Gene Ward Smith, BenFrantzDale, Lupin, Nayuki, LiDaobing, Zeimusu, CSTAR, Fintor, Andreas Kaufmann, Giuscarl~enwiki, Mormegil, Zowie, EugeneZelenko, Mazi, Bender235, LemRobotry, Bobo192, Touriste, Viriditas,

Rbj, Jumbuck, Jclaer, PAR, Ross Burgess, Cburnett, Yurivict, Oleg Alexandrov, Mwilde, Tbsmith, OwenX, Linas, Jftsang, BD2412, Maurice-JFox3, Andrei Polyanin, Vegaswikian, Chubby Chicken, Fred Bradstadt, RexNL, Fresheneesz, YurikBot, Borgx, RobotE, Lenthe, Grubber, Morwan, Tony1, Supten, Tachs, Mütze, LeonardoRob0t, Slehar, Unaiaia~enwiki, Bo Jacoby, Zvika, SmackBot, Freestyle~enwiki, KYN, Dingar, Chris the speller, Oli Filth, Silly rabbit, Jyossarian, Nbarth, Baa, Szidomingo, Bob K, JustUser, RFightmaster, Voyajer, Radagast83, Eliyak, Frade, NongBot~enwiki, Dicklyon, Mahlerite, CRGreathouse, HenningThielemann, Thrapper, Wikid77, Headbomb, Miguelgoldstein, Futurebird, Ste4k, Alphachimpbot, Qwerty Binary, Thenub314, Strangealibi, Parsecboy, Andykass, Ayonbd2000, MistyMorn, Idioma-bot, John-Blackburne, LokiClock, Philip Trueman, Anonymous Dissident, GcSwRhIc, Rmcguire, Marcosaedro, Don4of4, Cogburnd02, Oxymoron83, Jdaloner, Lamoidfl, Melcombe, Denisarona, Loren.wilton, ClueBot, Sepia tone, Geoeg, Sun Creator, Brews ohare, Razorflame, Sylvestersteele, Planb 89, Kman543210, CanadianLinuxUser, Uncia, Quercus solaris, Legobot, Yobot, Ht686rg90, JAIG, Citation bot, Ayda D, Srich32977, GT5162, Sławomir Biały, Citation bot 1, Boxplot, Ashok567, Davidmeo, Math.geek3.1415926, Helwr, EmausBot, Helptry, Syncategoremata, D.Lazard, Snotbot, MerlIwBot, Helpful Pixie Bot, Augiecalisi, Solomon7968, Robert the Devil, Brad7777, Hebert Peró, BrightStarSky, Lugia2453, Hamoudafg, JaconaFrere, Monkbot, Fourier1789 and Anonymous: 136

- **Gibbs phenomenon** *Source:* https://en.wikipedia.org/wiki/Gibbs_phenomenon?oldid=689808846 *Contributors:* The Anome, Waveguy, Michael Hardy, William M. Connolley, Charles Matthews, Reddi, Dysprosia, Jitse Niesen, Hyacinth, RedWolf, Giftlite, Jrdioko, Chris Howard, Eb.hoop, Gadykozma, Mecanismo, Teorth, Rbj, Oarih, PAR, YebisYa, Oleg Alexandrov, Linas, Ryan Reich, Marudubshinki, BD2412, R.e.b., Phrood~enwiki, Srleffler, Hadaso, Jpkotta, WriterHound, Genba, Siddhant, YurikBot, Crasshopper, SmackBot, Jushi, Master of Puppets, Oli Filth, Nbarth, Bob K, Joey-das-WBF, Euchiasmus, Tawkerbot2, Jfcorbett, D.H, Oreo Priest, Thenub314, MSBOT, Salih, Policron, Unexpect, Fyo, AlleborgoBot, Mr. PIM, Gluefoot, YohanN7, Cwkmail, JP.Martin-Flatin, Excirial, Brews ohare, Addbot, LaaknorBot, Victor.zamanian, Lightbot, Zorrobot, Yobot, Jordsan, Linket, It's Been Emotional, AnomieBOT, Citation bot, Bdmy, Aa77zz, FrescoBot, Shiki2, Pinethicket, RjwilmsiBot, JamesCrook, Peretuset, Ain92, ClueBot NG, Twb5, Bradleyjefferson, CitationCleanerBot, BattyBot, Clongo01, Pokajanje, Sol1, Monkbot and Anonymous: 44

- **Weak convergence (Hilbert space)** *Source:* https://en.wikipedia.org/wiki/Weak_convergence_(Hilbert_space)?oldid=686625116 *Contributors:* Michael Hardy, Tobias Bergemann, Nonick, Giftlite, Gauge, Oleg Alexandrov, Linas, Nsteinberg, RussBot, SmackBot, Lavaka, Mct mht, Cj67, Futurebird, Dingenis, Addbot, Calle, J04n, Erik9bot, Sławomir Biały, Weaky87, Master Lenman, Mgkrupa, Kamerondeckerharris, Geoffbourque and Anonymous: 22

- **Banach space** *Source:* https://en.wikipedia.org/wiki/Banach_space?oldid=668076056 *Contributors:* AxelBoldt, Zundark, Taw, Toby Bartels, Miguel~enwiki, Michael Hardy, Kku, TakuyaMurata, Looxix~enwiki, Vargenau, Jitse Niesen, Cjmnyc, Prumpf, Fibonacci, Robbot, Josh Cherry, R3m0t, RedWolf, Romanm, MathMartin, Tobias Bergemann, Giftlite, Mikez, BenFrantzDale, Lethe, Lupin, Hellisp, Barnaby dawson, TheObtuseAngleOfDoom, Rich Farmbrough, Guanabot, Luqui, Mecanismo, Quistnix, Bender235, El C, Kwamikagami, Crisófilax, Spoon!, Wood Thrush, Tsirel, Sligocki, Oleg Alexandrov, Japanese Searobin, Joriki, Linas, Jannex, Isnow, Graham87, Rjwilmsi, Salix alba, Chobot, Volunteer Marek, YurikBot, Wavelength, KSmrq, Robertvan1, Dtrebbien, Buster79, Crasshopper, Bota47, Reyk, GrinBot~enwiki, Janek Kozicki, Attilios, SmackBot, Reedy, Sviemeister, Chris the speller, SMP, Jprg1966, Silly rabbit, DHN-bot~enwiki, Hard Nut, Madmath789, Twipie, 345Kai, Myasuda, Mct mht, Ntsimp, Konradek, Headbomb, Rlupsa, Eleuther, AntiVandalBot, Salgueiro~enwiki, Magioladitis, Sullivan.t.j, Policron, Skou, Camrn86, LokiClock, BotMultichill, OsamaBinLogin, Anchor Link Bot, Aohara1986, Pernambuko, Aprock, Sabalka, Topology Expert, Tassedethe, JackS333, Zorrobot, Jarble, سمع, Legobot, Luckas-bot, Yobot, AnomieBOT, Ciphers, AdjustShift, LilHelpa, Xqbot, Bdmy, Clément Pillias, Kaoru Itou, Howard McCay, FrescoBot, Sławomir Biały, Kiefer.Wolfowitz, Rausch, Tuplanolla, Trappist the monk, Skakkle, EmausBot, Slawekb, Werieth, Ida Shaw, Fæ, Skalkaz, Berthold1954, ClueBot NG, Chogg, Helpful Pixie Bot, BG19bot, Beaumont877, AdventurousSquirrel, Shedoblyde, The1337gamer, BattyBot, ChrisGualtieri, Illia Connell, ToccattaAndFugue, StefanEckert, Jochen Burghardt, Eozhik, Mark viking, Holmesindiana, Gladtobeherenow, Lemnaminor, Jnhlnd, MetaSlaya, Mgkrupa, Banannaboy, KasparBot, Srednuas Lenoroc and Anonymous: 77

- **Tensor product of Hilbert spaces** *Source:* https://en.wikipedia.org/wiki/Tensor_product_of_Hilbert_spaces?oldid=675185383 *Contributors:* Michael Hardy, AJP, Linas, BD2412, R.e.b., Silly rabbit, Nbarth, Sstarr7774, Ksoileau, Ubermichael, Willow1729, MystBot, Addbot, Yobot, AnomieBOT, FrescoBot, Sławomir Biały, Trappist the monk, Slawekb, Solomon7968, Mark viking, Mgkrupa, TSBM and Anonymous: 4

- **Hilbert algebra** *Source:* https://en.wikipedia.org/wiki/Hilbert_algebra?oldid=557863534 *Contributors:* R.e.b., Mathsci, Yobot and Frostie Jack

- **Hilbert manifold** *Source:* https://en.wikipedia.org/wiki/Hilbert_manifold?oldid=607162426 *Contributors:* Michael Hardy, Silverfish, Charles Matthews, Tobias Bergemann, Giftlite, ScottDavis, Tabletop, JIP, Rjwilmsi, MarSch, YurikBot, Trovatore, Sullivan.t.j, Cardano~enwiki, Arcfrk, MystBot, Addbot, Yobot, Citation bot, Djcrowley, Helpful Pixie Bot, Brad7777 and Anonymous: 2

- **Rigged Hilbert space** *Source:* https://en.wikipedia.org/wiki/Rigged_Hilbert_space?oldid=667114180 *Contributors:* Erik Zachte, Charles Matthews, Giftlite, Lethe, Rick Block, CSTAR, Garrison, Mat cross, Gauge, Kusma, Linas, Grammarbot, R.e.b., Mathbot, Silly rabbit, Gala.martin, Khazar, Mct mht, Josemiotto, Dstewart7309, LokiClock, Rafaelo10, ManDay, Addbot, Ginosbot, AnomieBOT, Davidcarfi, Dr-davidasmith, Airwoz, Snarayr and Anonymous: 17

24.5.2 Images

- **File:Affine_subspace.svg** *Source:* https://upload.wikimedia.org/wikipedia/commons/8/8c/Affine_subspace.svg *License:* CC BY-SA 3.0 *Contributors:* Own work *Original artist:* Jakob.scholbach

- **File:Altitudes_and_orthic_triangle_SVG.svg** *Source:* https://upload.wikimedia.org/wikipedia/commons/5/52/Altitudes_and_orthic_triangle_SVG.svg *License:* CC BY-SA 3.0 *Contributors:* Own work *Original artist:* Limaner

- **File:Astroid.svg** *Source:* https://upload.wikimedia.org/wikipedia/commons/0/03/Astroid.svg *License:* Public domain *Contributors:* ? *Original artist:* ?

- **File:Bass_Guitar_Time_Signal_of_open_string_A_note_(55_Hz).png** *Source:* https://upload.wikimedia.org/wikipedia/commons/a/a6/Bass_Guitar_Time_Signal_of_open_string_A_note_%2855_Hz%29.png *License:* CC BY-SA 4.0 *Contributors:* Own work *Original artist:* Fourier1789

- **File:Sinfrequency.jpg** *Source:* https://upload.wikimedia.org/wikipedia/commons/7/7e/Sinfrequency.jpg *License:* CC BY-SA 3.0 *Contributors:* Own work *Original artist:* Futurebird

- **File:SquareWave.gif** *Source:* https://upload.wikimedia.org/wikipedia/commons/f/f8/SquareWave.gif *License:* CC BY 3.0 *Contributors:* Own work *Original artist:* Peretuset

- **File:Standing_waves_on_a_string.gif** *Source:* https://upload.wikimedia.org/wikipedia/commons/5/5c/Standing_waves_on_a_string.gif *License:* CC BY-SA 3.0 *Contributors:* Own work *Original artist:* Adjwilley

- **File:Superellipse_rounded_diamond.svg** *Source:* https://upload.wikimedia.org/wikipedia/commons/1/13/Superellipse_rounded_diamond.svg *License:* CC BY-SA 3.0 *Contributors:* Own work *Original artist:* Krishnavedala

- **File:Tensor-diagramB.jpg** *Source:* https://upload.wikimedia.org/wikipedia/commons/5/55/Tensor-diagramB.jpg *License:* CC BY-SA 3.0 *Contributors:* File:Tensor-diagram.jpg *Original artist:* modified by User:Bdmy from work of StefanEckert

- **File:Tetrahedron_vertfig.png** *Source:* https://upload.wikimedia.org/wikipedia/commons/6/68/Tetrahedron_vertfig.png *License:* Public domain *Contributors:* ? *Original artist:* ?

- **File:Triangle.Orthocenter.svg** *Source:* https://upload.wikimedia.org/wikipedia/commons/9/93/Triangle.Orthocenter.svg *License:* Public domain *Contributors:* ? *Original artist:* ?

- **File:Triangle.Scalene.svg** *Source:* https://upload.wikimedia.org/wikipedia/commons/9/93/Triangle.Scalene.svg *License:* Public domain *Contributors:* No machine-readable source provided. Own work assumed (based on copyright claims). *Original artist:* No machine-readable author provided. Syp assumed (based on copyright claims).

- **File:Triangle_inequality_in_a_metric_space.svg** *Source:* https://upload.wikimedia.org/wikipedia/en/c/ca/Triangle_inequality_in_a_metric_space.svg *License:* GFDL *Contributors:*
Based on File:Dreieck.svg. *Original artist:* , modified by <a href='//en.wikipedia.org/wiki/User:S%C5%82awomir_Bia%C5%82y' title='User:Sławomir Biały'>Sławomir Biały</a> (talk)

- **File:Triangle_sphérique.svg** *Source:* https://upload.wikimedia.org/wikipedia/commons/1/14/Triangle_sph%C3%A9rique.svg *License:* Public domain *Contributors:* Own work *Original artist:* Herve1729

- **File:Trig_Functions.PNG** *Source:* https://upload.wikimedia.org/wikipedia/commons/5/5a/Trig_Functions.PNG *License:* CC BY-SA 3.0 *Contributors:* Own work *Original artist:* Brews ohare

- **File:Tâbit_ibn_Qorra.PNG** *Source:* https://upload.wikimedia.org/wikipedia/commons/b/b0/T%C3%A2bit_ibn_Qorra.PNG *License:* CC BY-SA 3.0 *Contributors:* Own work *Original artist:* Brews ohare

- **File:Universal_tensor_prod.svg** *Source:* https://upload.wikimedia.org/wikipedia/commons/d/d3/Universal_tensor_prod.svg *License:* CC BY-SA 4.0 *Contributors:* Own work *Original artist:* IkamusumeFan

- **File:UniversumUNAM18.JPG** *Source:* https://upload.wikimedia.org/wikipedia/commons/2/20/UniversumUNAM18.JPG *License:* CC BY-SA 3.0 *Contributors:* Own work *Original artist:* AlejandroLinaresGarcia

- **File:Vector_add_scale.svg** *Source:* https://upload.wikimedia.org/wikipedia/commons/a/a6/Vector_add_scale.svg *License:* CC BY-SA 3.0 *Contributors:* Plot SVG using text editor. *Original artist:* IkamusumeFan

- **File:Vector_addition3.svg** *Source:* https://upload.wikimedia.org/wikipedia/commons/e/e6/Vector_addition3.svg *License:* CC BY-SA 3.0 *Contributors:* Own work *Original artist:* Jakob.scholbach

- **File:Vector_components.svg** *Source:* https://upload.wikimedia.org/wikipedia/commons/8/87/Vector_components.svg *License:* CC BY-SA 3.0 *Contributors:* Own work *Original artist:* Jakob.scholbach

- **File:Vector_components_and_base_change.svg** *Source:* https://upload.wikimedia.org/wikipedia/commons/a/a6/Vector_components_and_base_change.svg *License:* CC BY-SA 3.0 *Contributors:* Own work *Original artist:* Jakob.scholbach

- **File:Vector_norms.svg** *Source:* https://upload.wikimedia.org/wikipedia/commons/4/4d/Vector_norms.svg *License:* CC-BY-SA-3.0 *Contributors:* ? *Original artist:* ?

- **File:Vector_norms2.svg** *Source:* https://upload.wikimedia.org/wikipedia/commons/0/02/Vector_norms2.svg *License:* CC BY-SA 3.0 *Contributors:* Own work (Original text: *I created this work entirely by myself.*) *Original artist:* Jakob.scholbach (talk)

- **File:Wikibooks-logo-en-noslogan.svg** *Source:* https://upload.wikimedia.org/wikipedia/commons/d/df/Wikibooks-logo-en-noslogan.svg *License:* CC BY-SA 3.0 *Contributors:* Own work *Original artist:* User:Bastique, User:Ramac et al.

24.5.3 Content license

- Creative Commons Attribution-Share Alike 3.0